Universitext

Universitext

Universitext is a series of textbooks that presents material from a wide variety of mathematical disciplines at master's level and beyond. The books, often well class-tested by their author, may have an informal, personal even experimental approach to their subject matter. Some of the most successful and established books in the series have evolved through several editions, always following the evolution of teaching curricula, to very polished texts.

Thus as research topics trickle down into graduate-level teaching, first textbooks written for new, cutting-edge courses may make their way into *Universitext*.

More information about this series at http://www.springer.com/series/223

Nicolas Bergeron

The Spectrum of Hyperbolic Surfaces

 Springer

 sciences

Nicolas Bergeron
IMJ-PRG
Universite Pierre et Marie Curie
Paris
France

Translated by Farrell Brumley

Translation from the French language edition:
Le spectre des surfaces hyperboliques
by Nicolas Bergeron
Copyright © 2011 EDP Sciences, CNRS Editions, France.
http://www.edpsciences.org/
http://www.cnrseditions.fr/

ISSN 0172-5939   ISSN 2191-6675   (electronic)
Universitext
ISBN 978-3-319-27664-9   ISBN 978-3-319-27666-3   (eBook)
DOI 10.1007/978-3-319-27666-3

Library of Congress Control Number: 2015960755

Mathematics Subject Classification: 11F72, 11M36, 35P15, 35P20, 35R01, 58J50, 58J51

Springer Cham Heidelberg New York Dordrecht London

Printed on acid-free paper

Springer International Publishing AG Switzerland is part of Springer Science+Business Media (www.springer.com)

Preface

The Laplacian, like few other objects, is nearly ubiquitous in mathematics. The aim of this book is to provide an introduction to the study of the spectrum of the Laplacian on *hyperbolic surfaces*. These are the Riemannian surfaces of constant negative curvature -1.

In the more familiar context of the Euclidean circle the Laplacian is given simply by the second derivative, and the analysis of its spectrum is the subject of classical Fourier theory. Besides its intrinsic interest, Fourier analysis has long been one of the principal tools in analytic number theory. Indeed, one of the proofs of the analytic continuation of the Riemann zeta function is based on the Poisson summation formula. In the first chapter of this book, in order to get things rolling, we will recall this and other basic Fourier analytic facts.

In 1956 the Norwegian mathematician Atle Selberg proposed a vast generalization of the Poisson summation formula, now referred to as the *Selberg trace formula*. This formula, which we prove in Chap. 5, is reminiscent of the so-called explicit formulae in the analytic theory of L-functions. In the powerful analogy that the Selberg trace formula evokes, the closed geodesics on a compact hyperbolic surface S play the role of prime numbers. The trace formula establishes a relation between these closed geodesics and the Laplacian spectrum of S. These results were, upon their publication, immediately recognized as a new perspective on the Riemann hypothesis. While the latter remains an enigma, the trace formula has, since the late 1960s, played an increasingly central role in the ambitious program of Robert Langlands. The latter aims to link number theory with harmonic analysis on *locally symmetric spaces*, chief among which are the hyperbolic surfaces.

There are several textbooks already dedicated to the extension of the classical Fourier theory to hyperbolic surfaces; we mention in particular the now classic text of Iwaniec [63] (geared towards applications to analytic number theory) as well as the book by Buser [24], which has a more geometric outlook. In French, the closest work to the present one is probably the book by Kowalski [71] where a large part of the basic theory is presented.[1]

Why Another Book?

For well over 30 years now the Langlands program has made enormous advances, and spectral theory and number theory alike have harvested its fruits. But it has become difficult to penetrate such a larger and larger mathematical landscape. We wanted to write a book on the Langlands program in a more classical, and hopefully more accessible, language. This inclination has naturally led us to explore in more detail compact arithmetic hyperbolic surfaces. In particular, we present a proof of the first striking result in the Langlands program: the *Jacquet-Langlands correspondence*. The proof we give – due to Bolte and Johansson – should be more readily comprehensible to a reader unfamiliar with the language of adeles. With any luck, this expository simplification will encourage the reader to dive headlong into the original work of Jacquet and Langlands.

An added motivation for an updating of the literature was given by three other recent results: the lower bound on the Laplacian eigenvalues of arithmetic hyperbolic surfaces by the method Luo-Rudnick-Sarnak, the lovely proof of the existence of cusp forms for congruence subgroups of $SL(2, \mathbf{Z})$ due to Lindenstrauss and Venkatesh, and finally Lindenstrauss' proof, by purely ergodic theoretic arguments, of the *arithmetic* quantum unique ergodicity conjecture of Rudnick and Sarnak in the compact setting. We describe these in detail in the introduction, and we later give complete proofs of the first two results. The last chapter contains an introduction to the work of Lindenstrauss on the quantum unique ergodicity conjecture. Via this last chapter in particular, we hope to lead the reader to the heart of current research.

To Whom Is This Book Addressed?

We presuppose a basic knowledge of differential geometry and functional analysis. Our desire has been to make the entire text comprehensible to an ambitious first year graduate student – or possibly a colleague whose speciality lies elsewhere but whose curiosity is piqued by the subject. Each chapter has its own level of difficulty, however. In the first chapters, for example, we develop the spectral theory of the Laplacian on hyperbolic surfaces using only the basic algebraic and analytic tools of a first year graduate course (integration, Fourier analysis, Hilbert spaces). By contrast, the three last chapters, being more directly plugged into current research, would more naturally find their place in an advanced graduate level course.

---

[1]Strictly speaking, the closest book in French to the present one is, well,...

Nevertheless, with an eye on highlighting the most illustrative cases, we have tried to simplify certain proofs; we then give references for the stronger statements.

Finally, to the reader eager to learn more on the subject than is presented here, we could not do better than to recommend the beautiful article of Sarnak [112], entitled "Spectra of hyperbolic surfaces", from which this book has borrowed its own title as well as a large part of its structural organization.

Acknowledgments

We extend an immense thank you to Nalini Anantharaman who proofread the text in depth and to Farrell Brumley who translated it into English. This book would have contained many more errors without their numerous corrections.

I'm also grateful to a second anonymous reader whose constructive remarks allowed me to simplify certain proofs and to correct several mistakes.

A big thank you to Aurélien Alvarez, Jean-Pierre Otal, Gabriel Rivière, Claude Sabbah and Emmanuel Schenck for their proofreading and the numerous corrections they pointed out.

Finally, we are deeply indebted to Sébastien Gouëzel for granting us permission to reproduce a part of his notes on Host's proof of Rudolph's theorem presented in § 9.4 and, of course, Valentin Blomer and Farrell Brumley who were willing to write Appendix C on elementary estimates on hyper-Kloosterman sums and who explained to me how to simplify the proof of Theorem 7.38.

Paris, France   Nicolas Bergeron

Contents

Chapter 1
Introduction

1.1 Spectral Analysis on the Torus

Before entering into the heart of the subject, we begin with the motivating and familiar example of spectral analysis on the torus. Let $(\mathbf{R}^n, \langle\,,\,\rangle)$ be standard n-dimensional Euclidean space, where

$$\mathbf{R}^n = \{x = (x_1, \ldots, x_n) \mid x_1, \ldots, x_n \in \mathbf{R}\} \text{ and } \langle x, y \rangle = x_1 y_1 + \cdots + x_n y_n.$$

The torus $\mathbf{T}^n$ is the quotient of $\mathbf{R}^n$ by the group of translations $\mathbf{Z}^n$. One thinks of the latter as a discrete (torsion-free) subgroup of the real group $G = \mathbf{R}^n$ acting on Euclidean space by translations. We write $q : \mathbf{R}^n \to \mathbf{T}^n$ for the quotient map.

Let $\Delta = -\partial^2/\partial x_1^2 - \cdots - \partial^2/\partial x_n^2$ be the Laplacian[1] on $\mathbf{R}^n$. A C^∞- function f on the torus $\mathbf{T}^n$ lifts to a $\mathbf{Z}^n$-periodic C^∞-function $f \circ q$ on $\mathbf{R}^n$. By abuse of notation we identify these two functions. The Laplacian commutes with the action of G on Euclidean space. In particular, since f is $\mathbf{Z}^n$-periodic, the function Δf is also $\mathbf{Z}^n$-periodic and defines a function on the torus. The Laplacian Δ descends in this way to a differential operator on $\mathbf{T}^n$, which we continue to denote by Δ; this is the Laplacian on the torus.

It is clear that the exponential functions[2]

$$\varphi_\xi(x) = e(\langle x, \xi \rangle), \quad \xi \in \mathbf{R}^n,$$

---

[1] We consider the negative of the usual Laplacian in order for its spectrum to lie in $\mathbf{R}^+$. One sometimes calls this operator the "geometric Laplacian".

[2] Throughout this book we shall use the notation $e(z) = e^{2\pi i z}$.

© Springer International Publishing Switzerland 2016
N. Bergeron, *The Spectrum of Hyperbolic Surfaces*, Universitext,
DOI 10.1007/978-3-319-27666-3_1

are Δ eigenfunctions;

$$\Delta\varphi_\xi = \lambda\varphi_\xi, \quad \lambda = \lambda(\xi) = 4\pi^2|\xi|^2.$$

Here $|\cdot|$ denotes the norm associated to $\langle\,,\,\rangle$.

Let $\mathcal{S}(\mathbf{R}^n)$ be the space of *Schwartz class* functions. Recall that a function is in the Schwartz class if it and all of its derivatives decay rapidly. Let $f \in \mathcal{S}(\mathbf{R}^n)$. The classical Fourier inversion identities (see [106, Chap. 9])

$$\widehat{f}(\xi) = \int_{\mathbf{R}^n} f(x)e(-\langle x,\xi\rangle)\,dx$$
$$f(x) = \int_{\mathbf{R}^n} \widehat{f}(\xi)\varphi_\xi(x)d\xi \tag{1.1}$$

can be interpreted as the "spectral resolution of the Laplacian Δ". Formally, we can write $\widehat{f}(\xi) = (f, \varphi_\xi)$, but the functions φ_ξ do not belong to $L^2(\mathbf{R}^n)$, so this expression does not literally make sense. Moreover, for $f \in L^2(\mathbf{R}^n)$ the Fourier integral does not have a well-defined meaning in general and the Fourier inversion formula is only obtained by analytic continuation.

The functions φ_ξ for $\xi \in \mathbf{Z}^n$ descend to C^∞-functions of L^2 norm 1 on the torus $\mathbf{T}^n$. They form a Hilbert orthonormal basis for $L^2(\mathbf{T}^n)$. Since they are also Laplacian eigenfunctions, we say that they form a complete orthonormal system of solutions to the spectral problem

$$\Delta\varphi = \lambda\varphi.$$

In this way, we recover Fourier theory (see [145, Chap. IV]). We remark that analytic number theory profits enormously from harmonic analysis on the torus since it often exploits properties of periodic functions, notably via the Poisson summation formula (see [145, p. 94]).

Poisson's formula Let $f \in \mathcal{S}(\mathbf{R}^n)$. Then,

$$\sum_{m\in\mathbf{Z}^n} f(m) = \sum_{m\in\mathbf{Z}^n} \widehat{f}(m). \tag{1.2}$$

Proof Consider the function defined by

$$F(x) = \sum_{m\in\mathbf{Z}^n} f(x+m) \tag{1.3}$$

for $x \in \mathbf{R}$. From the fact that $f \in \mathcal{S}(\mathbf{R}^n)$, it follows that this series converges absolutely and uniformly on compact sets. The function F is therefore C^∞, and being $\mathbf{Z}^n$-periodic it comes from a function on the torus $\mathbf{T}^n$.

We can thus expand F with respect to the basis $(\varphi_\xi)_{\xi \in \mathbf{Z}^n}$:

$$F(x) = \sum_{\xi \in \mathbf{Z}^n} c_\xi \varphi_\xi(x). \tag{1.4}$$

According to Dirichlet's theorem [145, p. 86] this series again converges absolutely and uniformly. We now compute the coefficients c_ξ. By definition we have

$$
\begin{aligned}
c_\xi &= \int_{\mathbf{T}^n} F(x) e(-\langle x, \xi \rangle)\, dx \\
&= \int_{x \in [0,1]^n} \sum_{m \in \mathbf{Z}^n} f(x + m) e(-\langle x, \xi \rangle)\, dx \\
&= \sum_{m \in \mathbf{Z}^n} \int_{x \in [0,1]^n} f(x + m) e(-\langle x, \xi \rangle)\, dx \\
&= \sum_{m \in \mathbf{Z}^n} \int_{x \in m + [0,1]^n} f(x) e(-\langle x, \xi \rangle)\, dx \\
&= \int_{x \in \mathbf{R}^n} f(x) e(-\langle x, \xi \rangle)\, dx = \widehat{f}(\xi).
\end{aligned}
$$

Here all interchanging of sums and integrals is justified by absolute convergence. Setting $x = 0$ in (1.4) gives the desired formula. □

The analytic continuation of the Riemann zeta function is a classical application of the Poisson summation formula (in the case $n = 1$ above) to number theory; see Exercise 1.16. This is the precise proof that Riemann gave in the sole paper [102] that the famous geometer and analyst dedicated to arithmetic.

Let us then consider the case $n = 1$; we first recall the following classical lemma.

Lemma 1.1 Let $y > 0$ and $f_y(x) = e^{-\pi y x^2}$. Then for every ξ

$$\widehat{f}_y(\xi) = \frac{1}{\sqrt{y}} e^{-\pi \xi^2 / y} = \frac{1}{\sqrt{y}} f_{y^{-1}}(\xi).$$

Proof Put $f = f_1$. Changing variables, we have

$$\widehat{f}_y(\xi) = \frac{1}{\sqrt{y}} \widehat{f}\left(\frac{\xi}{\sqrt{y}}\right).$$

It suffices then to treat the case $y = 1$. We must show that $\widehat{f} = f$. Now one has

$$\widehat{f}(\xi) = \int_{\mathbf{R}} e^{-\pi x^2} e(-x\xi)\, dx.$$

Differentiating under the integral (a valid operation since f is Schwartz class), we obtain (via an integration by parts):

$$\widehat{f}'(\xi) = -2\pi \xi \widehat{f}(\xi).$$

This first order linear differential equation admits f as a solution as well. We deduce that there exists $\lambda \in \mathbf{C}$ such that $\widehat{f} = \lambda f$. Applying once again the Fourier transform we obtain (recall that f is even):

$$f = \widehat{\widehat{f}} = \lambda \widehat{f} = \lambda^2 f.$$

Thus $\lambda = \pm 1$ and since $\widehat{f}(0) > 0$, $\lambda = 1$. $\square$

The modern theory of automorphic forms in concerned with the spectral theory of quotients of more general (non-abelian) groups. In this book we concentrate exclusively on the case of the group $G = \mathrm{SL}(2, \mathbf{R})$ and the *hyperbolic plane* associated with it. This basic example is absolutely central to the entire automorphic theory. It has the advantage of being explicit yet sufficiently rich to serve as a model case; in fact we shall see that many important questions in this setting remain stubbornly open.

1.2   The Hyperbolic Plane

As a model for the hyperbolic plane we shall use the *Poincaré model* of the upper half-plane

$$\mathcal{H} = \{z = x + iy \in \mathbf{C} \mid y > 0\}.$$

This is endowed with the complex structure coming from $\mathbf{C}$.

To gain our footing we proceed in the spirit of Euclid, who put forward a notion of geometry based on axioms – although we shall end up excluding the famous fifth axiom. We begin by defining *hyperbolic lines* in $\mathcal{H}$ as

- the vertical half-lines orthogonal to the real axis $\mathbf{R}$, and
- the half-circles perpendicular to $\mathbf{R}$.

See Fig. 1.1. Given any two distinct points in $\mathcal{H}$ there is exactly one hyperbolic line passing through them. On the other hand, given any point there is in general an infinite number of lines passing through it which are parallel to – in the sense of not intersecting with – a given line.

A more modern point of view consists in viewing $\mathcal{H}$ as a Riemannian manifold. This point of view is not necessary for a good understanding of the rest of the text but it would be a pity not to review it, at least rapidly. The reader is encouraged to see [6, 24, 67, 92] for more detailed presentations.

Fig. 1.1 Hyperbolic lines

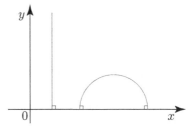

One can in fact endow the half-plane $\mathcal{H}$ with the Riemannian metric

$$ds^2 = \frac{dx^2 + dy^2}{y^2};$$ (1.5)

one calls this the *hyperbolic metric*. This defines a distance on $\mathcal{H}$, by taking the lim inf of the hyperbolic lengths

$$\ell(\gamma) = \int_\gamma ds = \int_0^1 \frac{\sqrt{x'(t)^2 + y'(t)^2}}{y(t)} \, dt$$

of piecewise C^1 curves joining z to w, $\gamma(t) = (x(t), y(t))$ with $t \in [0, 1]$.

The Riemannian metric allows us to define the angle[3] between two tangent vectors at a point $z \in \mathcal{H}$. As ds^2 is proportional to the Euclidean metric, we find that the hyperbolic angles are the same as the Euclidean angles.

The presence of a large group of isometries will allow us to calculate explicitly the distance function on $\mathcal{H}$.

For that we begin by recalling a few well known properties of homographic transformations

$$z \longmapsto \frac{az + b}{cz + d}, \quad a, b, c, d \in \mathbf{C}, \ ad - bc \neq 0$$ (1.6)

of the Riemann sphere $\widehat{\mathbf{C}} = \mathbf{C} \cup \{\infty\}$. We can always assume that

$$ad - bc = 1$$

so that the homography determines a matrix

$$\begin{pmatrix} a & b \\ c & d \end{pmatrix} \in \mathrm{SL}(2, \mathbf{C})$$

---

[3]Let (M, g) be a Riemannian manifold and x a point in M. Given two tangent vectors $v, w \in T_x(M)$ of norm 1, we have $g_x(v, w) = \cos \theta$ where θ is the angle between v and w.

up to sign. In particular, the matrices $I = \begin{pmatrix} 1 & \\ & 1 \end{pmatrix}$ and $-I = \begin{pmatrix} -1 & \\ & -1 \end{pmatrix}$ both define the identity transformation.

A homography g sends a Euclidean circle to another one – as long as Euclidean lines in $\widehat{\mathbf{C}}$ are deemed circles (these are the circles passing through infinity). Of course, the centers are not in general preserved since g is not a Euclidean isometry, apart from the case $g = \pm \begin{pmatrix} 1 & * \\ 0 & 1 \end{pmatrix}$ representing a translation.

Let $G = \mathrm{SL}(2, \mathbf{R})$ be the group of real 2×2 matrices of determinant 1. Given an element $g = \begin{pmatrix} * & * \\ c & d \end{pmatrix} \in G$ we write

$$j(g, z) = cz + d. \tag{1.7}$$

The function j satisfies the cocycle condition:

$$j(gh, z) = j(g, hz)j(h, z). \tag{1.8}$$

Taking into account the formula

$$|j(g, z)|^2 \operatorname{Im}(gz) = \operatorname{Im}(z) \tag{1.9}$$

the Riemann sphere $\widehat{\mathbf{C}}$ is the union of three G-orbits: the upper half-plane $\mathcal{H}$, the lower half-plane $\overline{\mathcal{H}}$ and the boundary separating them $\widehat{\mathbf{R}} = \mathbf{R} \cup \{\infty\}$. Moreover, we have

$$\operatorname{Im}(gz)^{-1}|d(gz)| = \operatorname{Im}(z)^{-1}|dz|,$$

which implies that the metric ds on $\mathcal{H}$ is invariant under the action of G. The homographies are therefore the isometries of the hyperbolic plane.

The action of the group G on $\mathcal{H}$ is, moreover, *transitive*, meaning that for any given pair $z, w \in \mathcal{H}$, there exists $g \in G$ such that $gz = w$. To see this, it suffices to take $z = i$ and $w = x + iy$ with $y > 0$. Then the matrix

$$g = \begin{pmatrix} y^{1/2} & xy^{-1/2} \\ 0 & y^{-1/2} \end{pmatrix}$$

does the job.

The group G preserves the orientation of $\mathcal{H}$ and the only homography acting trivially on $\mathcal{H}$ is the one associated to (plus or minus) the identity matrix. Besides these isometries there is also the orthogonal symmetry about the imaginary axis, $z \mapsto -\bar{z}$, which reverses orientation. One can show that the symmetry $z \mapsto -\bar{z}$ along with all real homographies generate the whole group of isometries of $\mathcal{H}$; we will not, however, have need of this fact in the rest of the text. For the proof one can refer to [6, §7.2], for example.

If $z = i$ and $w = iy$, $y > 0$, one sees immediately that

$$\ell(\gamma) \geq \left| \int_1^y \frac{dt}{t} \right| = |\log y|$$

for every curve adjoining z to w. This lower bound is realized by the vertical segment linking z to w, and so the hyperbolic distance $\rho(z, w)$ between z and w is given by $\rho(i, iy) = |\log y|$.

Now let z and w be arbitrary. Since the action of G on $\mathcal{H}$ is transitive there exists $g_0 \in G$ such that $g_0 z = i$. We can replace g_0 by $k(\theta)g_0$ for every

$$k(\theta) = \begin{pmatrix} \cos\theta & -\sin\theta \\ \sin\theta & \cos\theta \end{pmatrix} \in SO(2, \mathbf{R}) \subset SL(2, \mathbf{R})$$

for such a k fixes i. As $\mathrm{Re}(k(\pi/2)g_0 w)$ and $\mathrm{Re}(g_0 w)$ have opposite sign, by continuity there exists θ such that $\mathrm{Re}(k(\theta)g_0 w) = 0$. We have thus found an element $g \in G$ such that $gz = i$ and $gw = iy$ for some $y > 0$, allowing us to calculate the distance $\rho(z, w) = \rho(i, iy)$. Explicitly we have

$$\rho(z, w) = \log \frac{|z - \overline{w}| + |z - w|}{|z - \overline{w}| - |z - w|}. \tag{1.10}$$

Equivalently, one can write

$$\cosh \rho(z, w) = 1 + 2u(z, w), \tag{1.11}$$

where

$$u(z, w) = \frac{|z - w|^2}{4 \, \mathrm{Im}\, z \, \mathrm{Im}\, w}; \tag{1.12}$$

the latter function is often more practical to use than the true distance function $\rho(z, w)$.

By using the above method, we can show that the *geodesics* of $\mathcal{H}$ – the curves locally realizing the distance between two points – are precisely the hyperbolic lines introduced at the beginning of this section. In particular, the space $\mathcal{H}$ is complete (in the Riemannian sense of the word, or as a metric space relative to ρ). Finally, we isolate a few properties of these hyperbolic lines, see [24].

1. Distances are *globally* minimized by geodesics.
2. If γ is a geodesic and $p \notin \gamma$, there exists a unique geodesic passing through p and perpendicular to γ.
3. If γ and η are two geodesics at distance > 0 apart, there exists a unique geodesic perpendicular to γ and η.

The set $\mathcal{C}(i, \rho)$ of points at hyperbolic distance ρ from i coincides with the orbit of the point ie^{ρ} under the action of the group $SO(2, \mathbf{R})$. Since

$$k(\theta) = \frac{1}{2} \begin{pmatrix} 1 & 1 \\ i & -i \end{pmatrix} \begin{pmatrix} e^{i\theta} & 0 \\ 0 & e^{-i\theta} \end{pmatrix} \begin{pmatrix} 1 & -i \\ 1 & i \end{pmatrix}^{-1},$$

the set $\mathcal{C}(i, \rho)$ is the image by a homography of $\widehat{\mathbf{C}}$ of a Euclidean circle centered at 0, and is thus again a Euclidean circle. We note that the orbit is traced out twice: $k(\pi)$ is the identity transformation. The set $\mathcal{C}(i, \rho)$ is therefore the Euclidean circle of center $i \cosh(\rho)$ and radius $\sinh(\rho)$. More generally, since the action of G on $\mathcal{H}$ is transitive, every hyperbolic circle – the locus of points at fixed hyperbolic distance from a given point – is represented by a Euclidean circle in $\mathcal{H}$ (but of course with a different center).

Having at our disposal a notion of a line, we can now speak about hyperbolic triangles – and, more generally, polygons: these are closed arcwise geodesic Jordan curves. One defines the sides and internal angles of hyperbolic polygons in the usual way.

There are several cute and surprising relations in the hyperbolic plane. For example, an exercise in hyperbolic trigonometry shows (see for example [6, §7.7]) that in each hyperbolic triangle one has

$$\frac{\sin \alpha}{\sinh a} = \frac{\sin \beta}{\sinh b} = \frac{\sin \gamma}{\sinh c}, \tag{1.13}$$

$$\sin \alpha \sin \beta \cosh c = \cos \alpha \cos \beta + \cos \gamma, \tag{1.14}$$

as well as

$$\cosh c = \cosh a \cosh b - \sinh a \sinh b \cos \gamma, \tag{1.15}$$

where α, β, γ are the interior angles opposite the sides of length a, b, c, respectively. The relation (1.14) implies, contrary to the Euclidean case, that the length of a given side depends only on the interior angles of the triangle.

Even more counter-intuitive phenomena appear when we consider the notion of area. As for any Riemannian metric, one can define a measure starting from the hyperbolic metric. By using formulae valid for an arbitrary Riemannian metric (see for example [41, p. 140]), one finds that the hyperbolic measure is

$$d\mu(z) = \frac{dx \, dy}{y^2}. \tag{1.16}$$

According to the general theory, this measure is invariant under isometric motions and hence invariant under the action of the group G.

The following theorem reduces easily to the case of a triangle. We refer the reader to [92, Lem. 1.4.4] for a proof.

Theorem 1.2 *The area of an n-sided hyperbolic polygon P with internal angles* $\theta_1, \ldots, \theta_n$ *is given by the formula:*

$$\text{area}(P) = (n-2)\pi - (\theta_1 + \cdots + \theta_n).$$

There exist other models for hyperbolic geometry. For example, the Poincaré disk is obtained from the homography

$$z \longmapsto \frac{z-i}{z+i}, \quad z \in \mathcal{H}$$

sending the half-plane $\mathcal{H}$ biholomorphically onto the open unit disk **D**,

$$\mathbf{D} = \{w = x + iy \in \mathbf{C} \mid x^2 + y^2 < 1\}.$$

This map transports the hyperbolic metric on $\mathcal{H}$ to the Riemannian metric

$$ds^2 = \frac{4(dx^2 + dy^2)}{(1 - (x^2 + y^2))^2} \tag{1.17}$$

and the corresponding geodesics are the half-circles orthogonal to the boundary unit circle.

Although we will not often use the unit disk model, it nevertheless has its advantages. Being more symmetric,[4] it sometimes lends itself more easily to a qualitative understanding of hyperbolic geometry. The very accessible article by Ghys [44] adopts this point of view and gives a beautiful proof of Theorem 1.2 and also provides a rich bibliography that one can consult for more details. That article begins in particular with a quote of Poincaré which highlights a certain phenomenon of concentration, beautifully illustrating a fundamental difference between Euclidean and hyperbolic geometry. On one hand, the hyperbolic disk of radius ρ centered at i has hyperbolic area $4\pi(\sinh(\rho/2))^2$ and perimeter $2\pi \sinh \rho$. On the other, since the Euclidean center of this same disk is $i \cosh \rho$ with radius $\sinh \rho$, the Euclidean area is $\pi(\sinh \rho)^2$ while its perimeter is again $2\pi \sinh \rho$. Thus, despite having the same perimeter, the Euclidean area is much larger than the hyperbolic area as ρ gets large (and the ratio of the two areas tends to 1 as ρ tends toward 0). The bulk of the hyperbolic area is therefore concentrated close to the edge of the disk.

We conclude this section on hyperbolic geometry with a classification of the isometries of $\mathcal{H}$. We group them into three subsets according to the way in which they displace points. We begin by remarking that conjugate isometries act in the same way; the classification should therefore be invariant under conjugation. Given

---

[4]We in fact made indirect use of the unit disk model to show that the hyperbolic circles are Euclidean circles.

a homography $g \in \mathrm{PSL}(2, \mathbf{R})$, we let

$$[g] = \{hgh^{-1} \mid h \in \mathrm{PSL}(2, \mathbf{R})\}$$

denote its conjugacy class. The identity transformation forms a class by itself; there is not much more we can say about it.

Each

$$g = \begin{pmatrix} a\ b \\ c\ d \end{pmatrix} \neq \pm I$$

has one or two fixed points in $\widehat{\mathbf{C}}$. There are precisely three possibilities:

1. g has one fixed point in $\widehat{\mathbf{R}}$,
2. g has two distinct fixed points in $\widehat{\mathbf{R}}$,
3. g fixes a unique point in $\mathcal{H}$ and fixes the conjugate of this point in $\overline{\mathcal{H}}$.

The matrix (and the corresponding transformation) is said to be *parabolic*, *hyperbolic*, or *elliptic*, respectively, and each classification obviously applies to their entire conjugacy class. Each conjugacy class contains a representative which acts on $\mathcal{H}$ in one of the following three ways:

1. $z \mapsto z + v$, with $v \in \mathbf{R}$ (translation, fixed point ∞),
2. $z \mapsto pz$, with $p \in \mathbf{R}^+$ (homothety, fixed points 0 and ∞),
3. $z \mapsto k(\theta)z$, with $\theta \in \mathbf{R}$, $k(\theta) = \begin{pmatrix} \cos\theta\ -\sin\theta \\ \sin\theta\ \ \cos\theta \end{pmatrix}$ (rotation, fixed point i).

A parabolic transformation is always of finite order, and moves points along horocycles.

The geodesic line between the two points fixed by a hyperbolic transformation g is preserved by g. Of the two fixed points, one attracts, the other repels. The homothety factor p is called the *norm* of g. For every point z on this geodesic, $|\log p|$ is the hyperbolic distance between z and gz.

An elliptic transformation can be of finite or infinite order and moves points along circles centered at its fixed point in $\mathcal{H}$.

The trace – or more precisely its absolute value, since the transformation g determines the matrix $\begin{pmatrix} a\ b \\ c\ d \end{pmatrix}$ up to sign – is invariant under conjugation and the above classes (other than the identity transformation) are characterized in the following way:

1. g is parabolic if and only if $|a + d| = 2$,
2. g is hyperbolic if and only if $|a + d| > 2$,
3. g is elliptic if and only if $|a + d| < 2$.

The isometries of $\mathcal{H}$ do not in general commute, but it is clear that if $gh = hg$, then the fixed points of g are those of h. The converse is also true. More precisely, two non-trivial isometries of $\mathcal{H}$ commute if and only if they have the same set of fixed points. The centralizer $\{h \mid gh = hg\}$ of a parabolic (resp. hyperbolic, elliptic) isometry g is just the subgroup consisting of the identity transformation along with all the parabolic (resp. hyperbolic, elliptic) isometries fixing the same points as g.

1.3 Hyperbolic Surfaces

A *Fuchsian group* is a *discrete* subgroup $\Gamma \subset \mathrm{SL}(2, \mathbf{R})$. We are tacitly taking the topology to be the usual one on $\mathrm{SL}(2, \mathbf{R}) \subset \mathbf{R}^4$.

Fuchsian groups shall play the role of $\mathbf{Z}^n$ – a discrete subgroup of $\mathbf{R}^n$ – in the non-abelian version of the Fourier theory that we have in mind. One considers this time the quotient space $\Gamma \backslash \mathcal{H}$, endowed with the quotient topology rendering the projection $\mathcal{H} \to \Gamma \backslash \mathcal{H}$ continuous.

One can show (see [92, Cor. 1.5.3]) that the Fuchsian groups of $\mathrm{SL}(2, \mathbf{R})$ are precisely the subgroups of $\mathrm{SL}(2, \mathbf{R})$ – viewed as the isometry group of $\mathcal{H}$ – acting *properly discontinuously* on $\mathcal{H}$, in other words no orbit Γz, $z \in \mathcal{H}$, has an accumulation point in $\mathcal{H}$.

Let us fix a Fuchsian group $\Gamma \subset \mathrm{SL}(2, \mathbf{R})$. Since Γ acts properly discontinuously on $\mathcal{H}$ – a Hausdorff space – the quotient $\Gamma \backslash \mathcal{H}$ is also Hausdorff (see for example [92, Lem. 1.7.2]).

We write $\overline{\Gamma}$ for the image of Γ in $\mathrm{PSL}(2, \mathbf{R})$. If moreover $\overline{\Gamma}$ acts *freely* on $\mathcal{H}$, that is to say that for every $z \in \mathcal{H}$ the stabilizer

$$\Gamma_z := \{\gamma \in \Gamma \mid \gamma z = z\}$$

is contained in $\{\pm I\}$, then the projection $p : \mathcal{H} \to \Gamma \backslash \mathcal{H}$ is a covering map. (See for example [37, 91] for an introduction to covering theory.) One can use p to transport the C^∞, Riemannian or even hyperbolic structures. The quotient $S = \Gamma \backslash \mathcal{H}$ is naturally a *hyperbolic surface*; its universal cover is $\mathcal{H}$ and its fundamental group is isomorphic to $\overline{\Gamma}$.

We shall speak throughout this text of functions defined on the hyperbolic surfaces $S = \Gamma \backslash \mathcal{H}$, but this point of view is not strictly necessary: one could more simply think of such a function f as being defined on $\mathcal{H}$ while being periodic with respect to the action of the group Γ: $f(\gamma z) = f(z)$ for every $\gamma \in \Gamma$. The latter approach has the advantage of generalizing to all Fuchsian groups.

Finally, we shall say that a Fuchsian group $\Gamma \subset G$ is of the *first kind* if it is finitely generated and if for every $z \in \mathcal{H}$, the compactification of the real axis $\widehat{\mathbf{R}} \subset \widehat{\mathbf{C}}$ is contained in the closure (for the topology on $\widehat{\mathbf{C}}$) of the orbit Γz.

1.3.1   Fundamental Domains

One can visualize a Fuchsian group $\Gamma \subset G$ with the help of a fundamental domain. An open $D \subset \mathcal{H}$ is a *fundamental domain for Γ* if:

1. the sets γD, for $\gamma \in \overline{\Gamma}$, are pairwise disjoint; and
2. $\mathcal{H} = \bigcup_{\gamma \in \Gamma} \gamma \overline{D}$.

Every Fuchsian group $\Gamma \subset G$ admits a fundamental domain which is of course far from unique. Nevertheless, all of the "classical" fundamental domains – those which we are going to construct – have a boundary of measure zero and therefore have the same area

$$|D| = \int_D d\mu(z).$$

If $\Gamma \subset G$ is of the first kind, we can choose for a fundamental domain a convex polygon (in the sense of hyperbolic lines). Indeed, let $w \in \mathcal{H}$ be a point fixed by no element $\gamma \in \Gamma$ other than $\pm I$, then the set

$$D(w) = \{z \in \mathcal{H} \mid \rho(z, w) < \rho(z, \gamma w) \text{ for every } \gamma \in \Gamma, \ \gamma \neq \pm I\}$$

is a fundamental domain for Γ called a *Dirichlet domain*. One can show that $D(w)$ is a polygon (connected and convex) with an even number of sides. (We adopt the convention that if a side contains the fixed point of an elliptic element of order 2 of Γ, then this point is considered as a vertex and the hyperbolic line on which it lies counts for two sides.) The sides of $D(w)$ can be grouped into pairs in such a way that the isometries identifying each pair generate the group Γ.

From these properties of $D(w)$, one can show – it is a rather delicate theorem of Siegel [123, pp. 75–77], see also [92, Th. 1.9.1] – that a Fuchsian group is of the first kind if and only if it admits a a fundamental domain of finite area.

Fuchsian groups of the first kind break further into two classes according to whether the closure $\overline{D}$ in $\mathcal{H}$ of a fundamental domain D is compact or not. In the former case, we shall say that Γ is a *cocompact group*.

For Γ of the first kind, if $\Gamma \backslash \mathcal{H}$ is non-compact this fact can be read off from the fundamental domain by the presence of a finite number of *cusps x*. These are points where D – or rather its closure in $\mathcal{H} \cup \widehat{\mathbf{R}}$ – meets $\widehat{\mathbf{R}}$. Modifying D if necessary we can always assume that the two sides of D adjacent to a cusp are identified by an element of Γ fixing the cusp.

Lemma 1.3 *Let Γ be a Fuchsian group of the first kind, x a cusp and*

$$\Gamma_x = \{\gamma \in \Gamma \mid \gamma x = x\}$$

its stabilizer in Γ.

1. *Every element $\gamma \in \Gamma_x$ is parabolic.*
2. *The stabilizer Γ_x is isomorphic to $\mathbf{Z}$ or $\mathbf{Z}/2\mathbf{Z} \times \mathbf{Z}$ according to whether $-I \notin \Gamma$ or $-I \in \Gamma$, respectively; the image $\overline{\Gamma}_x$ of Γ_x in $\mathrm{PSL}(2, \mathbf{R})$ is always isomorphic to $\mathbf{Z}$.*
3. *Let γ be a generator of $\overline{\Gamma}_x$. There exists a matrix $\sigma_x \in \mathrm{SL}(2, \mathbf{R})$ such that*

$$\sigma_x \infty = x \quad and \quad \sigma_x^{-1} \gamma \sigma_x = \begin{pmatrix} 1 & 1 \\ 0 & 1 \end{pmatrix}; \tag{1.18}$$

moreover every matrix σ satisfying these two properties is of the form

$$\sigma = \pm \sigma_x \begin{pmatrix} 1 & t \\ 0 & 1 \end{pmatrix} \quad with\ t \in \mathbf{R}.$$

Proof Changing D if necessary, we can in fact assume that the cusps of D are pairwise nonequivalent. Let x be a cusp of D and γ the element of Γ identifying the two sides of D adjacent to x (and fixing x).

To see that γ is parabolic, we consider the two hyperbolic lines d_0 and d_1 bounding D in a neighborhood of x and such that $\gamma(d_0) = d_1$. Since γ fixes the point $x \in \widehat{\mathbf{R}}$, it is a parabolic or hyperbolic transformation. Arguing by contradiction we assume that it is hyperbolic and consider the position of its second fixed point $y \in \widehat{\mathbf{R}}$ relative to the two lines. Now the hyperbolic line d adjoining x to y cannot lie strictly between d_0 and d_1, for otherwise (as one sees by considering the germ of D along d at x) the intersection of the fundamental domain D with $\gamma(D)$ would be non-empty. Replacing γ by γ^{-1} if necessary, we can assume that x repels and y attracts, so that the sequence of geodesics $d_n = \gamma^n(d_0)$ tends towards d. The side of D abutted by d_0 is then sent by γ^n to a side of $D_n = \gamma^n(D)$ abutted by d_n. The tiling $\mathcal{H} = \bigcup_{\gamma \in \Gamma} \gamma(D)$ is therefore not locally finite in a neighborhood of d. Since Γ acts properly discontinuously on $\mathcal{H}$, we have found a contradiction.

The above argument in fact shows that the stabilizer Γ_x contains only parabolic elements. As such, it is contained in a one-parameter subgroup consisting entirely of parabolic elements. In fact there exists σ in $\mathrm{SL}(2, \mathbf{R})$ such that $\sigma \infty = x$ and thus

$$\sigma^{-1} \overline{\Gamma}_x \sigma \subset \left\{ \begin{pmatrix} 1 & t \\ 0 & 1 \end{pmatrix} \middle| t \in \mathbf{R} \right\}.$$

As $\sigma^{-1} \overline{\Gamma}_x \sigma$ is discrete, it must therefore be of the form

$$\begin{pmatrix} 1 & h\mathbf{Z} \\ 0 & 1 \end{pmatrix}.$$

We easily conclude the argument by using the formula

$$\begin{pmatrix} a & 0 \\ 0 & a^{-1} \end{pmatrix} \begin{pmatrix} 1 & t \\ 0 & 1 \end{pmatrix} \begin{pmatrix} a^{-1} & 0 \\ 0 & a \end{pmatrix} = \begin{pmatrix} 1 & a^2 t \\ 0 & 1 \end{pmatrix}.$$

□

Conversely, translating by Γ if necessary, the cusps are precisely the fixed points of parabolic elements of Γ. We deduce the following proposition.

Proposition 1.4 *A Fuchsian group of the first kind is cocompact if and only if it contains no parabolic element.*

Now a given fundamental domain D is not canonically associated with Γ, so for a rigorous definition of a *cusp* of Γ we take a point of $\widehat{\mathbf{R}}$ which is fixed by a parabolic element in Γ. The group Γ induces an equivalent relation on the set of cusps and each equivalence class has a cusp of D as a representative. In particular, it follows from Siegel's theorem mentioned earlier that there are only a finite number of cusps modulo Γ (again, the latter is always assumed to be of the first kind). We saw that if a Fuchsian group Γ acts freely on $\mathcal{H}$, the quotient $S = \Gamma \backslash \mathcal{H}$ is a hyperbolic Riemann surface. The surface S is obtained by gluing a fundamental domain of Γ along identified sides. The Riemannian measure on the quotient has a concrete interpretation as the restriction of μ to D. This measure is in particular of finite area if Γ is of the first kind. Finally, the surface S is compact if Γ is a cocompact group.

A general Fuchsian group Γ does not necessarily act freely on $\mathcal{H}$. If $z \in \mathcal{H}$ and $\overline{\Gamma}_z \neq \{I\}$ then all of the elements of Γ_z are elliptic. Since Γ is discrete and the stabilizer of z in G is compact, the group Γ_z is necessarily finite. If Γ is of the first kind, it admits a fundamental domain with a finite number of sides. We easily deduce from this that Γ contains only a finite number of elliptic elements. The cover $\mathcal{H} \to \Gamma \backslash \mathcal{H}$ therefore has ramification points precisely at the fixed points of elliptic elements in Γ.

Figure 1.2, illustrated by Frank Loray for the collective work [34], represents a quotient with two ramification points and one cusp.

1.3.2 Hyperbolic Laplacian

The Laplacian admits a natural generalization to any Riemannian manifold. Using the general formulae for a Riemannian metric (see for example [41, p. 183]), one finds that in the coordinates $z = x + iy$ of the Poincaré upper half-plane the hyperbolic Laplacian is the differential operator

$$\Delta = -y^2 \left(\frac{\partial^2}{\partial x^2} + \frac{\partial^2}{\partial y^2} \right) \tag{1.19}$$

acting on C^∞ functions on $\mathcal{H}$.

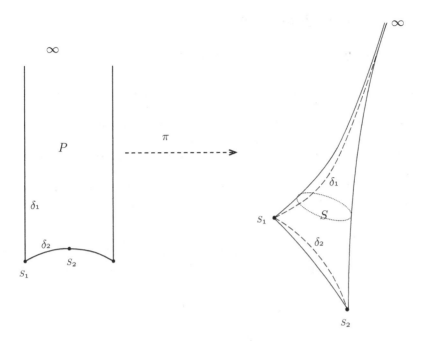

Fig. 1.2 A quotient with one cusp and two ramification points

According to the general theory, the Laplacian is an object intrinsic to the Riemannian metric, invariant under the action by isometries, and so in particular by G. In other words, if $L_g f(z) = f(g^{-1}z)$ then

$$L_g \Delta = \Delta L_g, \quad \text{for } g \in G. \tag{1.20}$$

In polar coordinates (r, θ) of the hyperbolic plane (see Appendix A), one has

$$\Delta = -\frac{\partial^2}{\partial r^2} - \frac{1}{\tanh r}\frac{\partial}{\partial r} - \frac{1}{(2\sinh r)^2}\frac{\partial^2}{\partial \theta^2}. \tag{1.21}$$

Property (1.20) is fundamental: one can show that it, in a sense, characterizes the Laplacian. Indeed, *the differential operators which commute with all isometries form a polynomial algebra in the Laplacian Δ* (see [56]).

Finally, we note that, as in the case of the torus, it follows from (1.20) that if f is a C^∞ function on the hyperbolic surface $S = \Gamma \backslash \mathcal{H}$, viewed as a C^∞ Γ-periodic function on $\mathcal{H}$, the function Δf is again Γ-periodic on $\mathcal{H}$. This function therefore comes from a function on S. The operator defined in this way on S is the Laplacian associated with the Riemannian metric on S; we shall continue to denote it by Δ.

1.4   Description of the Main Results

1.4.1   Examples of Hyperbolic Surfaces

There are several ways to construct hyperbolic surfaces. One can begin by drawing a convex hyperbolic polygon $D \subset \mathcal{H}$ of finite area having an even number of sides. The polygon D will not in general be the fundamental domain of a Fuchsian group of the first kind Γ; for that D must satisfy certain conditions which were completely explicated by Poincaré. We shall not enter into the details here. We only mention that it is easy to show using this construction (see [34]) that every surface of genus g at least 2 can be endowed with a hyperbolic structure and that the set of such structures is a real manifold of dimension $6g - 6$. There are therefore a great many hyperbolic surfaces (up to isometry).

From the number theoretic viewpoint, of all the hyperbolic surfaces of finite area, the most interesting are the so-called *arithmetic* hyperbolic surfaces; we shall construct them in Chap. 2. Among other results we give a proof of the following important theorem.

Theorem 1.5 *Given two positive integers a and b, the group*

$$\Gamma_{a,b} = \left\{ \begin{pmatrix} x_0 + x_1 \sqrt{a} & x_2 + x_3 \sqrt{a} \\ b(x_2 - x_3 \sqrt{a}) & x_0 - x_1 \sqrt{a} \end{pmatrix} \;\middle|\; \begin{array}{l} x_0, x_1, x_2, x_3 \in \mathbf{Z}, \\ x_0^2 - ax_1^2 - bx_2^2 + abx_3^2 = 1 \end{array} \right\}$$

is Fuchsian of the first kind. It is cocompact if and only if $(0,0,0)$ is the unique integral solution to the Diophantine equation $x^2 - ay^2 - bz^2 = 0$.

The most important example corresponds to the choice $a = b = 1$, where one obtain a finite index subgroup of the *modular group*

$$\Gamma(1) = \mathrm{SL}(2, \mathbf{Z}).$$

The associated surface

$$X(1) = \Gamma(1) \backslash \mathcal{H}$$

is called the *modular surface*. It is not strictly speaking a hyperbolic surface: the group $\Gamma(1)$ contains elliptic elements.

Let us determine the cusps of $\Gamma(1)$. The fixed points of parabolic elements are necessarily in $\mathbf{Q} \cup \{\infty\}$ since if

$$\frac{az + b}{cz + d} = z$$

has a unique solution, it is either rational or ∞. The point ∞ is indeed a cusp as is every rational number since they are Γ-equivalent: $\gamma\infty = a/c$ runs through all of $\mathbf{Q}$.

We take ∞ as a representative of the unique equivalence class of cusps for $\Gamma(1)$. Its stabilizer is the subgroup of integer translations (up to sign)

$$\Gamma(1)_\infty = \left\{ \pm \begin{pmatrix} 1 & n \\ 0 & 1 \end{pmatrix} \middle| n \in \mathbf{Z} \right\}.$$

In the notation of Lemma 1.3, a suitable choice of σ_∞ is I.

In this case there exists an explicit construction of a fundamental domain. Starting from the fundamental domain

$$D_\infty = \{z \in \mathcal{H} \mid |x| < 1/2\}$$

for $\Gamma(1)_\infty$ we consider the subset

$$D = \{z \in D_\infty \mid \mathrm{Im}(z) > \mathrm{Im}(\gamma z) \text{ for all } \gamma \in \Gamma(1), \ \gamma \notin \Gamma(1)_\infty\}.$$

For a proof that D is indeed a fundamental domain, see for example [118, Chap. VII, Th. 1]. This method of construction generalizes to all Fuchsian groups with at least one cusp. The general construction – different from that of Dirichlet – is due to Ford [39].

The polygon D is a hyperbolic triangle with angles $\pi/3$, $\pi/3$ and 0, see Fig. 1.3.[5] Its area is therefore equal to $\pi/3$. Thus, although the modular surface $X(1)$ is non-compact, one has

$$\mathrm{area}(X(1)) = \pi/3. \tag{1.22}$$

One can moreover show that i and $\zeta = (1 + i\sqrt{3})/2$ are representatives of the two classes of elliptic fixed points of Γ (of order 2 and 3, respectively). Topologically the modular surface is therefore a sphere minus a point (the cusp) with two conical points of respective angles π and $2\pi/3$. See Fig. 1.2.

The modular group is naturally accompanied by the *congruence subgroups* of $\mathrm{SL}(2, \mathbf{Z})$

$$\Gamma(N) = \{\gamma \in \mathrm{SL}(2, \mathbf{Z}) \mid \gamma \equiv I \pmod{N}\}.$$

They are of finite index. The associated quotients $X(N) = \Gamma(N)\backslash\mathcal{H}$ ($N \geqslant 1$) are finite covers of $X(1)$ called *congruence covers*. For $N \geqslant 2$, we shall see that the

---

[5]Figure 1.3 was realized by McMullen, see http://www.math.harvard.edu/~ctm/gallery/index.html.

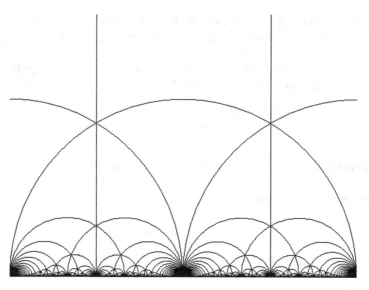

Fig. 1.3 Tesselation of $\mathcal{H}$ by translates of D

surfaces $X(N)$ are truly hyperbolic surfaces whose genus grows roughly like N^3 as N grows to infinity. Just like the modular group, the groups $\Gamma_{a,b}$ also contain congruence subgroups $\Gamma_{a,b}(N)$, each of finite index. Congruence subgroups are the heroes of this text.

1.4.2   The Spectral Theorem

In Chap. 3, which is completely independent of Chap. 2, we shall be interested in the spectral decomposition of the Laplacian on compact hyperbolic surfaces. More precisely, the goal of this chapter is to provide the proof of the following theorem.

Theorem 1.6 (Spectral theorem) *Let S be a compact hyperbolic surface. The spectral problem*

$$\Delta\varphi = \lambda\varphi$$

admits a complete orthonormal system of C^∞ eigenfunctions $\varphi_0, \varphi_1, \ldots$ in $L^2(S)$ with corresponding eigenvalues

$$0 = \lambda_0 < \lambda_1 \leqslant \lambda_2 \leqslant \ldots, \quad \lambda_n \longrightarrow \infty \text{ as } n \longrightarrow \infty.$$

This result is rather clearly the basis of the entire theory. The proof that we propose is far from original; in a sense it is intermediate between those given in the books of Buser [24] and Iwaniec [63], from which we once again borrow many ideas.

The spectral theorem allows one to formulate the conjecture – due to Selberg [116], see also Vignéras [134] – that will occupy us for a large part of this book:

Conjecture 1.7 *For every* $N \geqslant 1$,

$$\lambda_1(\Gamma_{a,b}(N)\backslash\mathcal{H}) \geqslant 1/4.$$

At the end of Chap. 3 we explain the overall geometric tenor of such lower bounds; what we find will be quite surprising. The explanation of the precise bound $1/4$ shall come later!

1.4.3 Maaß Forms

In the case of the non-compact surfaces $X(N)$, one can formulate the spectral problem in the following way. We are looking for non-zero solutions $\phi : \mathcal{H} \to \mathbf{C}$ to the spectral problem

$$\Delta\phi = \lambda\phi$$

$$\phi(\gamma z) = \phi(z), \ \gamma \in \Gamma(N) \tag{1.23}$$

$$\int_{X(N)} |\phi(z)|^2 d\mu(z) < +\infty.$$

Contrary to the case of the torus, the solutions to the problem (1.23) are not at all explicit. The same can be said of the associated eigenvalues. In the case of a compact surface, however, we do have an existence theorem. This follows essentially from the spectral theorem, and so it is not constructive.

In the case of a non-compact but finite area surface, Theorem 1.6 is no longer applicable. In Chap. 4 we begin by establishing its analog for the modular surface $X(1)$; a new feature here is the presence of a continuous spectrum. The eigenvalues appearing in the latter do not literally correspond to solutions to (1.23). We already encountered an example of a continuous spectrum when we interpreted Fourier inversion as the spectral resolution of the Laplacian on $\mathbf{R}^n$.

One cannot deduce from this spectral theorem the existence of non-constant solutions to the problem (1.23). It could very well happen that the entire spectrum is continuous. Nevertheless, a recent idea of Lindenstrauss and Venkatesh [83] allows one to deduce rather easily from the spectral theorem the existence of an infinite number of solutions of (1.23) for the modular surface $X(1)$. This theorem, proved for the first time by Selberg using more involved techniques, brings to the fore the arithmetic nature of the modular surface.

A solution to problem (1.23) is called a *Maaß form* in honor of Hans Maaß who was the first to construct explicit solutions to problem (1.23). By all appearances analytic objects, Maaß forms are every bit as arithmetic as their cousins the *holomorphic modular forms*.[6] On the other hand, at present we know of no single explicit Maaß form for $X(1)$. It is one of the fundamental problems in the theory; see [80] for some recent results.

We conclude Chap. 4 with a result – a special case of a more general theorem of Maaß – which sheds light on the arithmetic nature of certain Maaß forms. This result can be thought of as providing a construction of explicit solutions to problem (1.23) for the surface $X(8)$:

Theorem 1.8 *For every integer $k \geqslant 1$, there exists a solution to the spectral problem* (1.23) *for $N = 8$ and*

$$\lambda = \frac{1}{4} + \left(\frac{\pi k}{\log(1 + \sqrt{2})} \right)^2.$$

1.4.4 The Selberg Trace Formula

To prove his general theorem on the existence of Maaß forms mentioned earlier, Selberg developed an analog of the Poisson formula for the group $\mathrm{SL}(2, \mathbf{R})$: the famous Selberg trace formula. This is the main result of Chap. 5. We shall give the proof in the case of compact surfaces and then in the case of the modular surface. We deduce from it two applications, one of which is the *prime geodesic theorem*.

Theorem 1.9 *For a positive integer N let $\pi_N(x)$ be the number of prime closed geodesics of norm*[7] *at most x on $X(N)$. Then*

$$\pi_N(x) \sim \frac{x}{\log(x)}$$

as x goes to infinity.

A closed geodesic on $X(N)$ is the projection of a hyperbolic line $d \subset \mathcal{H}$ which is preserved by a non-trivial element $\gamma \in \Gamma(N)$. Such an element is necessarily

---

[6]There are quite a few works dedicated to the holomorphic theory [70, 92, 118]. Although many ideas in this text were originally developed in the context of holomorphic modular forms and are simply extended here to Maaß forms, we wanted to focus our attention on analytic questions and have therefore not recalled – even briefly – this beautiful theory.

[7]The norm of a closed geodesic is the exponential of its length. The prime closed geodesics are the closed geodesics which are traced out just once.

hyperbolic of axis d. Hence there exists $\sigma \in G$ such that

$$\sigma \gamma \sigma^{-1} \subset \left\{ \begin{pmatrix} t & 0 \\ 0 & t^{-1} \end{pmatrix} \,\Big|\, t \in \mathbf{R}^* \right\}.$$

As $\sigma \Gamma(N) \sigma^{-1}$ is discrete, it must be of the form

$$\left\{ \begin{pmatrix} t_0^m & 0 \\ 0 & t_0^{-m} \end{pmatrix} \,\Big|\, m \in \mathbf{Z} \right\}.$$

The subgroup of $\Gamma(N)$ made up of hyperbolic elements preserving d is therefore isomorphic to $\mathbf{Z}$ and generated by an element of norm t_0^2. Note that $2|\log t_0|$ is equal to the (hyperbolic) length of the projection of d on $X(N)$. Such an element is called *primitive*. In particular, γ is primitive if it cannot be written as a non-trivial power – different from ± 1 – of a hyperbolic element of $\Gamma(N)$.

Two conjugate hyperbolic elements in $\Gamma(N)$ have axes projecting onto the same closed geodesic in $X(N)$. The multiset of norms of prime closed geodesics in $X(N)$ is therefore equal to the multiset of norms $N(\gamma_0)$, where $\{\gamma_0\}$ runs through the set of conjugacy classes of primitive hyperbolic elements in $\Gamma(N)$.

The norms $N(\gamma_0)$ have the same asymptotic behavior as the prime numbers,[8] namely

$$\pi_N(x) = |\{\{\gamma_0\} \mid N(\gamma_0) \leqslant x\}| \sim \int_2^x \frac{dt}{\log t}.$$

Many mathematicians have had a hand in establishing a good error term for the asymptotic. Among them we mention just a few: Huber [59], Hejhal [53, 54], Venkov [131] and Kuznetsov [72] (sometimes for difference groups). Analogously to the Riemann hypothesis we expect that

$$\pi_N(x) = \int_2^x \frac{dt}{\log t} + O_\varepsilon(x^{1/2+\varepsilon}), \tag{1.24}$$

as $x \to +\infty$. The role of the Riemann zeta function is in this context played by the Selberg zeta function

$$Z_N(s) = \prod_{\gamma_0} \prod_{k=0}^{+\infty} \frac{1}{1 - N(\gamma_0)^{-s-k}}$$

---

[8]We sketch a proof of the prime number theorem in Chap. 4. We refer the reader to the book of Titchmarsh [129] for a complete proof of this theorem, its link with the Riemann zeta function and the Riemann hypothesis.

which converges absolutely in the half-plane Re(s) > 1; see Selberg [116] or Iwaniec [63]. It is well-known that the analogous asymptotic behavior to (1.24) for prime numbers follows from the Riemann hypothesis on the zeros of the Riemann zeta function. This feature does not continue to hold here (see [62] for a detailed discussion of the problems one encounters). We shall nevertheless show that the analog of the Riemann hypothesis for the Selberg zeta function would imply

$$\pi_N(x) = \int_2^x \frac{dt}{\log t} + O(x^{3/4}/\log x). \tag{1.25}$$

1.4.5 The Selberg Eigenvalue Conjecture

The Riemann hypothesis for the Selberg zeta function is in fact equivalent to the non-existence of small non-zero Laplacian eigenvalues. It is better known in the following form, and it is this statement which motivates a large part of this work.

Conjecture 1.10 (Selberg) *For every $N \geqslant 1$,*

$$\lambda_1(X(N)) \geqslant \frac{1}{4}.$$

Roelcke [105] has verified the conjecture for $N = 1$; see Theorem 3.38.

Selberg stated his conjecture in [116], motivated by questions on the cancellation in certain arithmetic exponential sums introduced earlier, and in another context, by Hendrik Kloosterman. Selberg showed that the absence of small eigenvalues gives rise to non-trivial cancellation in these sums.

These questions are at the heart of most arithmetic applications of the spectral theory of hyperbolic surfaces. In fact, every result in the direction of the Selberg conjecture is *useful* and is not simply to be viewed as abstract evidence in support of the conjecture. One can find several illustrations of this phenomenon in Sarnak's article [112].

Selberg himself showed [116] the following approximation to his conjecture.

Theorem 1.11 (Selberg) *For every $N \geqslant 1$,*

$$\lambda_1(X(N)) \geqslant \frac{3}{16}.$$

We shall see in Chap. 6 the following theorem. While it gives a weaker lower bound than that of Selberg, it has the conceptual advantage of being essentially geometric.

Theorem 1.12 *For every sufficiently large prime number p,*

$$\lambda_1(X(p)) \geqslant \frac{5}{36}.$$

The proof of the above theorem is based on an idea of Kazhdan. It was subsequently developed by Sarnak and Xue [113] in the case of arithmetic compact surfaces and then by Gamburd [42] for the covers $X(p)$ of the modular surface.

Given a positive integer N, we put

$$\Gamma_0(N) = \left\{ \begin{pmatrix} a & b \\ c & d \end{pmatrix} \in SL(2, \mathbf{Z}) \mid N \mid c \right\}.$$

In light of the link between arithmetic and Maaß forms, we shall come back to the Selberg conjecture in Chap. 7 via the study of L-functions associated to Maaß forms. We show:

Theorem 1.13 *For every square free integer $N \geqslant 1$, we have*

$$\lambda_1(\Gamma_0(N) \backslash \mathcal{H}) \geqslant \frac{5}{36}.$$

The proof, due to Luo, Rudnick and Sarnak [84], is analytic. We restrict ourselves to congruence subgroups of the above form (including the square free assumption on N) to simplify the exposition, but the argument of [84] is general enough to apply to any of the congruence surfaces $Y(N)$. It can moreover be adapted to the setting of automorphic representations of $GL(n)$. This method is at the heart of the more recent proof by Kim and Sarnak [68] of the best known bound towards the Selberg eigenvalue conjecture, namely $\lambda_1 \geqslant \frac{1}{4} - \left(\frac{7}{64}\right)^2$.

1.4.6 *The Jacquet-Langlands Correspondence*

Chapter 8 is dedicated to the proof of the surprising correspondence – due to Jacquet and Langlands [65] – between the spectra of different arithmetic surfaces. We shall present the proof in a special case and in classical language, following the work of Bolte and Johansson [12].

Let a and b be positive integers such that the quotient $\Gamma_{a,b} \backslash \mathcal{H}$ is compact. Consider a discrete subgroup $\tilde{\Gamma}_{a,b} \subset G$ containing $\Gamma_{a,b}$ and maximal for this property. The Jacquet-Langlands correspondence [65] implies, as a special case, the following theorem.

Theorem 1.14 *The set of Laplacian eigenvalues in $L^2(\tilde{\Gamma}_{a,b} \backslash \mathcal{H})$ is contained in the set of eigenvalues associated with Maaß forms for a group $\Gamma_0(N)$ for some positive square free integer N.*

In particular, this allows us to prove the following approximations to the Selberg eigenvalue conjecture.

Theorem 1.15 *We have*

$$\lambda_1(\tilde{\Gamma}_{a,b}\backslash\mathcal{H}) \geqslant \frac{5}{36}.$$

By working with adelic groups one can generalize Theorem 1.14 to an arbitrary congruence subgroup of $\Gamma_{a,b}$, see [43]. This allows one to transport the theorem of Kim and Sarnak to the groups $\Gamma_{a,b}(N)$.

1.4.7 Arithmetic Quantum Unique Ergodicity

The behavior of Maaß forms ϕ_λ as $\lambda \to +\infty$ is especially interesting, for one begins to broach questions in mathematical physics. We conclude this book with a description of the recent breakthroughs concerning this problem in the case of arithmetic surfaces.

1.5 Notation

Throughout this text we shall use the following notation: given a set X and two functions f and g on X, we write $f(x) = O(g(x))$ for $x \in Y \subset X$ if there exists a constant $c > 0$ such that, for every $x \in Y$, one has $|f(x)| \leqslant cg(x)$. If the set Y is not specified, it is taken to be X by default. If the constant c depends on auxiliary parameters, for example ε and τ, we indicate this as a subscript: $f(x) = O_{\varepsilon,\tau}(g(x))$.

If $x \to x_0$, the notation

$$f(x) \sim g(x)$$

means that the limit

$$\lim_{x \to x_0} \frac{f(x)}{g(x)}$$

is well-defined and equal to 1.

Given two integers n and m the notation $n|m$ means that n divides m. We write (n, m) for the greatest common divisor of n and m.

We adopt the usual notation $\mathcal{M}_n(K)$ for the space of $n \times n$ matrices over a field K and I_n (or simply I if $n = 2$) for the identity matrix in $\mathcal{M}_n(K)$. The symbols $\mathrm{GL}(n, K)$ and $\mathrm{SL}(n, K)$ have their standard meaning.

Finally, we denote by $|A|$ the cardinality of a set A.

1.6 Remarks and References

Most of the references are grouped at the end of each chapter in a section entitled "Remarks and References". When a result is stated without proof in the body of the text, the references where one can find the proof are given in the same section. One will also find there more specialized commentary and general statements of theorems which are stated and proved in special cases in the body of the text.

We note finally that certain proofs – sometimes entire paragraphs – are typeset in small letters. These passages are not strictly necessary for the rest of the text and are more difficult in nature.

1.7 Exercises

Exercise 1.16 (Poisson summation formula and the Riemann zeta function)
The zeta function $\zeta(s)$ is defined on $\mathrm{Re}(s) > 1$ by

$$\zeta(s) = \sum_{n=1}^{\infty} n^{-s} = \prod_{p} (1 - p^{-s})^{-1}.$$

Here the product runs over all prime numbers and the equality is equivalent to the uniqueness of the prime factorization of an integer.

1. By applying the Poisson summation formula to f_y show that for every $y > 0$

$$\sum_{n \in \mathbf{Z}} e^{-\pi n^2 y} = \frac{1}{\sqrt{y}} \sum_{n \in \mathbf{Z}} e^{-\pi n^2/y}.$$

2. Show that for $\mathrm{Re}(s) > 1$,

$$\int_0^{+\infty} \left(\sum_{n \in \mathbf{Z}} e^{-\pi n^2 y} \right) y^{s/2} \frac{dy}{y} = \xi(s) := \pi^{-s/2} \Gamma(s/2) \zeta(s),$$

where Γ denotes the Euler *Gamma function*:

$$\Gamma(s) = \int_0^{+\infty} e^{-x} x^s \frac{dx}{x}. \tag{1.26}$$

The latter is first defined on $\mathrm{Re}(s) > 0$ but admits a meromorphic continuation to all of **C**. See Appendix B for a brief summary of these basic properties.

3. Deduce that the function $s \mapsto \xi(s)$ admits a meromorphic continuation to all of $\mathbf{C}$ with two simple poles at $s = 0$ and $s = 1$ and that it satisfies the functional equation

$$\xi(s) = \xi(1 - s). \tag{1.27}$$

This last exercise is non-trivial, but it is worthwhile to try solving it without help. In case of difficulty, the reader can consult the proof of Theorem 4.2, which contains the answer.

Exercise 1.17 (The hyperbolic disk) Let $\mathbf{D} = \{z \in \mathbf{C} \mid |z| < 1\}$ be the open unit disk in $\mathbf{C}$. We endow the disc $\mathbf{D}$ with the *hyperbolic* Riemannian metric

$$ds = \frac{|dz|}{1 - |z|^2}.$$

As in the case of the half-plane, this defines a distance dist_{hyp}, that we again call *hyperbolic*.

1. If $\alpha \in \mathbf{R}$ and $a \in \mathbf{D}$, show that the transformation

$$f_{\alpha,a} : z \longmapsto \exp(i\alpha) \frac{z - a}{1 - \bar{a}z}$$

preserves the disk $\mathbf{D}$.
2. Show that the set of transformations $f_{\alpha,a}$ forms a group acting transitively on $\mathbf{D}$.
3. Check that

$$\frac{|df_{\alpha,a}(z)|}{1 - |f_{\alpha,a}(z)|^2} = \frac{|dz|}{1 - |z|^2};$$

in other words, check that the hyperbolic metric is invariant under the group of transformations $f_{\alpha,a}$.

Exercise 1.18 (Geodesics in the disk model)

1. Let $\gamma : [0, 1] \to \mathbf{D}$ be a piecewise C^1 curve joining $0 \in \mathbf{D}$ to a point $r \in [0, 1[\subset \mathbf{D}$. Show that the radial projection $|\gamma| : [0, 1] \to [0, 1[\subset \mathbf{D}$ is again a piecewise C^1 curve joining 0 to r and that its hyperbolic length is no greater than that of γ.
2. Deduce from the first question that the unique curve minimizing the hyperbolic length between 0 and r is the ray $[0, r]$ of length $\text{Argth}(r) = \frac{1}{2} \log \left(\frac{1+r}{1-r} \right)$.
3. Recall that a homography

$$z \in \mathbf{C} \cup \{\infty\} \longmapsto \frac{az + b}{cz + d} \in \mathbf{C} \cup \{\infty\}, \quad ad - bc \neq 0,$$

sends a circle to a circle in the Riemann sphere $\mathbf{C} \cup \{\infty\}$ and preserves the intersection angle between two circles. Deduce that the geodesics of $\mathbf{D}$ are precisely the arcs of circles orthogonal to the unit circle or the diameters.

Exercise 1.19 (Hyperbolic distance and cross-ratio) Recall that the cross-ratio of four distinct point x, y, z, t in $\mathbf{C} \cup \{\infty\}$ is defined as

$$[x : y : z : t] = \frac{z - x}{z - y} \frac{t - y}{t - x}$$

and that for any homography f, we have

$$[f(x) : f(y) : f(z) : f(t)] = [x : y : z : t].$$

1. Deduce from the previous exercise that

$$\mathrm{dist}_{\mathrm{hyp}}(0, r) = \frac{1}{2} |\log[-1 : 1 : 0 : r]|.$$

2. More generally, show that if z_0 and z_1 are two distinct points in $\mathbf{D}$, there exists a unique geodesic passing through z_0 and z_1, and that if u and v are the endpoints of this geodesic on the unit circle we have

$$\mathrm{dist}_{\mathrm{hyp}}(z_0, z_1) = \frac{1}{2} |\log[u : v : z_0 : z_1]|.$$

Exercise 1.20 (From the disk to the half-plane) Show that the transformation $z \mapsto \frac{1}{z-i}$ sends the disk $\mathbf{D}$ biholomorphically to the half-plane $\mathcal{H}$, that it transports the hyperbolic metric on $\mathbf{D}$ to the hyperbolic metric on $\mathcal{H}$, and that it conjugates (in the group of all homographies) the group of transformations $f_{\alpha,a}$ to the group $\mathrm{PSL}(2, \mathbf{R})$.

Exercise 1.21 Show that the group generated by the symmetry $z \mapsto -\bar{z}$ and all of the real homographies is the group of all isometries of $\mathcal{H}$.

Exercise 1.22 Show that the geodesics in $\mathcal{H}$ are precisely the hyperbolic lines introduced at the beginning of § 1.2.

Exercise 1.23

1. Check directly, with the help of (1.9), that the measure $d\mu$ is invariant under the action of the group G.
2. Show that such an invariant measure is unique up to scalar multiples.

Exercise 1.24 (Area of hyperbolic triangles) By a triangle in $\mathcal{H}$ or $\mathbf{D}$ we understand three distinct points joined by three geodesic segments. Since the two models $\mathcal{H}$ and $\mathbf{D}$ of the hyperbolic plane are *conformal* with the Euclidean metric $|dz|$, the hyperbolic angles are the same as the Euclidean angles.

1. Show that every *ideal* triangle in $\mathcal{H}$ or $\mathbf{D}$ (those having all three vertices on $\mathbf{R} \cup \{\infty\}$ or on the unit circle) has area π.

2. Consider a triangle $T(\alpha)$ having one angle equal to $\alpha \in [0, \pi]$ and whose two other vertices are at infinity (lying in $\mathbf{R} \cup \{\infty\}$ or on the unit circle). Show that

$$\text{area}(T(\alpha + \beta)) = \text{area}(T(\alpha)) + \text{area}(T(\beta)) - \pi.$$

3. Let $F(\alpha) = \pi - \text{area}(T(\alpha))$. Show that there exists a constant c such that $F(\alpha) = c\alpha$ ($\alpha \in [0, \pi]$). Deduce from the value of $F(\pi)$ that for every $\alpha \in [0, \pi]$ one has

$$\text{area}(T(\alpha)) = \pi - \alpha.$$

4. Now consider a triangle $T(\alpha, \beta, \gamma)$ whose three vertices are properly in the hyperbolic plane $\mathcal{H}$ or $\mathbf{D}$. By extending the sides to the boundary at infinity ($\mathbf{R} \cup \{\infty\}$ or the unit circle), one obtains an ideal hexagon. Each vertex of the triangle determines two isometric copies of $T(\alpha)$, $T(\beta)$, $T(\gamma)$, respectively. Deduce that the area of the hexagon is

$$2[(\pi - \alpha) + (\pi - \beta) + (\pi - \gamma)] - 2\,\text{area}(T(\alpha, \beta, \gamma)).$$

By cutting up this hexagon into ideal triangles, deduce finally that

$$\text{area}(T(\alpha, \beta, \gamma)) = \pi - (\alpha + \beta + \gamma).$$

Exercise 1.25 (Area of hyperbolic polygons) If P is a polygon in $\mathcal{H}$ with n sides and angles $\alpha_1, \ldots, \alpha_n$ we put

$$\mathcal{A}(P) = (n - 2)\pi - \sum_{i=1}^{n} \alpha_i.$$

1. Show that $\mathcal{A}$ is invariant under isometries and that if we cut up P into two polygons P_1 and P_2 along a geodesic, we have

$$\mathcal{A}(P) = \mathcal{A}(P_1) + \mathcal{A}(P_2).$$

2. Show that $\mathcal{A}(P)$ goes to 0 as the diameter of P goes to 0.
3. By mimicking the construction of the Lebesgue measure in the plane, show that $\mathcal{A}$ defines a measure in the hyperbolic plane which is invariant under isometries.
4. From the uniqueness of such a measure show that there is a constant $c > 0$ such that

$$\text{area}(P) = c\left(((n - 2)\pi - \sum_{i=1}^{n} \alpha_i)\right).$$

Exercise 1.26 (Discrete subgroups of $PSL(2, \mathbf{R})$**)** Show that a subgroup $\Gamma \subset$ $PSL(2, \mathbf{R})$ is discrete if and only if $\gamma_n \to I$ $(\gamma_n \in \Gamma)$ implies $\gamma_n = I$ for sufficiently large n.

Exercise 1.27 (Proper discontinuous actions) Let Γ be a subgroup of $PSL(2, \mathbf{R})$. Show that the following statements are equivalent.

1. The group Γ acts properly discontinuously on $\mathcal{H}$.
2. For every compact set $K \subset \mathcal{H}$, the set $\{\gamma \in \Gamma \mid \gamma(K) \cap K \neq \varnothing\}$ is finite.
3. Every point x in $\mathcal{H}$ admits a neighborhood V such that if $\gamma(V) \cap V \neq \varnothing$ then $\gamma(x) = x$.

Exercise 1.28 (Fuchsian cyclic groups)

1. Show that every cyclic group generated by a hyperbolic or parabolic transformation is Fuchsian.
2. Show that a cyclic group generated by an elliptic element is Fuchsian if and only if it is finite.

Exercise 1.29 Show that if Γ is a Fuchsian group then it acts properly discontinuously on $\mathcal{H}$.

Exercise 1.30 Show that the polygon D in Fig. 1.3 is a fundamental domain for the group $PSL(2, \mathbf{Z})$, that i is the unique elliptic fixed point of order 2 and that $\zeta = (1 + i\sqrt{3})/2$ and $\zeta + 1$ are the only elliptic fixed points of order 3.

Chapter 2
Arithmetic Hyperbolic Surfaces

As we recalled in the introduction, there are several ways to construct Fuchsian groups of the first kind. Of all of these groups, the most important from the number theoretic viewpoint are the arithmetic groups. The general definition of these groups is a bit technical so we content ourselves for the moment with describing a family of examples: the arithmetic groups coming from a quaternion algebra over $\mathbf{Q}$. Before describing them, we begin by proving several general results concerning the space of lattices.

2.1 The Space of Lattices

A *lattice* in $\mathbf{R}^n$ is a discrete subgroup isomorphic to $\mathbf{Z}^n$. In this section we study the set $\mathcal{L}_n$ of lattices in $\mathbf{R}^n$. We endow $\mathbf{R}^n$ with the Euclidean norm $|\cdot|$ for which the canonical basis $(e_1, \ldots, e_n)$ is orthonormal.

We define two fundamental invariants of a lattice $L \in \mathcal{L}_n$:

1. its *height* $H = H(L) = \min_{v \in L - \{0\}}\{|v|\}$, and
2. its *volume* $V = V(L)$, i.e., the Euclidean volume of the paralleletope subtended by a basis of L.

The group $\mathrm{GL}(n, \mathbf{R})$ acts transitively (on the left) on the set $\mathcal{L}_n$. Denote by $L_0 = \mathbf{Z}^n$ the standard lattice in $\mathbf{R}^n$ generated by the canonical basis. A lattice L generated by n (linearly independent) vectors $v_1, \ldots, v_n$ is the image of L_0 under the action of the invertible matrix whose columns are the v_i. The stabilizer of the $\mathbf{Z}$-module L_0 in $\mathrm{GL}(n, \mathbf{R})$ is the subgroup $\mathrm{GL}(n, \mathbf{Z})$ consisting of matrices that are invertible in $\mathcal{M}_n(\mathbf{Z})$. Note that the determinant of such a matrix is necessarily invertible in $\mathbf{Z}$ and is therefore equal to ± 1. Finally, the set $\mathcal{L}_n$ is naturally identified with the quotient $\mathrm{GL}(n, \mathbf{R})/\mathrm{GL}(n, \mathbf{Z})$. This identification induces a topology on $\mathcal{L}_n$: the quotient topology. By definition, a sequence L_m of lattices in $\mathbf{R}^n$ converges, in the quotient

© Springer International Publishing Switzerland 2016
N. Bergeron, *The Spectrum of Hyperbolic Surfaces*, Universitext,
DOI 10.1007/978-3-319-27666-3_2

topology, towards a lattice L of $\mathbf{R}^n$ if and only if there exists a basis $(f_1^m, \ldots, f_n^m)$ of L_m which converges towards a basis $(f_1, \ldots, f_n)$ of L. The lattice L_0 corresponds to the identity class in $\mathrm{GL}(n, \mathbf{R})/\mathrm{GL}(n, \mathbf{Z})$; it will serve as a convenient base point for the space $\mathcal{L}_n$. Recall that the determinant of n vectors $v_1, \ldots, v_n$ relative to the canonical basis is equal to the Euclidean volume of the paralleletope they subtend. Thus, if $L = g(L_0)$ with $g \in \mathrm{GL}(n, \mathbf{R})$, we have $V(L) = |\det g|$.

The set $\mathcal{L}_n$ is not compact: the volume can explode or the height can tend towards 0. The following theorem tells us that these are the only two ways that a family of lattices can degenerate.

Theorem 2.1 (Hermite-Mahler criterion) *A subset $M \subset \mathcal{L}_n$ is relatively compact if and only if there exist constants $\varepsilon > 0$ and $C > 0$ such that*

$$\begin{cases} H > \varepsilon \\ V < C \end{cases}$$

on M.

The main idea of the proof of Theorem 2.1 is to bring M into a fundamental domain for the action of $\mathrm{GL}(n, \mathbf{Z})$ on $\mathrm{GL}(n, \mathbf{R})$, and to study the ways of degenerating in this fundamental domain. When $n = 2$ the construction of such a fundamental domain can be deduced from that of the modular group acting on $\mathcal{H}$. In higher dimensions, the explicit knowledge of a fundamental domain is replaced by the Siegel sets that we now construct.

For g in $\mathrm{GL}(n, \mathbf{R})$, we write g_{ij} for the matrix entries of g. Let

$$K := O(n) = \{g \in \mathrm{GL}(n, \mathbf{R}) \mid {}^t g g = 1\},$$

$$A := \{g \in \mathrm{GL}(n, \mathbf{R}) \mid g \text{ is diagonal with positive entries}\},$$

$$A_s := \{a \in A \mid a_{ii} \leqslant s a_{i+1, i+1} \text{ for } i = 1, \ldots, n-1\} \quad \text{with } s \geqslant 1,$$

$$N := \{g \in \mathrm{GL}(n, \mathbf{R}) \mid g - I_n \text{ is strictly upper triangular}\} \text{ and,}$$

$$N_t := \{n \in N \mid |n_{ij}| \leqslant t \text{ for } 1 \leqslant i < j \leqslant n\} \quad \text{with } t \geqslant 0.$$

According to the *Iwasawa decomposition* of $\mathrm{GL}(n, \mathbf{R})$ – which boils down to the orthonormalization procedure of Schmidt – the map

$$(k, a, n) \longmapsto kan$$

is a homeomorphism of $K \times A \times N$ onto $\mathrm{GL}(n, \mathbf{R})$. We denote by $S_{s,t}$ the *Siegel domain*

$$S_{s,t} = KA_s N_t$$

and

$$\Gamma = \text{SL}(n, \mathbf{Z}).$$

One knows that N is a closed subgroup of $\text{GL}(n, \mathbf{R})$, homeomorphic to $\mathbf{R}^m$ ($m = n(n-1)/2$) by the map $\theta : n \mapsto (n_{ij})_{1 \leqslant i < j \leqslant n}$; as a result, N_t is compact.

We would like to prove the following

Lemma 2.2 *One has*

$$\text{GL}(n, \mathbf{R}) = S_{s,t}\Gamma$$

as soon as $s \geqslant 2/\sqrt{3}, t \geqslant 1/2$.

Proof Let g be in $\text{GL}(n, \mathbf{R})$ and $L = g(L_0)$. We shall proceed by induction on n.

We begin by remarking that given a linear subspace V of Euclidean space $\mathbf{R}^n$, the quotient space $\mathbf{R}^n/V$ can be identified with the orthogonal of V in $\mathbf{R}^n$ and, as such, is itself naturally a Euclidean vector space. One can then define, by induction on n, the notion of an admissible family of vectors in L. This is a family $(f_1, \ldots, f_n)$ of vectors in L such that

- f_1 is a vector in $L - \{0\}$ of minimal norm,
- the images $\dot{f}_2, \ldots, \dot{f}_n$ of $f_2, \ldots, f_n$ in the lattice $\dot{L} := L/\mathbf{Z}f_1$ of the Euclidean space $\mathbf{R}^n/\mathbf{R}f_1$ form an admissible family of $\dot{L}$, and
- the vectors f_i are of minimal norm among the vectors of L whose image in $\dot{L}$ is $\dot{f}_i$.

(When $n = 1$, we pay no attention to the last two conditions.)

It is a standard fact that the $\mathbf{Z}$-module L admits an admissible family $(f_1, \ldots, f_n)$ and that this family is in fact a $\mathbf{Z}$-basis. Multiplying g on the right by an element of Γ, if necessary, we can assume that for every $i = 1, \ldots, n$ we have $ge_i = f_i$.

We show by induction on n that if $g \in \text{GL}(n, \mathbf{R})$ sends the canonical basis of $\mathbf{R}^n$ to an admissible family of vectors of the lattice $g(L_0)$ then $g \in S_{2/\sqrt{3}, 1/2}$.

Write $g = kan$. As $(k^{-1}f_1, \ldots, k^{-1}f_n)$ is an admissible basis of $k^{-1}L$, we can assume that $k = I_n$. One then has $g = an$, so that

$$f_1 = a_{11}e_1$$
$$f_2 = a_{22}e_2 + a_{11}n_{12}e_1$$
$$\cdots$$
$$f_i = a_{ii}e_i + a_{i-1,i-1}n_{i-1,i}e_{i-1} + \cdots + a_{11}n_{1i}e_1.$$

Denote by $\mathbf{R}^{n-1}$ the subspace of $\mathbf{R}^n$ orthogonal to $\mathbf{R}f_1 = \mathbf{R}e_1$. This is the Euclidean subspace of $\mathbf{R}^n$ generated by $e_2, \ldots, e_n$. The linear transformation $g : \mathbf{R}^n \to \mathbf{R}^n$ defines on the quotient by $\mathbf{R}f_1$ a linear transformation $\bar{g} : \mathbf{R}^{n-1} \to \mathbf{R}^{n-1}$ which, by the definition of an admissible family, sends the canonical basis $(e_2, \ldots, e_n)$ to an admissible family of the quotient lattice $L/\mathbf{Z}f_1$ obtained by the orthogonal projection of L onto $\mathbf{R}^{n-1}$. By the induction hypothesis, we therefore have

$$|n_{ij}| \leqslant 1/2 \quad \text{for } 2 \leqslant i < j \leqslant n$$

and

$$a_{ii} \leqslant 2/\sqrt{3}\, a_{i+1,i+1} \quad \text{for } 2 \leqslant i \leqslant n-1.$$

It remains to show

$$|n_{1j}| \leqslant 1/2 \quad \text{for } 2 \leqslant j \leqslant n \text{ and } a_{11} \leqslant 2/\sqrt{3}a_{22}.$$

The first inequality is a consequence of

$$|f_j|^2 \leqslant |f_j + p f_1|^2, \quad \forall p \in \mathbf{Z}.$$

By subtracting the left-hand side from the expanded right-hand side and factoring out by a_{11}^2 (which is non-zero!), we deduce that for all $p \in \mathbf{Z}$,

$$p^2 + 2n_{1,j}p \geqslant 0.$$

This forces n_{1j} to be less than $1/2$ in absolute value.

The second inequality comes from $|f_1|^2 \leqslant |f_2|^2$, as this, when written out, is $a_{11}^2 \leqslant a_{22}^2 + a_{11}^2 n_{12}^2 \leqslant a_{22}^2 + 1/4 a_{11}^2$. $\square$

Proof of Theorem 2.1 Fix $s \geqslant 2/\sqrt{3}$ and $t \geqslant 1/2$. It is clear that a subset M of $\mathcal{R}$ is relatively compact if and only if there exists a compact subset S of $S_{s,t}$ such that $M \subset S \cdot L_0 := \{g L_0 \mid g \in S\}$.

We first show the forward direction of the equivalence stated in Theorem 2.1. Let us fix $0 < r < R$ such that for every $g = kan$ in S and for every $i = 1, \ldots, n$ we have $r \leqslant a_{ii} \leqslant R$. Then

$$|\det g| = \prod_{i=1}^{n} a_{ii} \leqslant R^d$$

and

$$\min_{v \in \mathbf{Z}^n - \{0\}} |gv| \geqslant r,$$

for if $v = \sum_{i=1}^{\ell} m_i e_i \in \mathbf{Z}^n$ with $m_\ell \neq 0$, then

$$|gv| \geqslant |\langle k e_\ell, gv \rangle| = |\langle e_\ell, anv \rangle| = a_{\ell\ell}|m_\ell| \geqslant r.$$

We now show the converse direction. Let $S = \{g \in S_{s,t} \mid gL_0 \in \overline{M}\}$. It follows from Lemma 2.2 that $\overline{M}$ is contained in $S \cdot L_0$. But for every $g = kan$ in S we have

- $a_{11} = |ge_1| \geqslant \varepsilon$,
- $a_{ii} \leqslant sa_{i+1,i+1}$ for $i = 1, \ldots, n-1$,
- $\prod_{i=1}^{n} a_{ii} \leqslant C$.

We deduce that there exists $R > r > 0$ such that, for every $g = kan$ in S and for all $i = 1, \ldots, n$, we have $r \leqslant a_{ii} \leqslant R$. Thus S is compact, as is $\overline{M}$.   □

2.2 Quaternion Algebras and Arithmetic Groups

Let F be an arbitrary field and let a and b be two non-zero elements of F. The corresponding *quaternion algebra* over F is the ring

$$D_{a,b}(F) = \{x_0 + x_1 i + x_2 j + x_3 k \mid x_0, \ldots, x_3 \in F\},$$

where

- addition is defined in the obvious way, so as to form a vector space of dimension 4 over F;
- the map $x \mapsto x + 0i + 0j + 0k$ is an (injective) ring homomorphism from F into $D_{a,b}(F)$ whose image – again written F – is contained in the center of $D_{a,b}(F)$. In other words,

$$x\alpha = \alpha x, \quad \text{for all } x \in F, \ \alpha \in D_{a,b}(F);$$

- multiplication is determined by the relations

$$i^2 = a, \ j^2 = b, \ ij = k = -ji.$$

The *reduced norm* of $\alpha = x_0 + x_1 i + x_2 j + x_3 k \in D_{a,b}(F)$ is

$$N_{\mathrm{red}}(\alpha) = x_0^2 - ax_1^2 - bx_2^2 + abx_3^2. \tag{2.1}$$

The *conjugate* of α is

$$\overline{\alpha} = x_0 - x_1 i - x_2 j - x_3 k,$$

so that $N_{\mathrm{red}}(\alpha) = \alpha\overline{\alpha} = \overline{\alpha}\alpha$. One defines the *trace* of α by

$$\mathrm{tr}(\alpha) = \alpha + \overline{\alpha} = 2x_0. \tag{2.2}$$

The terminology naturally comes from the famous example $D_{-1,-1}(\mathbf{R})$ of Hamilton's quaternions. On the other hand the map $D_{1,1}(\mathbf{R}) \rightarrow \mathcal{M}_2(\mathbf{R})$, which on basis vectors is defined by

$$1 \longmapsto I_2, \quad i \longmapsto \begin{pmatrix} 1 & 0 \\ 0 & -1 \end{pmatrix}, \quad j \longmapsto \begin{pmatrix} 0 & 1 \\ 1 & 0 \end{pmatrix}, \quad k \longmapsto \begin{pmatrix} 0 & 1 \\ -1 & 0 \end{pmatrix},$$

extends in a unique way to a ring isomorphism. This will be justified at the beginning of the proof of Theorem 2.3.

One calls a quaternion algebra D a *division algebra* if every non-zero element $\alpha \in D$ admits an inverse. This happens if and only if $\mathrm{N}_{\mathrm{red}}(\alpha) \neq 0$ for all $\alpha \neq 0$, in which case $\alpha^{-1} = \overline{\alpha}/\mathrm{N}_{\mathrm{red}}(\alpha)$.

Henceforth we fix two positive integers a and b. We can then naturally speak of the subring $\mathcal{O} := D_{a,b}(\mathbf{Z})$ in $D_{a,b}(\mathbf{Q})$.

Our goal is to prove the following theorem.

Theorem 2.3

1. *There exists a group isomorphism from*

$$D_{a,b}(\mathbf{R})^1 := \mathrm{SL}(1, D_{a,b}(\mathbf{R})) = \{g \in D_{a,b}(\mathbf{R}) \mid \mathrm{N}_{\mathrm{red}}(g) = 1\}$$

   to $G = \mathrm{SL}(2, \mathbf{R})$.
2. *The image $\Gamma_{a,b}$ in G of the group $\mathcal{O}^1 = \mathrm{SL}(1, D_{a,b}(\mathbf{Z}))$ via this isomorphism is a Fuchsian group of the first kind.*
3. *The following statements are equivalent:*

   a. *$\Gamma_{a,b}$ is cocompact in G;*
   b. *$(0,0,0)$ is the unique solution in integers of the Diophantine equation $x^2 - ay^2 - bz^2 = 0$;*
   c. *$D_{a,b}(\mathbf{Q})$ is a division algebra.*

4. *Two subgroups $\Gamma_{a,b}$ and $\Gamma_{a',b'}$ are commensurable in G if and only if the quadratic forms $x^2 - ay^2 - bz^2$ and $x^2 - a'y^2 - b'z^2$ are similar over $\mathbf{Q}$.*

Two subgroups Γ and Λ in G are said to be *commensurable* in G if, conjugating Γ in G if necessary, the group $\Gamma \cap \Lambda$ is of finite index in both Γ and Λ. Recall furthermore that two quadratic forms over a field of characteristics $\neq 2$ are said to be *equivalent* if a change of basis over the field transforms one into the other and are said to be *similar* if they are equivalent up to a non-zero scalar multiple (in the base field).

According to Property 4 of Theorem 2.3, there exist infinitely many commensurability classes of discrete cocompact subgroups in G. Indeed, if p and q are two distinct prime numbers congruent to -1 modulo 4, the groups $\Gamma_{p,p}$ and $\Gamma_{q,q}$ are not commensurable, for otherwise there would exist $M \in \mathrm{GL}(3, \mathbf{Q})$ and $\lambda \in \mathbf{Q}$ such that $\Delta' = \lambda\,{}^t\!M \Delta M$, with $\Delta = \mathrm{diag}\{1, -p, -p\}$, $\Delta' = \mathrm{diag}\{1, -q, -q\}$. We can therefore assume that $\det M = \pm q/p$ (and $\lambda = 1$). Let $p^\alpha n$ be the least common multiple of the denominators of the coefficients of M; we have $p^{\alpha} n M = (m_{ij})$ with $m_{ij} \in \mathbf{Z}$, and $\alpha \geq 1$. The first equation obtained in identifying the coefficients of Δ' and $\lambda \, {}^t M \Delta M$ is

$$(np^{\alpha})^2 = m_{11}^2 - pm_{21}^2 - pm_{31}^2.$$

Thus $m_{11} \equiv 0 \pmod{p}$ and $m_{21}^2 + m_{31}^2 \equiv 0 \pmod{p}$. Since -1 is not a square modulo p, we necessarily have $m_{21} \equiv m_{31} \equiv 0 \pmod{p}$. Likewise, all of the coefficients of $p^{\alpha} n M$ are divisible by p. Therefore $p^{\alpha-1} n M \in \mathcal{M}_3(\mathbf{Z})$, in contradiction with the choice of $p^{\alpha} n$.

Proof of Theorem 2.3 The linear map $\Phi : D_{a,b}(\mathbf{R}) \to \mathcal{M}_2(\mathbf{R})$ defined on basis vectors by

$$\Phi(1) = I_2, \ \Phi(i) = \begin{pmatrix} \sqrt{a} & 0 \\ 0 & -\sqrt{a} \end{pmatrix}, \ \Phi(j) = \begin{pmatrix} 0 & 1 \\ b & 0 \end{pmatrix}, \ \Phi(k) = \begin{pmatrix} 0 & \sqrt{a} \\ -b\sqrt{a} & 0 \end{pmatrix}$$

is bijective. Moreover, Φ preserves multiplication, making it a ring isomorphism. Finally, if $\alpha = x_0 + x_1 i + x_2 j + x_3 k \in D_{a,b}(\mathbf{R})$, we have

$$\det(\Phi(\alpha)) = (x_0 + x_1\sqrt{a})(x_0 - x_1\sqrt{a}) - (x_2 + x_3\sqrt{a})(bx_2 - bx_3\sqrt{a})$$
$$= x_0^2 - ax_1^2 - bx_2^2 + abx_3^2 = \mathrm{N}_{\mathrm{red}}(\alpha).$$

Thus $\Phi(D_{a,b}(\mathbf{R})^1) = \mathrm{SL}(2, \mathbf{R})$, the first claim of the theorem is proved, and

$$\Gamma_{a,b} = \left\{ \begin{pmatrix} x_0 + x_1\sqrt{a} & x_2 + x_3\sqrt{a} \\ b(x_2 - x_3\sqrt{a}) & x_0 - x_1\sqrt{a} \end{pmatrix} \ \middle| \ \begin{array}{l} x_0, x_1, x_2, x_3 \in \mathbf{Z}, \\ x_0^2 - ax_1^2 - bx_2^2 + abx_3^2 = 1 \end{array} \right\}.$$

The proofs of the other claims will be taken up in the following subsections. $\quad\square$

2.2.1 An Exceptional Isomorphism

From now on we denote by $D_{a,b}$ the real quaternion algebra $D_{a,b}(\mathbf{R})$, remembering that this algebra is defined over $\mathbf{Q}$ (and even over $\mathbf{Z}$) and that we can therefore speak of its rational (or integral) points. Let $P = \{\alpha \in D_{a,b} \mid \mathrm{tr}(\alpha) = 0\}$ be the set of pure quaternions in $D_{a,b}$. This is a subspace of $D_{a,b}$ defined over $\mathbf{Q}$ and isomorphic to $\mathbf{R}^3$. The reduced norm restricts to P as the quadratic form $q = -ax_1^2 - bx_2^2 + abx_3^2$.

Lemma 2.4 *There is a rational isomorphism*

$$D_{a,b}(\mathbf{Q})^* / \mathbf{Q}^* \xrightarrow{\ \cong\ } \mathrm{SO}(q, \mathbf{Q})$$

which induces an isomorphism on the real groups[1]

$$D^1_{a,b}/\{\pm 1\} \longrightarrow SO_0(q).$$

Proof It suffices to construct the first isomorphism. For that we consider the action of the group $D_{a,b}(\mathbf{Q})^*$ on $D_{a,b}(\mathbf{Q})$ by interior automorphisms. For $\alpha \in D_{a,b}(\mathbf{Q})^*$ and $\beta \in D_{a,b}(\mathbf{Q})$ we write

$$S_\alpha(\beta) = \alpha\beta\alpha^{-1}.$$

We verify immediately the following properties:

- S_α preserves the reduced norm,
- $(S_\alpha)_{|\mathbf{Q}} = \mathrm{Id}_\mathbf{Q}$, and
- $S_\alpha(P) = P$.

Let $s_\alpha = (S_\alpha)_{|P}$. Then $s : D_{a,b}(\mathbf{Q})^* \to O(q)$ is a homomorphism whose kernel is $\mathbf{Q} \cap D_{a,b}(\mathbf{Q})^* = \mathbf{Q}^*$.

Before continuing we make a few preliminary calculations. For $\beta \in P$ such that $q(\beta) \neq 0$, the reflection τ_β associated with β is given by the formula

$$\tau_\beta(x) = x - 2\frac{q(x,\beta)}{q(\beta,\beta)}\beta \quad (x \in P).$$

From the definition of $q = (\mathrm{N_{red}})_{|P}$ we deduce that

$$\tau_\beta(x) = x - \frac{x\overline{\beta} + \beta\overline{x}}{\beta\overline{\beta}}\beta = -\beta\overline{x}\overline{\beta}^{-1},$$

and since x and β are both pure quaternions we find

$$\tau_\beta(x) = -\beta x\beta^{-1} \quad (x \in P).$$

Already we see that if σ_β is the rotation of angle π with axis β, we have $\sigma_\beta = -\tau_\beta = s_\beta$.

Let $u \in O(q, \mathbf{Q})$. We can write u as a product of reflections, $u = \tau_{\beta_1} \cdots \tau_{\beta_r}$, for $\beta_1, \ldots, \beta_r \in P$, with $q(\beta_i) \neq 0$, the integer r being even if u is in SO and odd otherwise. We therefore have

$$u(x) = (-1)^r\beta_1 \cdots \beta_r x(\beta_1 \cdots \beta_r)^{-1} \quad \text{for all } x \in P.$$

---

[1]We denote by $SO_0(q) \cong SO_0(2,1)$ the connected component of the identity of the special orthogonal group of the quadratic form q on P.

We now continue with the argument and show that s takes values in $SO(q)$. Let $\alpha \in D_{a,b}(\mathbf{Q})^*$ and, arguing by contradiction, suppose that $s_\alpha \in O_3^-(q)$. Under this hypothesis we would then have, for every $x \in P$,

$$s_\alpha(x) = \alpha x \alpha^{-1} = -\alpha' x (\alpha')^{-1}, \quad \text{where } \alpha' = \beta_1 \cdots \beta_r.$$

In other words, $-x = \beta x \beta^{-1}$, for every $x \in P$, with $\beta = \alpha^{-1}\alpha'$. As $\beta x \beta^{-1} = x$ for every $x \in \mathbf{Q}$, we would then have, for every x in $D_{a,b}(\mathbf{Q})$: $\bar{x} = \beta x \beta^{-1}$, with $\beta \in D_{a,b}(\mathbf{Q})^*$. But $x \mapsto \bar{x}$ is an anti-automorphism and $x \mapsto \beta x \beta^{-1}$ an automorphism – contradiction.

Summarizing, we have a homomorphism $s : D_{a,b}(\mathbf{Q})^* \to SO(q, \mathbf{Q})$, and since the rotations of angle π are in the image and these generate the special orthogonal group, s is surjective. $\qquad \square$

Henceforth we denote the isomorphism of Lemma 2.4[2] by

$$\Psi : D_{a,b}^1/\{\pm 1\} \longrightarrow SO_0(q).$$

The group $SO_0(q)$ is naturally embedded in $GL(3, \mathbf{R})$ (take the natural coordinates (x_1, x_2, x_3) of P). We let $SO_0(q, \mathbf{Z})$ be the intersection $SO_0(q) \cap GL(3, \mathbf{Z})$. The proof of Lemma 2.4 implies that

$$\mathcal{O}^1/\{\pm 1\} = \Psi^{-1}(SO_0(q, \mathbf{Z})). \tag{2.3}$$

The subgroup $\mathcal{O}^1$ is therefore discrete in $D_{a,b}^1$ and its image, $\Gamma_{a,b}$, by Φ is equally a discrete subgroup of G.

Let $g_{ij}, 1 \leq i, j \leq 3$, be the matrix entries of an element $g \in O(q) \subset GL(3, \mathbf{R})$. By a *rational representation* of the linear group $O(q)$ on $\mathbf{R}^n$ we mean a homomorphism $\rho : O(q) \to GL(n, \mathbf{R})$ such that for every pair of integers $(\mu, \nu) \in [1, n]^2$, there exists a polynomial $P_{\mu,\nu} \in \mathbf{Q}[X_{ij} | 1 \leq i, j \leq 3]$ such that for every $g \in O(q)$,

$$\rho(g)_{\mu,\nu} = P_{\mu,\nu}(g_{ij}),$$

where $\rho(g)_{\mu,\nu}$ denotes the (μ, ν) matrix entry of $\rho(g) \in GL(n, \mathbf{R})$.

Lemma 2.5 *Let ρ be a rational representation of the linear group $O(q)$ on $\mathbf{R}^n$. Then there exists a finite index subgroup of $O(q, \mathbf{Z})$ which, via the action induced by ρ, leaves invariant the standard lattice $\mathbf{Z}^n$.*

---

[2]For $a = b = 1$ the map obtained in composing Ψ and Φ^{-1} induces an isomorphism between $PSL(2, \mathbf{R})$ and $SO_0(2, 1)$. This is one of the isomorphisms – called *exceptional* – which exist only between certain "small" Lie groups (see [57, pp. 518–520]).

Proof The coefficients of the polynomials $P_{\mu,\nu}$ defined by $\rho(g)_{\mu,\nu} = P_{\mu,\nu}(g_{ij})$ are rational. The same is true for the polynomials $Q_{\mu,\nu}$ defined by

$$\rho(g)_{\mu,\nu} - \delta_{\mu,\nu} = Q_{\mu,\nu}(g_{ij} - \delta_{ij}).$$

But the latter have vanishing constant term, for $\rho(I_3) = I_n$. Thus if m is the common denominator of all their coefficients, we have $Q_{\mu,\nu}(g_{ij} - \delta_{ij}) \in \mathbf{Z}$ as soon as $g_{ij} - \delta_{ij} \equiv 0 \pmod{m}$, i.e., $g \equiv I_3 \pmod{m}$. For such a choice of g the coefficients $\rho(g)_{\mu,\nu}$ are integral, which assures the invariance of $\mathbf{Z}^n$ under $\rho(g)$. Finally, these elements form a finite index subgroup of $O(q, \mathbf{Z})$ since the quotient injects in $GL(n, \mathbf{Z}/m\mathbf{Z})$.   □

2.2.2   A Subset of the Space of Lattices

The map Ψ passes to the quotient as an embedding (continuous for the quotient topologies) of $D_{a,b}^1/\mathcal{O}^1$ into the space of lattices $\mathcal{L}_3 = GL(3, \mathbf{R})/GL(3, \mathbf{Z})$. Denote by M the image of this embedding. The subset M is also the orbit of the standard lattice L_0 under the action by left multiplication of the subgroup $SO_0(q) \subset GL(3, \mathbf{R})$ on $\mathcal{L}_3$.

Lemma 2.6 *The subset M is closed in $\mathcal{L}_3$.*

Proof Let L be a lattice in $\mathcal{L}_3$ and assume that there exists a sequence $\{L_i\}$ of lattices in M which converge to L. Each L_i is the image by an element of the group $SO_0(q)$ of the standard lattice $L_0 = \mathbf{Z}^3$. The quadratic form q therefore takes integer values on each one. Since it is continuous, the form q must also take integer values on L.

Consider now a basis (e_1, e_2, e_3) of L. Since each convergent sequence of integers is stationary, there must exist a lattice L_{i_0} in the sequence $\{L_i\}$ with a basis (f_1, f_2, f_3) sufficiently close to the basis (e_1, e_2, e_3) such that

- $q(f_i) = q(e_i)$ for $i = 1, 2, 3$, and
- $q(f_i \pm f_j) = q(e_i \pm e_j)$ for $i, j = 1, 2, 3$.

We thus have $B(e_i, e_j) = B(f_i, f_j)$, where B is the underlying symmetric bilinear form of the quadratic form q. Taking $g \in GL(3, \mathbf{R})$ to satisfy $g(e_i) = f_i$, we see that the lattice L_{i_0} can be sent to L by a matrix in $GL(3, \mathbf{R})$ which leaves the quadratic form q invariant. It follows that the lattice L is contained in the orbit of the standard lattice under the action of the group $O(q)$. The group $SO_0(q)$ being of finite index (equal to 4) in $O(q)$, we conclude that $L \in M$.   □

2.2.3   The Cocompactness Criterion

The reduced norm being multiplicative, the algebra $D_{a,b}(\mathbf{Q})$ is a division algebra if and only if $(0, 0, 0, 0)$ is the unique integral solution of the Diophantine equation

$x_0^2 - ax_1^2 - bx_2^2 + abx_3^2 = 0$. In this case we shall show that the subset M of $\mathcal{L}_3$ is compact.

The Subset M Is Compact

According to the preceding subsection, it suffices to show that M is relatively compact. For this we shall apply the Hermite-Mahler criterion.

Let us begin by noticing that every lattice L in M is of volume 1. It suffices therefore to verify that the height is bounded uniformly from below on M. Now by hypothesis $(0,0,0)$ is the only integral solution to the Diophantine equation $q(x_1, x_2, x_3) = 0$. The set

$$U = \{x \in \mathbf{R}^3 \mid |q(x)| < 1\}$$

is therefore an open neighborhood of 0 in $\mathbf{R}^3$ intersecting the standard lattice $L_0 = \mathbf{Z}^3$ only at the origin. But M is equal to the orbit of L_0 in $\mathcal{L}_3$ under the action of the group $SO_0(q)$, and since $SO_0(q)$ preserves the quadratic form q, the latter takes the same set of values on each of the lattices contained in M. The intersection of the open set U with any lattice belonging to M is therefore always reduced to the point $0 \in \mathbf{R}^3$. In other words, the height stays uniformly bounded from below on M. We may then apply the criterion of Hermite-Mahler to deduce that the subset M of $\mathcal{L}_3$ is relatively compact.

We have thus shown that the lattice $\Gamma_{a,b}$ is cocompact in G whenever $D_{a,b}(\mathbf{Q})$ is a division algebra.

Proof of the converse direction Suppose now that there exists an integral solution $\neq (0,0,0)$ to the Diophantine equation $q(x_1, x_2, x_3) = -ax_1^2 - bx_2^2 + abx_3^2 = 0$. Let us show that the quadratic form q is similar over $\mathbf{Q}$ to the quadratic form $-x_1^2 - x_2^2 + x_3^2$.

Indeed, there exists $w \in P(\mathbf{Q})$ – a vector of P with rational coordinates – such that the orthogonal complement $w^\perp$ of w with respect to q contains a non-zero q-isotropic rational vector u. Write $\lambda = -1/q(w)$ and $q' = \lambda q$. We have $q'(w) = -1$ and $q'(u) = 0$. Since $q_{|w^\perp}$ is non-degenerate, there exists a rational vector $v \in w^\perp$ such that $\mu := q'(u, v) \neq 0$. Note that the expression

$$q'(v + tu) = q'(v) + 2tq'(u, v)$$

is linear in t and has a zero $t \in \mathbf{Q}$. Replacing v by $v + tu$ if necessary, we can suppose that $q(v) = 0$. If x and y are in $\mathbf{Q}$,

$$q'(xu + yv) = 2\mu xy = \left(\frac{x + 2\mu y}{2}\right)^2 - \left(\frac{x - 2\mu y}{2}\right)^2.$$

In the basis $(w, u - v/(2\mu), u + v/(2\mu))$ the quadratic form q' is therefore equal to $-x_1^2 - x_2^2 + x_3^2$.

Lemma 2.7 *If $D_{a,b}(\mathbf{Q})$ is not a division algebra, the group $\Gamma_{a,b}$ is commensurable with the group* $\mathrm{SL}(2,\mathbf{Z})$. *In particular, it is Fuchsian of the first kind and non-cocompact.*

Proof Assume that $D_{a,b}(\mathbf{Q})$ is not a division algebra. There exists an integral solution $(x_0, x_1, x_2, x_3) \neq (0,0,0,0)$ to the Diophantine equation $x_0^2 - ax_1^2 - bx_2^2 + abx_3^2 = 0$. We first prove that there exists an integral solution $\neq (0,0,0)$ to $-ax_1^2 - bx_2^2 + abx_3^2 = 0$. If $x_0 = 0$, there is nothing to show. Suppose therefore that $x_0 \neq 0$. The integers ax_1^2 and bx_2^2 cannot be both the zero. Assume for example that ax_1^2 is non-zero. Since $x_0^2 - ax_1^2 - bx_2^2 + abx_3^2 = 0$, we obtain $x_0^2 - bx_2^2 = a(x_1^2 - bx_3^2)$. Then a simple calculation shows that

$$(b(x_0 x_3 + x_1 x_2), a(x_1^2 - bx_3^2), x_0 x_1 + bx_2 x_3)$$

is an integer solution to $-ax_1^2 - bx_2^2 + abx_3^2 = 0$. If the latter is zero, then $-ax_1^2 + abx_3^2 = 0$ and $(x_1, 0, x_3)$ is a non-zero integer solution (since $x_1 \neq 0$) to $-ax_1^2 - bx_2^2 + abx_3^2 = 0$.

According to the paragraph preceding Lemma 2.7, the quadratic forms $q = -ax_1^2 - bx_2^2 + abx_3^2$ and $q' = -x_1^2 - x_2^2 + x_3^2$ are similar over $\mathbf{Q}$. Since multiplying a quadratic form by a non-zero scalar does not change the associated orthogonal group, the groups $\mathrm{O}(q)$ and $\mathrm{O}(q')$ are conjugate by a matrix belonging to $\mathrm{GL}(3,\mathbf{Q})$. Conjugation by a rational invertible matrix being a rational representation, Lemma 2.5 implies that the image of the group $\mathrm{O}(q,\mathbf{Z})$ in $\mathrm{O}(q')$ is commensurable with the group $\mathrm{O}(q',\mathbf{Z})$. The isomorphism between $\mathrm{SO}_0(q')$ and $\mathrm{PSL}(2,\mathbf{R})$ then implies that the groups $\Gamma_{a,b}$ and $\Gamma_{1,1}$ are commensurable in G. The result follows by observing that $\Gamma_{1,1}$ is commensurable with the group $\mathrm{SL}(2,\mathbf{Z})$. □

The group $\Gamma_{a,b}$ is therefore commensurable with $\mathrm{SL}(2,\mathbf{Z})$ if and only if $-ax^2 - by^2 + abz^2$ is similar to $-x^2 - y^2 + z^2$. On the other hand

$$ab(-ax^2 - by^2 + abz^2) = -b(ax)^2 - a(by)^2 + (abz)^2,$$

so that the quadratic forms $-ax^2 - by^2 + abz^2$ and $x^2 - ay^2 - bz^2$ are themselves similar. The three first claims of Theorem 2.3 are thus proved.

2.2.4 Commensurability Classes

Although not directly used in the rest of the text, the fourth and final claim of Theorem 2.3 will be established in this subsection. In order to be brief, we freely use more advanced algebraic concepts and results; appropriate references are quoted at the end of this chapter. Proofs requiring more background have been put in small typeface.

We begin by remarking that it follows from the proof of Lemma 2.5 that if the quadratic forms $x^2 - ay^2 - bz^2$ and $x^2 - a'y^2 - b'z^2$ (and hence $-ax^2 - by^2 + abz^2$ and $-a'x^2 - b'y^2 + a'b'z^2$) are similar, the groups $\Gamma_{a,b}$ and $\Gamma_{a',b'}$ are commensurable. We must now prove the converse.

Assume then that the groups $\Gamma_{a,b}$ and $\Gamma_{a',b'}$ are commensurable. Then there exists an $\mathbf{R}$-linear map $\alpha : \mathbf{R}^3 \to \mathbf{R}^3$ which sends the quadratic form $q = -ax^2 - by^2 + abz^2$ to $q' = -a'x^2 - b'y^2 + a'b'z^2$ and, by conjugation, a finite index subgroup $\Gamma \subset SO(q, \mathbf{Z})$ to a finite index subgroup $\Gamma' \subset SO(q', \mathbf{Z})$. Let $\overline{\Gamma} \subset O(q)$ (resp. $\overline{\Gamma'} \subset O(q')$) be the extension by $\{\pm I_3\}$ of the group Γ (resp. Γ').

Lemma 2.8 *The $\mathbf{Q}$-vector subspace generated by $\overline{\Gamma}$ in $\mathcal{M}_3(\mathbf{R})$ is equal to $\mathcal{M}_3(\mathbf{Q})$.*

Proof The $\mathbf{Q}$-vector subspace generated by $\overline{\Gamma}$ is a subalgebra of $\mathcal{M}_3(\mathbf{Q})$. According to the Burnside lemma [76, Cor. 3.4] it suffices to show that the space $\mathbf{Q}^3$ is a simple $\mathbf{Q}\overline{\Gamma}$-module.

We first show that the group $\overline{\Gamma}$ is dense in $O(q) \cong O(1, 2)$ with respect to the *Zariski topology*. In other words, every polynomial in the matrix entries of matrices in $\mathcal{M}_3(\mathbf{R})$ which vanishes on $\overline{\Gamma}$ should also vanish on $O(q)$. This can be deduced from the following facts:

1. The closure (in the Zariski topology) of $\overline{\Gamma}$ in $O(q)$ is a subgroup of $O(1, 2)$ which is not abelian.
2. The group $SO(q)$ is minimal among the non-abelian subgroups of $O(q)$ which are closed with respect to the Zariski topology.

The first statement is immediate. The second statement follows from the fact that the group $SO(q)$ is a connected algebraic group. Its Lie algebra is isomorphic to the Lie algebra of $SL(2, \mathbf{R})$ which does not contain any proper non-abelian sub Lie algebra.

Since $\overline{\Gamma}$ contains $\{\pm I_3\}$ it follows from the preceding facts that $\overline{\Gamma}$ is dense in $O(q) \cong O(1, 2)$ with respect to the Zariski topology.

Now, since the natural representation of $O(1, 2)$ in $\mathbf{R}^3$ is irreducible over $\mathbf{C}$, the representation ρ of $\overline{\Gamma}$ in $GL(3, \mathbf{C})$ is also irreducible. □

End of the proof of Theorem 2.3 Since α sends $\overline{\Gamma}$ onto $\overline{\Gamma'}$, α sends $\mathbf{Q}\overline{\Gamma}$ onto $\mathbf{Q}\overline{\Gamma'}$. But according to Lemma 2.8, we have $\mathbf{Q}\overline{\Gamma} = \mathcal{M}_3(\mathbf{Q})$, which implies that $\alpha = \lambda\alpha_0$, where α_0 is defined over $\mathbf{Q}$ and $\lambda \in \mathbf{R}^*$. Thus α_0 sends q to $\lambda^{-2}q'$ and, since $q \neq 0$, the scalar λ^{-2} is in $\mathbf{Q}$. This shows that the forms q and q' are similar over $\mathbf{Q}$, completing the proof of the proof of Theorem 2.3. □

Remark 2.9 Theorem 2.3 implies in particular that there always exists an infinity of solutions to the Diophantine equation $x_0^2 - ax_1^2 - bx_2^2 + abx_3^2 = 1$.

2.3 Arithmetic Hyperbolic Surfaces

Hyperbolic surfaces are associated with discrete subgroups of $SL(2, \mathbf{R})$ that are torsion free – those containing no non-trivial element of finite order. We now explain how to eliminate torsion if we have it.

2.3.1 Eliminating Torsion

The process of eliminating torsion is based on the following theorem of Minkowski.

Theorem 2.10 *Let n be an integer $\geqslant 2$ and F a finite subgroup of $\mathrm{GL}(n, \mathbf{Z})$. Then F injects in $\mathrm{GL}(n, \mathbf{Z}/\ell\mathbf{Z})$ for every integer $\ell \geqslant 3$.*

Proof We consider the standard embedding $\mathbf{Z}^n \subset \mathbf{R}^n$, so that the group F identifies as a subgroup of $\mathrm{GL}(n, \mathbf{R})$. By averaging any choice of positive definite quadratic form on $\mathbf{R}^n$ over F we obtain a norm invariant under the action of F. We may normalize so that the non-zero points of $\mathbf{Z}^n$ closest to 0 are at distance 1; let $V \subset \mathbf{R}^n$ be the subspace generated by these points. By renormalizing the orthogonal complement $V^\perp$, we may again assume that the points of the lattice $\mathbf{Z}^n$ in $\mathbf{R}^n - V$ closest to the origin are at distance 1. Since F preserves V and $V^\perp$, the resulting quadratic form is still F-invariant. Continuing in this way we obtain an F-invariant norm on $\mathbf{R}^n$ such that

1. every vector in $\mathbf{Z}^n$ is of norm greater than or equal to 1, and
2. the points in $\mathbf{Z}^n$ of norm 1 generate the space $\mathbf{R}^n$.

Now let $\gamma \in F$ be such that γ is sent to the identity in $\mathrm{GL}(n, \mathbf{Z}/\ell\mathbf{Z})$ for some integer $\ell \geqslant 1$. Viewed as a matrix in $\mathcal{M}_n(\mathbf{Z})$, $\gamma - I_n \in \ell\mathcal{M}_n(\mathbf{Z})$. If $\gamma \neq I_n$, there exists an element $x \in \mathbf{Z}^n$ of norm 1 such that $\gamma x \neq x$. But $\gamma x - x$ is then a non-zero vector $\in \ell\mathbf{Z}^n$ and is thus of norm $\geqslant \ell$. On the other hand, γx and x both belong to the sphere[3] of radius 1. They are thus at distance at most 2 from one another and ℓ is necessarily less than or equal to 2. $\square$

Let a and b be two positive integers. Given a positive integer N we define the N-th *principal congruence subgroup* of $\Gamma_{a,b}$ as the subgroup $\Gamma_{a,b}(N) \subset \Gamma_{a,b}$ given by the image in $\Gamma_{a,b}$ of the subgroup

$$\{x \in \mathcal{O}^1 \mid x - 1 \in N D_{a,b}(\mathbf{Z})\} \subset \mathcal{O}^1.$$

Note that $\Gamma_{a,b}(N)$ is contained in the image in $\mathrm{SL}(2, \mathbf{R})$ of the subgroup

$$\mathrm{SO}_0(q, \mathbf{Z}) \cap \mathrm{Ker}\left(\mathrm{GL}(3, \mathbf{Z}) \longrightarrow \mathrm{GL}(3, \mathbf{Z}/N\mathbf{Z})\right)$$

under the map $\Phi \circ \Psi^{-1}$. We keep the notation from the proof of Theorem 2.3: the quadratic form q is equal to $-ax^2 - by^2 + abz^2$.

Corollary 2.11 *For every pair of integers $(a, b) \geqslant 1$ and for every integer $N \geqslant 3$, the subgroup $\Gamma_{a,b}(N) \subset \mathrm{SL}(2, \mathbf{R})$ is torsion free. The quotient*

$$X_{a,b}(N) := \Gamma_{a,b}(N) \backslash \mathcal{H}$$

---

[3]Recall that the norm is F-invariant.

is therefore a hyperbolic surface of finite area. It is compact if and only if the quadratic form $x^2 - ay^2 - bz^2$ does not represent 0 over $\mathbf{Q}$.

Proof The subgroup $SO_0(q, \mathbf{Z}) \cap \mathrm{Ker}\left(GL(3, \mathbf{Z}) \to GL(3, \mathbf{Z}/N\mathbf{Z})\right)$ is torsion free since, according to Theorem 2.10, the kernel

$$\mathrm{Ker}\left(GL(3, \mathbf{Z}) \longrightarrow GL(3, \mathbf{Z}/N\mathbf{Z})\right) \quad (N \geqslant 3)$$

contains no finite subgroup. The group $\Gamma_{a,b}(N)$ $(N \geqslant 3)$ is therefore itself torsion free and is clearly of finite index in $\Gamma_{a,b}$. Theorem 2.3 then implies Corollary 2.11. □

An *arithmetic surface* is any surface admitting a cover isometric to a finite cover of one of the surfaces $X_{a,b}(N)$ defined in Corollary 2.11. Note that the fourth claim of Theorem 2.3 implies that $X_{a,b}(N)$ and $X_{a',b'}(N')$ admit two finite isometric covers if and only if $x^2 - ay^2 - bz^2$ and $x^2 - a'y^2 - b'z^2$ are similar as quadratic forms over $\mathbf{Q}$.

2.3.2 The Modular Surface and Its Covers

The fundamental example of an arithmetic surface is the modular surface. As we recalled in the introduction, the subset

$$D = \{z = x + iy \in \mathcal{H} \mid |x| < 1/2 \text{ and } |z| > 1\}$$

of the upper half-plane $\mathcal{H}$ is a fundamental domain for the action of the modular group $\Gamma(1) = SL(2, \mathbf{Z})$. Moreover, i is an elliptic vertex of order 2, $\zeta = (1 + i\sqrt{3})/2$ is an elliptic vertex of order 3, ∞ is the only cusp (up to equivalence) and the genus of $X(1)$ is 0.

We may add to the above example all of the congruence covers of the modular surface. Let N be an integer $\geqslant 1$. The *principal congruence subgroup of level N*, denoted $\Gamma(N)$, is the subgroup of the modular group made up of matrices congruent to the identity modulo N, i.e.,

$$\Gamma(N) = \left\{ \gamma \in SL(2, \mathbf{Z}) \,\middle|\, \gamma \equiv \begin{pmatrix} 1 & 0 \\ 0 & 1 \end{pmatrix} \pmod{N} \right\}. \tag{2.4}$$

Theorem 2.12 *The group $\Gamma(N)$ is normal in $\Gamma(1) = SL(2, \mathbf{Z})$ and of index*

$$\mu_N = [\Gamma(1) : \Gamma(N)] = N^3 \prod_{p|N} (1 - p^{-2}).$$

The group $\overline{\Gamma(N)}$ is of index

$$\overline{\mu_N} = \begin{cases} \mu_N/2 & \text{if } N > 2 \\ \mu_N & \text{if } N = 2 \end{cases}$$

in $\mathrm{PSL}(2, \mathbf{Z})$. *It is torsion free as soon as $N > 1$ and the number of cuspidal equivalence classes is*

$$h_N = \frac{\overline{\mu_N}}{N}.$$

Proof Let $N \geqslant 1$. We begin with the following result.

Lemma 2.13 *The homomorphism*

$$\psi : \mathrm{SL}(2, \mathbf{Z}) \longrightarrow \mathrm{SL}(2, \mathbf{Z}/N\mathbf{Z})$$

is surjective.

Proof Let us show more generally that the homomorphism

$$\psi : \mathrm{SL}(m, \mathbf{Z}) \longrightarrow \mathrm{SL}(m, \mathbf{Z}/N\mathbf{Z})$$

is surjective.

We first assume that N is a prime power p^s. It suffices to show that the group $\mathrm{SL}(m, \mathbf{Z}/p^s\mathbf{Z})$ is generated by the transvection matrices $I_m + E_{ij}$, $i \neq j$, where E_{ij} is the $m \times m$ matrix all of whose coefficients are zero except in the i-th row and j-th column, where it is 1. Indeed, the transvections are clearly in the image of ψ.

We prove this by induction on m. The case $m = 1$ is trivial. Assume then that the statement is true for $m - 1$ where $m > 1$. Let $A \in \mathcal{M}_m(\mathbf{Z})$ be such that $\det(A) \equiv 1 \pmod{p^s}$. The first column of A contains at least one coefficient c which is invertible mod p^s; we shall use it as a pivot for row operations – these are realized by multiplying A by transvection matrices – until we are brought to a matrix whose first column is the vector $(1, 0, \ldots, 0)$. By operating next on the columns of this new matrix we are led to assume that

$$A = \left(\begin{array}{c|c} 1 & 0 \\ \hline 0 & A' \end{array} \right)$$

with $A' \in \mathcal{M}_{m-1}(\mathbf{Z})$ and $\det(A') \equiv 1 \pmod{p^s}$. The result easily follows by induction.

We deduce the case for arbitrary N from the Chinese Remainder Theorem. If $N = \prod_k p_k^{s_k}$ is the prime power decomposition of N for distinct primes p_k, we have

$$\mathbf{Z}/N\mathbf{Z} \cong \prod_k (\mathbf{Z}/p_k^{s_k}\mathbf{Z}),$$

$$\mathrm{GL}(2, \mathbf{Z}/N\mathbf{Z}) \cong \prod_k \mathrm{GL}(2, \mathbf{Z}/p_k^{s_k}\mathbf{Z}),$$

$$\mathrm{SL}(2, \mathbf{Z}/N\mathbf{Z}) \cong \prod_k \mathrm{SL}(2, \mathbf{Z}/p_k^{s_k}\mathbf{Z}).$$

From this it follows that the transvection $I_m + (\prod_{k \neq \ell} p_k^{s_k}) E_{ij}$ is congruent to the identity modulo $p_k^{s_k}$ for k different from ℓ and congruent to the transvection $I_m + E_{ij}$ modulo $p_\ell^{s_\ell}$. □

Consider now the kernel K of the surjective morphism

$$\mathrm{GL}(2, \mathbf{Z}/p_k^{s_k}\mathbf{Z}) \longrightarrow \mathrm{GL}(2, \mathbf{Z}/p_k\mathbf{Z}).$$

Since K is made up of matrices of $\mathcal{M}_2(\mathbf{Z}/p_k^{s_k}\mathbf{Z})$ which are congruent to the identity matrix $I_2 \bmod p_k$, the cardinality of K is $p_k^{4(s_k-1)}$.

Let us calculate the cardinality of $\mathrm{GL}(2, \mathbf{Z}/p_k\mathbf{Z})$. This group is made up of matrices with coefficients in $\mathbf{Z}/p_k\mathbf{Z}$, such that the column vectors are linearly independent. There are $p_k^2 - 1$ choices for the first column vector and, once the first has been fixed, there are $p_k^2 - p_k$ remaining choices for the second (we exclude the p_k multiples of the first vector). The cardinality of $\mathrm{GL}(2, \mathbf{Z}/p_k\mathbf{Z})$ is therefore equal to $(p_k^2 - 1)(p_k^2 - p_k)$.

The cardinality of $\mathrm{GL}(2, \mathbf{Z}/p_k^{s_k}\mathbf{Z})$ is therefore equal to $p_k^{4(s_k-1)}(p_k^2 - p_k)(p_k^2 - 1) = p_k^{4s_k}(1 - p_k^{-1})(1 - p_k^{-2})$. Now $\mathrm{SL}(2, \mathbf{Z}/p_k^s\mathbf{Z})$ is the kernel of the surjective morphism $\det : \mathrm{GL}(2, \mathbf{Z}/p_k^{s_k}\mathbf{Z}) \to (\mathbf{Z}/p_k^{s_k}\mathbf{Z})^*$. The cardinality of $\mathrm{SL}(2, \mathbf{Z}/p_k^{s_k}/p_k^{s_k}\mathbf{Z})$ is then equal to $p_k^{3s_k}(1 - p_k^{-2})$. As a consequence

$$[\Gamma(1) : \Gamma(N)] = N^3 \prod_{p|N} (1 - p^{-2}).$$

The matrix $-I_2$ belongs to the group $\Gamma(N)$ if and only if $N = 2$. We deduce

$$[\overline{\Gamma(1)} : \overline{\Gamma(N)}] = \begin{cases} \mu_N/2, & \text{if } N > 2, \\ \mu_N, & \text{if } N = 2. \end{cases}$$

Let us now show that if $N > 1$ the group $\overline{\Gamma(N)}$ has no elliptic elements. The only fixed points of $\overline{\Gamma(1)} = \mathrm{PSL}(2, \mathbf{Z})$ in the fundamental hyperbolic triangle D are the two vertices of angle $\pi/3$ and the point i. The elliptic conjugacy classes in

$\mathrm{PSL}(2, \mathbf{Z})$ are therefore represented by

$$\begin{pmatrix} 0 & -1 \\ 1 & 0 \end{pmatrix}, \quad \begin{pmatrix} 0 & -1 \\ 1 & -1 \end{pmatrix}, \quad \begin{pmatrix} -1 & 1 \\ -1 & 0 \end{pmatrix}.$$

None of these elements is congruent to the identity modulo N if $N > 1$. Since $\overline{\Gamma(N)}$ is normal in $\overline{\Gamma(1)}$ we obtain, as claimed, that $\overline{\Gamma(N)}$ is torsion free.

Now, if s is a cusp, s is $\overline{\Gamma(1)}$-equivalent to ∞. But

$$\Gamma(1)_\infty = \left\{ \pm \begin{pmatrix} 1 & m \\ 0 & 1 \end{pmatrix} \,\middle|\, m \in \mathbf{Z} \right\},$$

$$\Gamma(N)_\infty = \Gamma(N) \cap \Gamma(1)_\infty = \left\{ \pm \begin{pmatrix} 1 & mN \\ 0 & 1 \end{pmatrix} \,\middle|\, m \in \mathbf{Z} \right\},$$

so that $[\Gamma(1)_\infty : \Gamma(N)_\infty] = N$. In this way $\Gamma(N)$ has exactly $\overline{\mu_N}/N$ equivalence classes of cusps.   □

The surfaces $X(N) = \Gamma(N) \backslash \mathcal{H}$ are therefore "true" hyperbolic surfaces as soon as $N > 1$. These are the covers – ramified over the elliptic points – of the modular surface $X(1)$. The degree of this cover is equal to $\overline{\mu_N}$. The area of $X(N)$ is then equal to $\overline{\mu_N}$ times the area of $X(1)$, namely

$$\mathrm{area}(X(N)) = \begin{cases} \dfrac{\pi N^3}{6} \prod_{p|N}(1 - p^{-2}), & \text{if } N > 2, \\ 2\pi, & \text{if } N = 2. \end{cases} \tag{2.5}$$

The surface $X(N)$ $(N > 1)$ has $h_N = \overline{\mu_N}/N$ cusps. We obtain a fundamental domain for the action of $\Gamma(N)$ on $\mathcal{H}$ by taking the union of $\overline{\mu_N}$ translates of the triangle D. This induces a triangulation of $X(N)$ by $\overline{\mu_N}$ triangles. Each of these triangles has two vertices of angle $\pi/3$ around which are glued 6 triangles and then another vertex of angle 0 around which are glued N triangles. There are therefore $\overline{\mu_N}(\frac{1}{3} + \frac{1}{N})$ vertices, $\frac{3}{2}\overline{\mu_N}$ sides and $\overline{\mu_N}$ faces. The Euler characteristic of $X(N)$ is equal to

$$\overline{\mu_N}\left(\frac{1}{N} - \frac{1}{6}\right).$$

Now the Euler characteristic is also equal to $2 - 2g_N$, where g_N is the genus of $X(N)$. So we obtain

$$g_N = \begin{cases} 1 + \mu_N \dfrac{N - 6}{24N}, & \text{if } N > 2, \\ 0, & \text{if } N = 2. \end{cases} \tag{2.6}$$

Every subgroup of the modular group containing $\Gamma(N)$ is called a *congruence subgroup*. The following example is particularly important,

$$\Gamma_0(N) = \left\{ \gamma \in \mathrm{SL}(2, \mathbf{Z}) \;\middle|\; \gamma \equiv \begin{pmatrix} * & * \\ 0 & * \end{pmatrix} \pmod{N} \right\} .$$

2.4 Commentary and References

§ 2.1

The Iwasawa decomposition is discussed in detail in [93, p. 44].

§ 2.2

In general there are two possibilities for a quaternion algebra D over a field F of characteristic zero: either D is a division algebra, or $D \cong \mathcal{M}_2(F)$, the algebra of 2×2 matrices over F. See for example [88, Th. 2.1.7] for a proof of this result which we do not use here.

The general notion of arithmetic group is a bit technical; the reader can consult [99]. The classification of arithmetic groups described in this reference implies, in particular, that every arithmetic subgroup of $\mathrm{SL}(2, \mathbf{R})$ defined over $\mathbf{Q}$ is commensurable to one of the subgroups constructed in Theorem 2.3. One can more generally replace $\mathbf{Q}$ by a totally real number field. The construction is similar and treated in [67, 88].

The proof of Theorem 2.3 that we give here is slightly different from that which one usually finds in the literature, as for example in [67]. We reduce the argument to the study of arithmetic subgroups of $\mathrm{SO}(2, 1)$ via the exceptional isomorphism between $\mathrm{PSL}(2, \mathbf{R})$ and $\mathrm{SO}_0(2, 1)$. This way of constructing Fuchsian groups is essentially the same used by Poincaré in his article "Les fonctions fuchsiennes et l'arithmétique," which appeared in 1887. The groups obtained in this way are in fact the first examples of Fuchsian groups that Poincaré succeeded in constructing. He also used the geometry of the space of lattices $\mathcal{L}_3$ previously studied by Hermite, to whom we owe the proof of Theorem 2.1 when $n = 3$. Before that the geometry of the space $\mathcal{L}_2$ had been completely understood by Gauß in his *Disquisitiones arithmeticae* which contains a very clear exposition of the action of the modular group on the upper half-plane $\mathcal{H}$.

The construction of infinitely many commensurability classes of discrete cocompact subgroups in $\mathrm{SL}(2, \mathbf{R})$ is borrowed from [88, pp. 87–88].

Through the exceptional isomorphism the geometry of the orthogonal group of a non-degenerate quadratic form comes into play. A convenient reference on this subject is the book [96]. In particular, [96, Th. 2.6] gives a proof of the fact that the rotations of angle π generate the special orthogonal group.

The classification of commensurability classes relies on notions in algebraic geometry and algebraic groups. The books [97] and [16] can serve as good introductions to these notions.

§ 2.3

Our study of congruence subgroups of the modular group follows Shimura's exposition of the topic in [121, §1.6]. One can also have a look at the excellent book of Miyake [92]. These two references contain general computations on the cardinality of equivalence classes of elliptic fixed points of order 2 and 3 for the groups $\Gamma_0(N)$.

We have assumed familiarity with the Euler characteristic of a surface and the Euler phi function. For an introduction to these notions the reader can refer to [92, 121] and Serre's Course in Arithmetic [118], respectively.

2.5 Exercises

Exercise 2.14 Show that the group $SL(2, \mathbf{Z}[\sqrt{2}])$ is not a Fuchsian group.

Exercise 2.15

1. Show that the quotient $SL(2, \mathbf{Z}) \backslash SL(2, \mathbf{R})$, endowed with the quotient topology, is homeomorphic to the unit tangent bundle of the modular surface $SL(2, \mathbf{Z}) \backslash \mathcal{H}$. (This question is answered in § 9.2.)
2. Using the explicit description of the fundamental domain for $SL(2, \mathbf{Z})$ given in § 1.4.1, deduce from the preceding question that a subset of volume 1 lattices in $\mathbf{R}^2$ is relatively compact if and only if the height is uniformly bounded from below on this set by a positive constant.

Exercise 2.16 When $n = 2$, compare the set D of § 1.4.1 and the orbit $S_{s,t}^{-1} i$ of the point $i \in \mathcal{H}$ under the action of a Siegel set. Deduce that $D(1) \subset S_{s,t}^{-1} i$ as soon as $s \geqslant 2/\sqrt{3}, t \geqslant 1/2$.

Exercise 2.17 Taking inspiration from the proof of Theorem 2.3, show that there exist infinitely many integer solutions (m, n) to the "Pell-Fermat" Diophantine equation

$$m^2 - an^2 = 1,$$

where a is an integer such that $\sqrt{a} \notin \mathbf{N}$.

Exercise 2.18

1. Let $\alpha = \begin{pmatrix} N & 0 \\ 0 & 1 \end{pmatrix}$. Show that $-I_2 \in \Gamma_0(N)$ and that

$$\Gamma_0(N) = \alpha^{-1} \Gamma(1) \alpha \cap \Gamma(1).$$

Fig. 2.1 Degree 6 cover of the modular surface

2. Show that the group $\Gamma_0(N)/\Gamma(N)$ is isomorphic to the subgroup of upper triangular matrices in $\mathrm{SL}(2, \mathbf{Z}/N\mathbf{Z})$. Deduce that it is of order $N\phi(N)$, where ϕ is the *Euler totient function* which associates with a positive integer N the cardinality of the group of invertible elements in $\mathbf{Z}/N\mathbf{Z}$.
3. Deduce from Question 2 that the index of $\Gamma_0(N)$ in $\Gamma(1)$ is given by

$$[\Gamma(1) : \Gamma_0(N)] = N \prod_{p|N}(1 + p^{-1}).$$

4. Show that the number h of equivalence classes of cusps for $\Gamma_0(N)$ is equal to the number of double classes in $\Gamma_0(N)\backslash\Gamma(1)/\Gamma(1)_\infty$.
5. Let M_N be the set of elements $(a, c) \in (\mathbf{Z}/N\mathbf{Z})^2$ of order N. Show that the map from $\Gamma(1)$ to the set M_N defined by

$$\begin{pmatrix} a & b \\ c & d \end{pmatrix} \longmapsto (a, c)$$

induces a bijective map from $\Gamma_0(N)\backslash\Gamma(1)/\Gamma(1)_\infty$ onto $M_N/\sim$, where $\sim$ is the equivalence relation on M_N given by $(a, c) \sim (a', c')$ if

$$(a', c') \equiv \pm(ma + nc, m^{-1}c) \quad (m \in (\mathbf{Z}/N\mathbf{Z})^*, \ n \in \mathbf{Z}/N\mathbf{Z}).$$

6. Deduce from the two preceding questions that

$$h = \sum_{d|N,\ d>0} \phi\left(\gcd\left(d, (N/d)\right)\right),$$

where ϕ is the Euler phi function.

Exercise 2.19

1. Let $\begin{pmatrix} a & b \\ c & d \end{pmatrix} \in \Gamma_0(8)$.

   a. Show that $a + d \neq 0$ (use the fact that -1 is not a square modulo 8).
   b. Show that a and d are odd and thus that $a + d \neq \pm 1$.
   c. Deduce that $|a + d| \geq 2$ and thus that the image of the group $\Gamma_0(8)$ in $\text{PSL}(2, \mathbf{R})$ is a torsion free subgroup.

2. Deduce from the first question and the preceding exercise that the quotient $\Gamma_0(8) \backslash \mathcal{H}$ is a non-singular hyperbolic surface of genus 0 with 4 cusps, which is a degree 12 cover of the modular surface. The union of two adjacent polygons in Fig. 2.1, generated by Arnaud Chéritat, forms a fundamental domain for the group $\Gamma_0(8)$.

Chapter 3
Spectral Decomposition

The goal of this chapter is to prove the Spectral Theorem (Theorem 1.6).

3.1 The Laplacian

Every element $g \in G = \mathrm{SL}(2, \mathbf{R})$ defines a linear operator L_g on functions $f : \mathcal{H} \to \mathbf{C}$ by the formula $(L_g f)(z) = f(g^{-1}z)$. A linear operator T on functions $f : \mathcal{H} \to \mathbf{C}$ is said to be *invariant* if it commutes with the G-action:

$$L_g T = T L_g, \quad \text{for all } g \in G.$$

According to (1.20), the Laplacian Δ is invariant. All of the invariant differential operators are important but chief among them is the Laplacian, which plays a particular role: a diffeomorphism of $\mathcal{H}$ is an isometry if and only if it preserves the Laplacian. In the $z = x + iy$ coordinates of the upper half-plane, the hyperbolic Laplacian is given by

$$\Delta = -y^2 \left(\frac{\partial^2}{\partial x^2} + \frac{\partial^2}{\partial y^2} \right). \tag{3.1}$$

According to (1.21), the Laplacian is expressed in the (r, θ) polar coordinates of the upper half-plane as

$$\Delta = -\frac{\partial^2}{\partial r^2} - \frac{1}{\tanh r} \frac{\partial}{\partial r} - \frac{1}{(2 \sinh r)^2} \frac{\partial^2}{\partial \theta^2}. \tag{3.2}$$

Harmonic analysis on $\mathcal{H}$ – a hyperbolic analog of the Euclidean Fourier theory – consists in spectrally decomposing the Laplacian.

© Springer International Publishing Switzerland 2016
N. Bergeron, *The Spectrum of Hyperbolic Surfaces*, Universitext,
DOI 10.1007/978-3-319-27666-3_3

The space of functions in question is the Hilbert space $L^2(\mathcal{H})$ of (classes of) square-integrable functions $f : \mathcal{H} \to \mathbf{C}$ equipped with the Hilbert space inner product

$$\langle f, g \rangle = \int_{\mathcal{H}} f(z)\overline{g(z)} \, d\mu(z).$$

Here μ is the hyperbolic area measure, see (1.16).

Let $\mathbf{H}$ be a *separable* Hilbert space, i.e., one which admits a countable orthonormal basis. Notice that, if $D \subset \mathcal{H}$ is a fundamental domain for Γ whose boundary is of zero measure, then the Hilbert space $L^2(\Gamma\backslash\mathcal{H})$ is isomorphic to $L^2(D)$; in particular it is separable.

We denote by $|\cdot|$ the Hilbert space norm on $\mathbf{H}$. A linear map $T : \mathbf{H} \to \mathbf{H}$ is a *bounded operator* if there exists a constant C such that $|Tx| \leq C|x|$ for all $x \in \mathbf{H}$. The smallest possible constant C is called the *operator norm* of T, and we write it as $|T|$. The linear map T is a bounded operator if and only if it is continuous.

Since the L^2 functions on $\mathcal{H}$ are not all differentiable, a differential operator is only defined on a dense subspace of the Hilbert space $L^2(\mathcal{H})$. We thus need a generalization of the notion of differential operator.

We agree to refer to any linear transformation defined on a dense subspace of $\mathbf{H}$ as an *operator*. In other words, an operator is a pair (T, D_T) consisting of a dense vector subspace $D_T \subset \mathbf{H}$, called the *domain* of T, and a linear transformation $T : D_T \to \mathbf{H}$. The operator is *closed* if its graph $\{(f, Tf) \mid f \in D_T\}$ is a closed subspace of the product $\mathbf{H} \times \mathbf{H}$. It is *unbounded* if it is not continuous (with respect to the topology on D_T induced by that of $\mathbf{H}$).[1] In practice one often speaks of "the operator" T, dropping the dependence on the domain of definition; if done in context, this should not lead to any confusion.

The Laplacian Δ is well-defined on the subspace $C_c^\infty(\mathcal{H})$ (dense in $L^2(\mathcal{H})$) consisting of C^∞ function of compact support in $\mathcal{H}$, although it is not continuous with respect to the L^2 topology: the Laplacian is an unbounded operator.

An operator (T, D_T) is said to be *symmetric* if

$$\langle Tf, g \rangle = \langle f, Tg \rangle \tag{3.3}$$

for all $f, g \in D_T$, where $\langle \, , \, \rangle$ is the Hilbert space inner product on $\mathbf{H}$.

Given an operator (T, D_T) on $\mathbf{H}$, we write D_{T^*} for the space of all $g \in \mathbf{H}$ such that $f \mapsto \langle Tf, g \rangle$ is a continuous linear form on D_T. For such a g, the corresponding linear form can be extended by continuity to all of $\mathbf{H}$, and one obtains in this way a bounded linear form on $\mathbf{H}$. A classical theorem of Riesz (see, for example, [18, p. 61]) states that every continuous linear form on a Hilbert space $\mathbf{H}$ is of the form $f \mapsto \langle f, h \rangle$ for a certain element $h \in \mathbf{H}$. Thus, if $g \in D_{T^*}$, there exists a unique element $T^*g \in \mathbf{H}$ such that $\langle Tf, g \rangle = \langle f, T^*g \rangle$. The operator (T^*, D_{T^*}) is called the *adjoint* of T. If T is symmetric then $D_T \subset D_{T^*}$.

---

[1] A continuous operator (T, D_T) on D_T naturally extends to a bounded operator of $\mathbf{H}$ to itself.

The operator T is said to be *self-adjoint* if $D_T = D_{T*}$ and $T = T^*$. A self-adjoint operator is in particular symmetric; it is also closed. For a bounded operator, it is one and the same to be symmetric or to be self-adjoint.

An *eigenvector* of T for the *eigenvalue* λ is a non-zero vector $f \in \mathbf{H}$ such that $Tf = \lambda f$. The scalar λ being fixed, the set of eigenvectors of eigenvalue λ is called the *λ-eigenspace*. If T is a symmetric operator, all of its eigenvalues are real and the eigenspaces corresponding to distinct eigenvalues are orthogonal.

From now on we denote by Δ the unbounded operator $(\Delta, C_c^\infty(\mathcal{H}))$ whose underlying Hilbert space is $L^2(\mathcal{H})$.

Let

$$\Delta^e = \frac{\partial^2}{\partial x^2} + \frac{\partial^2}{\partial y^2}$$

be the usual Euclidean Laplacian.[2] Let d be the exterior derivative which sends differential 1-forms to 2-forms. Let f and g be two C^∞ functions defined on a bounded region $\Omega \subset \mathbf{C}$ with boundary a C^∞ (or piecewise C^∞) curve $\partial\Omega$. Define

$$\nabla f = \begin{pmatrix} \partial f/\partial x \\ \partial f/\partial y \end{pmatrix}.$$

Then we have

$$d\left(g\left(\frac{\partial f}{\partial x} dy - \frac{\partial f}{\partial y} dx \right) \right) = (g\Delta^e f + \nabla f \cdot \nabla g)\, dx \wedge dy \tag{3.4}$$

so that

$$d\left(g\left(\frac{\partial f}{\partial x} dy - \frac{\partial f}{\partial y} dx \right) - f\left(\frac{\partial g}{\partial x} dy - \frac{\partial g}{\partial y} dx \right) \right) = (g\Delta^e f - f\Delta^e g)\, dx \wedge dy.$$

Stokes' Theorem then implies *Green's formula*:

$$\int_\Omega (g\Delta^e f - f\Delta^e g)\, dx \wedge dy$$
$$= \int_{\partial\Omega} \left(g\left(\frac{\partial f}{\partial x} dy - \frac{\partial f}{\partial y} dx \right) - f\left(\frac{\partial g}{\partial x} dy - \frac{\partial g}{\partial y} dx \right) \right), \tag{3.5}$$

where the integral along $\partial\Omega$ is taken in the clockwise direction.[3]

---

[2] The usual Euclidean Laplacian is negative definite. Throughout this text the hyperbolic Laplacian that we consider is the geometric Laplacian, which is positive definite. This accounts for the difference in sign here.

[3] The two interior parentheses on the right-hand side of (3.5) are the "normal derivatives" which arise in the general Green's formula on a Riemannian manifold.

Proposition 3.1 *The Laplacian Δ is a symmetric operator on $L^2(\mathcal{H})$ with domain* $C_c^\infty(\mathcal{H})$.

Proof Let f and g be two C^∞ functions with support in $\mathcal{H}$. According to Green's formula (3.5),

$$\int_\mathcal{H} (\bar{g}\Delta^e f - f\Delta^e \bar{g})\, dx \wedge dy = \int_C \left(\bar{g}\left(\frac{\partial f}{\partial x}dy - \frac{\partial f}{\partial y}dx\right) - f\left(\frac{\partial \bar{g}}{\partial x}dy - \frac{\partial \bar{g}}{\partial y}dx\right) \right),$$

where C is any contour which completely encircles the union of the supports of f and g. The integral on the right-hand side being clearly zero, we find

$$\int_\mathcal{H} \bar{g}\Delta^e f\, dx \wedge dy = \int_\mathcal{H} f\Delta^e \bar{g}\, dx \wedge dy. \tag{3.6}$$

Now one has

$$\langle \Delta f, g \rangle = \int_\mathcal{H} (\Delta f)\bar{g}\, \frac{dx \wedge dy}{y^2} = -\int_\mathcal{H} \bar{g}\Delta^e f\, dx \wedge dy,$$

and it therefore follows from (3.6) that Δ is symmetric.   □

The aim of this chapter is to decompose the Laplacian acting on the Hilbert space $L^2(\Gamma \backslash \mathcal{H})$, where Γ is a Fuchsian group of the first kind. Before we get to that we exhibit – in the next section – a few important eigenfunctions Δ on $\mathcal{H}$.

3.2   Eigenfunctions of the Laplacian on $\mathcal{H}$

Although it is not directly useful for what follows, it is important to point out that the Laplacian is an elliptic operator and hence a weak solution f, locally in L^2, of the equation

$$\Delta f = \lambda f \quad (\lambda \in \mathbf{C}) \tag{3.7}$$

is necessarily C^∞ – and moreover is real analytic – see, for example, [107, p. 212] for an introduction to the theory of elliptic operators and a proof of their regularizing nature. It is natural therefore to look for solutions to (3.7) among the functions $f \in C^\infty(\mathcal{H})$.

In the $z = x + iy$ coordinates of the upper half-plane, we begin by looking for a solution f of (3.7) which we require to be a function of y only, i.e., constant in x. For $s \neq 1/2$, we land immediately on the following two linearly independent solutions:

$$\frac{1}{2}(y^s + y^{1-s}) \quad \text{and} \quad \frac{1}{2s-1}(y^s - y^{1-s}), \tag{3.8}$$

where $s(1 - s) = \lambda$. When $s = 1/2$, so that $\lambda = 1/4$, we can take the limiting values of the above expressions to obtain

$$y^{1/2} \quad \text{and} \quad y^{1/2} \log y, \tag{3.9}$$

respectively. Of course if $s \neq 1/2$, one can obviously take the simpler pair (y^s, y^{1-s}).

If one just wants f to be 1-periodic in x, the change of variables $f(z) = e(nx)F(2\pi ny)$ ($n \neq 0$) reduces the study of the Eq. (3.7) to that of the ordinary differential equation in F

$$F''(y) + (\lambda y^{-2} - 1)F(y) = 0. \tag{3.10}$$

There are two linearly independent solutions to this equation, namely

$$(2\pi^{-1}y)^{1/2}K_{s-1/2}(y) \sim e^{-y}$$

and

$$(2\pi y)^{1/2}I_{s-1/2}(y) \sim e^y,$$

as $y \to +\infty$, where $K_\nu(y)$ and $I_\nu(y)$ are the standard Bessel functions, see Appendix B. If we assume that f doesn't grow too fast in the vertical strip

$$\{z \in \mathcal{H} \mid |\operatorname{Re}(z)| \leq 1/2\},$$

– more precisely, that $f(z) = o(e^{2\pi y})$ as $y \to \infty$ – then the second solution is excluded and $f(z)$ must be a scalar multiple of $W_s(nz)$. Here W_s denotes the *Whittaker function*:

$$W_s(z) = 2y^{1/2}K_{s-1/2}(2\pi y)\,e(x). \tag{3.11}$$

Note that these functions are *cuspidal*, meaning that for every $y > 0$, the average

$$\int_0^1 W_s(z)\,dx$$

is zero. Whittaker functions play the role of the exponential functions in the hyperbolic analog of Fourier inversion. We will not elaborate upon this since the harmonic analysis of interest to us will take place on quotients of the hyperbolic plane by discrete groups of symmetries Γ under which the Whittaker functions are not in general invariant.

3.2.1   Radial Functions

Let $F : \mathcal{H} \to \mathbf{C}$ be a *radial* function, i.e., a function depending only on the distance from a fixed point $z \in \mathcal{H}$. Assume furthermore that F satisfies the equation $\Delta F = \lambda F$ for a certain complex number λ. Since, in polar coordinates (r, θ) about z, the function F does not depend on θ, it follows from (3.2) that F is a solution to the second-order ordinary differential equation

$$F''(r) + \coth r F'(r) + \lambda F(r) = 0. \tag{3.12}$$

Given a function f on $\mathcal{H}$, one defines the *radialization* of f about a point $z \in \mathcal{H}$, written f_z^{rad}, by:

$$f_z^{\mathrm{rad}}(w) = \int_{S_z} f(Tw)\, dT,$$

where S_z is the isotropy subgroup of z in G, and dT is the normalized Haar measure on S_z. We remark that the groups S_z are pairwise conjugate and can be identified with the special orthogonal group $SO(2)$.

It is clear that the function f_z^{rad} is radial – about z – and satisfies $f_z^{\mathrm{rad}}(z) = f(z)$. Moreover, Δ being an invariant operator, it commutes with the radialization. In particular, the radialization of an eigenfunction is again an eigenfunction.

The function $y^s = \mathrm{Im}(z)^s$ is an eigenfunction. Let us calculate its radialization F_s in polar coordinates (r, θ) about the point i and relative to the vector pointing towards infinity. The point $z(r, \theta)$ with coordinates (r, θ) being $(\cos \theta e^{-r} i + \sin \theta)/(-\sin \theta e^{-r} i + \cos \theta)$, we find:

$$\begin{aligned} F_s(r) &= \frac{1}{\pi} \int_0^\pi (\mathrm{Im}\, z(r, \theta))^s d\theta \\ &= \frac{1}{\pi} \int_0^\pi (\cosh r + \sinh r \cos 2\theta)^{-s} d\theta. \end{aligned} \tag{3.13}$$

By construction, the function $z \mapsto F_s(\rho(z, i))$ is a solution to (3.7) for $\lambda = s(1 - s)$. Since in polar coordinates the function F_s is independent of θ, it is a solution to (3.12).

Equation (3.12) is completely solved – for λ real or complex – by the classical theory. From it we may extract the following proposition.[4]

Proposition 3.2 *Let $\lambda = s(1 - s) \in \mathbf{C}$. The space of solutions of (3.12) on $\mathbf{R}_+^*$ is 2-dimensional and F_s is the unique solution extending continuously to 0 by 1.*

Proof The 2-dimensionality of the solution space is classical. Moreover, by construction F_s is a solution of (3.12), and it is clear that F_s extends continuously to 0 by 1. To prove the uniqueness, we construct a second solution and show that it does not extend continuously to 0.

It will be useful to transform the radial coordinate r by the change of variables $u = (\cosh r - 1)/2$. In these new coordinates, Eq. (3.12) becomes

$$u(u + 1)G''(u) + (2u + 1)G'(u) + s(1 - s)G(u) = 0. \tag{3.14}$$

We have

$$(-\Delta + s(1 - s))\xi^{s-1}(1 - \xi)^{s-1}(\xi + u)^{-s} = s\frac{d}{d\xi}\xi^s(1 - \xi)^s(\xi + u)^{-s-1}, \tag{3.15}$$

where Δ is the Laplacian applied to the u variable (the θ variable being constant). Upon integrating (3.15) by parts, we obtain another solution to Eq. (3.12) (or, rather, (3.14)):

$$G_s(u) = \frac{1}{4\pi} \int_0^1 (\xi(1 - \xi))^{s-1}(\xi + u)^{-s}d\xi. \tag{3.16}$$

It follows from the following lemma that the function G_s tends toward infinity as u tends toward 0. We obtain in particular a solution which is linearly independent of F_s, and the proposition is proved. $\square$

Lemma 3.3 *The integral (3.16) converges absolutely for s in the half-plane $\mathrm{Re}(s) > 0$. It therefore defines a function $u \mapsto G_s(u)$ on $\mathbf{R}_+^*$ satisfying Eq. (3.14). Moreover,*

$$G_s(u) = \frac{1}{4\pi} \log \frac{1}{u} + O(1),$$

as $u \to 0$.

---

[4]Proposition 3.2 implies in particular that the functions F_s and F_{1-s} are equal for $s \in [0, 1]$; we shall in general assume that $s \leqslant 1/2$.

Proof The fact that $G_s(u)$ satisfies Eq. (3.14) follows immediately from (3.15). Put $v = (|s| + 1)u, \eta = (|s| + 1)^{-1}$ and $\sigma = \mathrm{Re}(s)$. We have

$$4\pi G_s(u) = \int_0^1 \left(\frac{\xi(1-\xi)}{\xi + u} \right)^{s-1} \frac{d\xi}{\xi + u} = \int_0^v + \int_v^\eta + \int_\eta^1$$

where

$$\int_0^v = O\left(u^{-\sigma} \int_0^v \xi^{\sigma-1} d\xi \right) = O(1)$$

and

$$\int_\eta^1 = O\left(\int_\eta^1 (1 - \xi)^{\sigma-1} d\xi \right) = O(1).$$

For the remaining integral we use the expansion

$$\left(\frac{\xi(1-\xi)}{\xi + u} \right)^{s-1} = \left(1 - \frac{u + \xi^2}{u + \xi} \right)^{s-1} = 1 + O\left(\frac{u + \xi^2}{u + \xi} \right)$$

to obtain

$$\int_v^\eta = \int_v^\eta \frac{d\xi}{\xi + u} + O\left(\int_v^\eta \frac{u + \xi^2}{(u + \xi)^2} d\xi \right)$$

$$= \log \frac{u + \eta}{u + v} + O(1) = \log \frac{1}{u} + O(1). \qquad \square$$

3.3 Invariant Integral Operators on $\mathcal{H}$

Let $k :]-\infty, +\infty[\to \mathbf{C}$ be an even C^∞-function. By letting[5] $k(z, w) = k(\rho(z, w))$, with $z, w \in \mathcal{H}$, we obtain a C^∞-function of z and w which depends only on the distance between them. One calls such a function a *point-pair invariant*.[6]

Lemma 3.4 *Let $k(z, w)$ be a point-pair invariant. Then*

$$\Delta_z k(z, w) = \Delta_w k(z, w). \tag{3.17}$$

Here the subscript indicates the variable on which one applies the Laplacian.

---

[5]Recall that $\rho(z, w)$ is the hyperbolic distance between two points z and w in $\mathcal{H}$.

[6]Here the term "invariant" comes from the fact that for every $g \in G$ and for all $z, w \in \mathcal{H}$ we have $k(gz, gw) = k(z, w)$.

Proof In polar coordinates about the point w, we obtain, since $k(z, w) = k(r)$,

$$\Delta_z k(z, w) = -k''(r) - \coth r\, k'(r).$$

But in the polar coordinates centered about z, we obtain the same expression for $\Delta_w k(z, w)$. □

A point-pair invariant k defines an *invariant integral operator*:

$$T_k : f \longmapsto \int_{\mathcal{H}} k(\cdot, w)f(w)\, d\mu(w), \tag{3.18}$$

where $d\mu$ is the Riemannian measure. In what follows we assume that the functions k and f are C^∞ and are chosen in a way so that the integral in (3.18) converges absolutely; concerning this latter assumption, one loses very little if one simply takes k to be of compact support.

Theorem 3.5 *The invariant integral operators commute with the action of the Laplacian:*

$$T_k \Delta = \Delta T_k.$$

Proof Let $f \in C_c^\infty(\mathcal{H})$ (the domain of Δ). According to Proposition 3.1,

$$\int_{\mathcal{H}} k(z, w)(\Delta_w f(w))\, d\mu(w) = \int_{\mathcal{H}} (\Delta_w k(z, w))f(w)\, d\mu(w).$$

Then, from Lemma 3.4, the integral on the right is equal to

$$\int_{\mathcal{H}} (\Delta_z k(z, w))f(w)\, d\mu(w). \qquad \square$$

Lemma 3.6 *For $\lambda \in \mathbf{C}$ and $z \in \mathcal{H}$, there exists a unique function $\omega_\lambda(z, w)$ in w, radial about the point z and such that*

1. $\omega_\lambda(z, z) = 1$, and
2. $\Delta_w \omega_\lambda(z, w) = \lambda \omega_\lambda(z, w)$.

In fact, one has

$$\omega_\lambda(z, w) = F_s(\rho(z, w)),$$

where $\lambda = s(1 - s)$.

Proof Let $F(r) = \omega_\lambda(z, w)$ where $r = \rho(z, w)$. The function F satisfies Eq. (3.12). According to Proposition 3.2 the normalization $F(0) = \omega_\lambda(z, z) = 1$ forces $F = F_s$. □

We remark that if $\lambda \in [0, 1/4]$, then $s \in [0, 1]$. Since, according to Proposition 3.2, we have $F_s = F_{1-s}$, we can assume that $s \leq 1/2$. When $s < 1/2$ the integral

$$\int_0^\pi (1 + \cos(2\theta))^{-s} d\theta$$

is absolutely convergent. It then follows from the explicit expression (3.13) of F_s that the function $\omega_\lambda(z, w)$ grows as

$$(\text{const}) e^{-s\rho(z,w)} = (\text{const}) e^{-\frac{1}{2}(1-\sqrt{1-4\lambda})\rho(z,w)}. \tag{3.19}$$

Theorem 3.7 *Every eigenfunction of Δ is also an eigenfunction of the invariant integral operators. More precisely, for every compactly supported C^∞ point-pair invariant k, there exists a function $h : \mathbf{C} \to \mathbf{C}$ such that if $f \in C^\infty(\mathcal{H})$ is a Δ-eigenfunction with eigenvalue $\lambda \in \mathbf{C}$ then $T_k f = h(\lambda) f$, i.e.,*

$$\int_{\mathcal{H}} k(z, w) f(w) \, d\mu(w) = h(\lambda) f(z). \tag{3.20}$$

Proof Let $f \in C^\infty(\mathcal{H})$ be an eigenfunction of the Laplacian with eigenvalue λ. For all $z \in \mathcal{H}$, the function f_z^{rad} is radial about z and satisfies $f_z^{\mathrm{rad}}(z) = f(z)$. It then follows from Lemma 3.6 that

$$f_z^{\mathrm{rad}}(w) = \omega_\lambda(z, w) f(z).$$

Let us first show that

$$\int_{\mathcal{H}} k(z, w) f(w) \, d\mu(w) = \int_{\mathcal{H}} k(z, w) f_z^{\mathrm{rad}}(w) \, d\mu(w). \tag{3.21}$$

Indeed,

$$\int_{\mathcal{H}} k(z, w) f_z^{\mathrm{rad}}(w) \, d\mu(w) = \int_{\mathcal{H}} k(z, w) \, d\mu(w) \int_{S_z} f(Tw) dT$$

$$= \int_{S_z} dT \int_{\mathcal{H}} k(z, w) f(Tw) \, d\mu(w)$$

$$= \int_{S_z} dT \int_{\mathcal{H}} k(z, T^{-1}w) f(w) \, d\mu(w)$$

$$= \int_{S_z} dT \int_{\mathcal{H}} k(z, w) f(w) \, d\mu(w)$$

$$= \int_{\mathcal{H}} k(z, w) f(w) \, d\mu(w).$$

Here we have written S_z for the isotropy subgroup of z in G and dT for the normalized Haar measure on S_z.

We therefore have

$$\int_{\mathcal{H}} k(z, w) f(w)\, d\mu(w) = \int_{\mathcal{H}} k(z, w) f_z^{\mathrm{rad}}(w)\, d\mu(w)$$

$$= f(z) \int_{\mathcal{H}} k(z, w) \omega_\lambda(z, w)\, d\mu(w).$$

Since k is of compact support, the integral

$$h(\lambda) := \int_{\mathcal{H}} k(z, w) \omega_\lambda(z, w)\, d\mu(w) \tag{3.22}$$

is well-defined. Finally, for a given λ, this integral doesn't depend on z: this follows from the transitivity of the G-action on $\mathcal{H}$ and the fact that for $g \in G$,

$$\int_{\mathcal{H}} k(gz, w) \omega_\lambda(gz, w)\, d\mu(w) = \int_{\mathcal{H}} k(z, g^{-1}w) \omega_\lambda(z, g^{-1}w)\, d\mu(w)$$

$$= \int_{\mathcal{H}} k(z, w) \omega_\lambda(z, w)\, d\mu(w). \qquad \square$$

The converse of Theorem 3.7 is also true, in the follows precise sense.

Theorem 3.8 *If f is an eigenfunction for all of the invariant integral operators for which the kernel k belongs to $C_c^\infty(\mathbf{R})$, then f is an eigenfunction for Δ.*

Proof Let f be an eigenfunction for all of the invariant integral operators whose kernel k belongs to $C_c^\infty(\mathbf{R})$. (Then f is necessarily C^∞.)

If f is constantly equal to 0 there is nothing to show. Assume then that f is non-zero at at least one point. It is not difficult to show – for example, by choosing a sequence of smooth approximations to the delta function [18, p. 70][7] – that there exists a point-pair invariant $k \in C_c^\infty(\mathbf{R})$ such that Eq. (3.20) is satisfied with $h(\lambda) \neq 0$. By applying Δ to Eq. (3.20), we find

$$\int_{\mathcal{H}} \Delta_z k(z, w) f(w)\, d\mu(w) = h(\lambda) \Delta f(z).$$

But $\Delta_z k(z, w)$ is again the compactly supported C^∞ kernel of an invariant integral operator, and by the hypothesis on f the above integral is therefore equal (as a function of z) to a scalar multiple of f. Since $h(\lambda) \neq 0$ Theorem 3.8 is proved. $\square$

---

[7]One could just as well refer to the proof of Lemma 3.29.

The condition in Theorem 3.7 that k be of compact support is not essential. It is nevertheless necessary to impose a decay at infinity condition. For example, it will suffice to assume that there is a real number $\delta > 0$ such that

$$|k(\rho)| = O(e^{-\rho(1+\delta)}). \qquad (3.23)$$

This is a natural condition: one wants the integral

$$\int_{\mathcal{H}} k(z, w) \, d\mu(w) = \int_0^{+\infty} k(\rho) \sinh \rho \, d\rho$$

to converge absolutely.

3.4   The Selberg Transform

In this section we seek to calculate the function h of Theorem 3.7. To this end, it is convenient to introduce the function U defined by

$$U(\cosh \rho) = k(\rho), \qquad (3.24)$$

so that

$$U\left(1 + \frac{|z - w|^2}{2 \operatorname{Im} z \operatorname{Im} w}\right) = k(z, w). \qquad (3.25)$$

The function $U : [1, +\infty[\to \mathbf{C}$ then satisfies: $U(t) = O(t^{-(1+\delta)})$.

Since $\lambda = s(1 - s)$ takes on all complex values as s runs through $\mathbf{C}$, we can view the function h as a function of a complex parameter r related to s by the equation $s = 1/2 + ir$. Note that $y = \operatorname{Im} z$. Recall that the function $y^{1/2+ir}$ $(r \in \mathbf{C})$ is an eigenfunction of the Laplacian Δ on $\mathcal{H}$ with eigenvalue $(r^2 + 1/4)$. By abuse of notation we write again

$$h(r) = h(\lambda) = h(r^2 + 1/4);$$

we're looking to calculate $h(r)$ as a function of U.

The formula (3.20) applied to the eigenfunction $z \mapsto y^{1/2+ir}$ implies that

$$h(r) = \int_{\mathcal{H}} k(z, i) y^{1/2+ir} d\mu(z). \qquad (3.26)$$

The function h is the *Selberg transform* $S(f)$ of the function $f : z \mapsto k(z, i)$. With the help of (3.22) it may be rewritten as

$$S(f)(r) = \int_{\mathcal{H}} f(z) \omega_{r^2+1/4}(i, z) \, d\mu(z). \qquad (3.27)$$

The Selberg transform (3.27) is well-defined if $f \in C_c^\infty(\mathcal{H})$. Nevertheless, on the hyperbolic circle $\{z \in \mathcal{H} \mid \rho(z, i) = \rho\}$ we have[8] $y = \operatorname{Im} z \leqslant e^\rho$. Under the rapid decay condition (3.23) of k, the integral (3.26) is therefore absolutely convergent as long as

$$|\operatorname{Im} r| < \frac{1}{2} + \delta', \tag{3.28}$$

where δ' is a constant such that $0 < \delta' < \delta$. Under this hypothesis, by a repeated change of variables $(1 + x^2 + y^2)/2y = t(x) = t$ (with $dx = y\,dt/x$), $y = e^u$ and $t = \cosh \rho$, we obtain the following lemma.

Lemma 3.9 *We have*

$$h(r) = \int_{-\infty}^{+\infty} \int_0^{+\infty} y^{1/2+ir} U\left(\frac{1 + x^2 + y^2}{2y}\right) \frac{dx\,dy}{y^2}$$

$$= \sqrt{2} \int_{-\infty}^{+\infty} e^{iru} \int_{|u|}^{+\infty} \frac{k(\rho) \sinh \rho}{\sqrt{\cosh \rho - \cosh u}}\,d\rho\,du.$$

Let

$$g(u) = \sqrt{2} \int_{|u|}^{+\infty} \frac{k(\rho) \sinh \rho}{\sqrt{\cosh \rho - \cosh u}}\,d\rho. \tag{3.29}$$

The function h is therefore equal to the Fourier transform[9] $\int_{-\infty}^{+\infty} e^{iru} g(u)\,du$ of g. We remark that the change of variables $x = \cosh u$ and $t = \cosh \rho$ transforms the expression (3.29) into

$$g(\operatorname{argch} x) = \sqrt{2} \int_x^{+\infty} \frac{k(\operatorname{argch} t)}{\sqrt{t - x}}\,dt. \tag{3.30}$$

One says that the function $G(x) = g(\operatorname{argch} x)$ is the *Abel transform* of the function $\Phi(t) = k(\operatorname{argch} t)$.

Lemma 3.10 *Let* $\Phi : [1, +\infty[\to \mathbf{C}$ *be a compactly supported* C^∞-*function. Then the Abel transform of* Φ,

$$Q(x) = \int_x^{+\infty} \frac{\Phi(t)}{\sqrt{t - x}}\,dt, \quad x \geqslant 1$$

---

[8]This can be easily seen if one recalls that this particular hyperbolic circle is a Euclidean circle whose maximal ordinate is on the imaginary axis.

[9]Note the change in convention relative to the introduction.

defines a compactly supported C^∞-function and this transformation is invertible via the formula:

$$\Phi(x) = -\frac{1}{\pi}\int_x^{+\infty}\frac{dQ(t)}{\sqrt{t-x}}.$$

Proof First note that

$$Q(x) = \int_0^{+\infty}\frac{\Phi(x+\xi^2)}{\xi}2\xi d\xi = 2\int_0^{+\infty}\Phi(x+\xi^2)d\xi.$$

The function Q is then clearly C^∞ and of compact support, with derivative

$$Q'(x) = 2\int_0^{+\infty}\Phi'(x+\xi^2)d\xi. \tag{3.31}$$

We then find that

$$\begin{aligned}
\Phi(x) &= -\int_0^{+\infty}d[\Phi(x+r^2)]\\
&= -\int_0^{+\infty}2\Phi'(x+r^2)\,rdr\\
&= -\frac{4}{\pi}\int_0^{\pi/2}\int_0^{+\infty}\Phi'(x+r^2)\,rdr\,d\theta\\
&= -\frac{4}{\pi}\int_0^{+\infty}\int_0^{+\infty}\Phi'(x+\eta^2+\xi^2)\,d\eta\,d\xi\\
&= -\frac{2}{\pi}\int_0^{+\infty}Q'(x+\eta^2)\,d\eta\quad\text{(d'après (3.31))}\\
&= -\frac{1}{\pi}\int_0^{+\infty}\frac{d(Q(x+\eta^2))}{\eta}\\
&= -\frac{1}{\pi}\int_x^{+\infty}\frac{dQ(t)}{\sqrt{t-x}}. \qquad\qquad\qquad\square
\end{aligned}$$

In particular,

$$k(\operatorname{argch} t) = -\frac{1}{\sqrt{2\pi}}\int_t^{+\infty}\frac{dg(\operatorname{argch} x)}{\sqrt{x-t}},$$

and the formula (3.29) inverts to

$$k(\rho) = -\frac{1}{\sqrt{2}\,\pi} \int_{\cosh\rho}^{+\infty} \frac{dg(\operatorname{argch}x)}{\sqrt{x - \cosh\rho}}$$

$$= -\frac{1}{\sqrt{2}\,\pi} \int_{\rho}^{+\infty} \frac{dg(u)}{\sqrt{\cosh u - \cosh\rho}}.$$

We admit the following proposition, which we include for completeness but which will not be used in the text. Let[10] $PW(\mathbf{C})$ be the space of even holomorphic functions $r \mapsto h(r)$ on $\mathbf{C}$ such that there exists a real number R (the *type* of h) with the property that for every $N \in \mathbf{N}$,

$$h(r) = O_N\left(\frac{e^{R|\operatorname{Im}(r)|}}{(1 + |r|)^N}\right).$$

Proposition 3.11 *The Selberg transform S is a bijection of $C_c^\infty(\mathcal{H})$ onto $PW(\mathbf{C})$ whose inverse is given by*

$$(S^{-1}h)(z) = \frac{1}{2\pi} \int_0^{+\infty} h(r)\omega_{r^2+1/4}(i, z) r \tanh(\pi r)\, dr.$$

Moreover, if $h \in PW(\mathbf{C})$ is of type R, then $f = S^{-1}h \in C^\infty(\mathcal{H})$ has support in the hyperbolic disc of center i and radius R.

We already remarked that it is not necessary to assume that k is of compact support in Theorems 3.7 and 3.8. The hypothesis (3.23) is not optimal; it is more convenient to give a condition on h. The most natural one is to ask for the existence of an $\varepsilon > 0$ such that h is an even holomorphic function defined on the strip $\mathcal{B}_\varepsilon = \{r \in \mathbf{C} \mid |\operatorname{Im}(r)| < \frac{1}{2} + \varepsilon\}$ and satisfying

$$h(r) = O((1 + |r|^2)^{-1-\varepsilon}).$$

3.5 A Family of Examples: The Heat Kernel

Let $p : \mathcal{H} \times \mathcal{H} \times (0, +\infty) \to \mathbf{R}$ be a continuous function of class C^2 in the first two variables (z, w) and of class C^1 in the last variable t. One calls p a *fundamental solution to the heat equation* on $\mathcal{H}$ if

1. $\partial p/\partial t = -\Delta_z p$,
2. $p(z, w, t) = p(w, z, t)$, and
3. $\lim_{t\downarrow 0} \int_{\mathcal{H}} p(z, w, t)f(w)\, d\mu(w) = f(z)$, locally uniformly in z and for every compactly supported continuous function f.

---

[10]The notation "PW" is in reference to the classical Paley-Wiener theorem, see [106].

Our goal now is to construct a fundamental solution to the heat equation on $\mathcal{H}$. This amounts to constructing the kernel – a function of t – which to a pair of points (z, w) in $\mathcal{H}$ associates the value $p(z, w, t)$. Denote this kernel by k. We shall first proceed formally as a way of guessing an expression for k; then we verify rigorously that this kernel furnishes a fundamental solution to the heat equation on $\mathcal{H}$.

As a function of t we can differentiate k. We denote by $\dot{k} = \partial k / \partial t$ the differentiated kernel. We would like for k to satisfy the equation

$$\dot{k} = -\Delta_z k. \tag{3.32}$$

Upon integrating (3.32) against an eigenfunction of the Laplacian with eigenvalue $(1/4 + r^2)$, and using the fact that Δ is symmetric, one deduces from Theorem 3.7 that the Selberg transform h of k satisfies the ordinary differential equation

$$\dot{h}(r) = -\left(\frac{1}{4} + r^2\right)h(r). \tag{3.33}$$

Moreover, condition 3 forces h to tend towards the constant function 1 as t tends toward 0. We then expect to have (at least formally)

$$h(r) = e^{-t/4 - tr^2}.$$

One would then like to take $k(z, w) = (\mathcal{S}^{-1}h)(z)$. Having done so, we continue to write g for the function (3.29). We then expect that $h(r) = \int_{-\infty}^{+\infty} e^{iru} g(u) \, du$ and thus – by Fourier inversion – that g should be of the form

$$\begin{aligned}
g(u) &= \frac{1}{2\pi} \int_{-\infty}^{+\infty} e^{-iru} h(r) \, dr \\
&= \frac{e^{-t/4}}{2\sqrt{\pi}} \int_{\mathbf{R}} e^{-\pi t(r/\sqrt{\pi})^2} e(-(r/\sqrt{\pi})(u/2\sqrt{\pi})) \, d(r/\sqrt{\pi}) \\
&= \frac{e^{-t/4}}{2\sqrt{\pi}} \int_{\mathbf{R}} e^{-\pi tr^2} e(-r(u/2\sqrt{\pi})) \, dr.
\end{aligned}$$

It then follows from Lemma 1.1 that

$$g(u) = \frac{e^{-t/4}}{2\sqrt{\pi t}} e^{-u^2/4t}.$$

By Abel inversion one expects that

$$\begin{aligned}
k(z, w) &= -\frac{1}{\sqrt{2}\pi} \frac{e^{-t/4}}{2\sqrt{\pi t}} \int_{\rho(z,w)}^{+\infty} \frac{d(e^{-u^2/4t})}{\sqrt{\cosh u - \cosh \rho(z, w)}} \\
&= \frac{\sqrt{2}}{(4\pi t)^{3/2}} e^{-t/4} \int_{\rho(z,w)}^{+\infty} \frac{u e^{-u^2/4t}}{\sqrt{\cosh u - \cosh \rho(z, w)}} du.
\end{aligned}$$

All of the above computations are formal but they allow us to "guess" the following theorem.

Theorem 3.12 *The function* $p_{\mathcal{H}} : \mathcal{H} \times \mathcal{H} \times (0, +\infty) \to \mathbf{R}$ *given by*

$$p_{\mathcal{H}}(z, w, t) = \frac{\sqrt{2}}{(4\pi t)^{3/2}} e^{-t/4} \int_{\rho(z,w)}^{+\infty} \frac{u e^{-u^2/4t}}{\sqrt{\cosh u - \cosh \rho(z, w)}} \, du$$

is a fundamental solution to the heat equation.

Proof It suffices to verify the expression that we "guessed" indeed provides a solution. We write $p_t(z, w) = p_{\mathcal{H}}(z, w, t)$. This is a point-pair invariant which corresponds to an even C^∞-function, which we again denote by $p_t : (-\infty, +\infty) \to \mathbf{C}$.

Lemma 3.13

$$p_t(\rho) = O\!\left(t^{-1} e^{-\rho^2/8t}\right).$$

Proof Let us fix $\rho > 0$. Then

$$\int_{\rho}^{2\rho} \frac{r e^{-u^2/4t}}{\sqrt{\cosh u - \cosh \rho}} \, du \leqslant 2\rho e^{-\rho^2/4t} \int_{\rho}^{2\rho} \frac{du}{\sqrt{(u - \rho)\sinh \rho}} \leqslant 4\rho e^{-\rho^2/4t}.$$

The inequality $x e^{-x^2} \leqslant e^{-x^2/2}$ implies

$$4\rho e^{-\rho^2/4t} = O\!\left(\sqrt{t} e^{-\rho^2/8t}\right).$$

Next, by using $\cosh u - \cosh \rho \geqslant \frac{1}{2}(u - \rho)^2$ and writing $r = u - 2\rho$, we find

$$\int_{2\rho}^{+\infty} \frac{u e^{-u^2/4t}}{\sqrt{\cosh u - \cosh \rho}} \, dr \leqslant 2\sqrt{2} \int_{2\rho}^{+\infty} e^{-u^2/4t} du$$

$$\leqslant 2\sqrt{2} \, e^{-\rho^2/t} \int_{0}^{+\infty} e^{-r^2/4t} dr$$

$$= O\!\left(\sqrt{t} e^{-\rho^2/t}\right). \qquad \square$$

This lemma allows us to justify all of the formal computations above. One verifies in particular that $p_{\mathcal{H}} \in C^\infty(\mathcal{H} \times \mathcal{H} \times (0, +\infty))$ is a solution to the equation[11]

$$\left(\Delta_z + \frac{\partial}{\partial t} \right) p_{\mathcal{H}}(z, w, t) = 0 \tag{3.34}$$

and that

$$\int_0^{+\infty} p_t(\rho) \sinh \rho \, d\rho = 1. \tag{3.35}$$

Equation (3.34) implies that $p_{\mathcal{H}}$ satisfies the first condition in the definition of a solution to the heat equation. The second condition is satisfied by construction. And finally, since p_t is positive and since, according to Lemma 3.13,

$$p_t \longrightarrow 0 \quad \text{as } t \longrightarrow 0,$$

uniformly on compacta avoiding 0, condition 3 follows from (3.35). □

We emphasize the fact that for for every $t > 0$, the kernel $p_t(z, w) = p_{\mathcal{H}}(z, w, t)$ is a point-pair invariant which satisfies the decay property (3.23); the results of the preceding sections therefore apply to it.

3.6   The Laplacian on $\Gamma \backslash \mathcal{H}$

We are setting our sites on the spectral decomposition of $L^2(\Gamma \backslash \mathcal{H})$, where Γ is a cocompact Fuchsian group or at least of the first kind. We fix then a Fuchsian group of the first kind Γ. The Laplacian Δ defines an unbounded operator on $L^2(\Gamma \backslash \mathcal{H})$. In this section we prove that this operator is symmetric and positive.

The Riemannian metric on $\mathcal{H}$ induces a norm on each tangent space and thus on the space of 1-forms on $\mathcal{H}$. One can therefore speak of a bounded 1-form on $\mathcal{H}$.

Lemma 3.14 *Let ω be a C^∞ differential 1-form, bounded and Γ-invariant on $\mathcal{H}$. Then*

$$\int_{\Gamma \backslash \mathcal{H}} d\omega = 0.$$

Proof First assume that $S = \Gamma \backslash \mathcal{H}$ is compact. Since ω is Γ-invariant it comes from a 1-form on S. Stokes' theorem implies that $\int_S d\omega = \int_{\partial S} \omega = 0$. If Γ is cocompact, the boundary of S is empty and the lemma is proved.

---

[11]One can just as well verify this directly by using the formula for $p_{\mathcal{H}}(z, w, t)$, but it is a bit tedious.

The lemma remains valid when $\Gamma\backslash\mathcal{H}$ is of finite area since ω is bounded whereas the length of a closed horocycle in $\Gamma\backslash\mathcal{H}$ tends toward 0 as it approaches infinity. $\quad\square$

Lemma 3.15 *Let f and g be two functions in $C^\infty(\Gamma\backslash\mathcal{H})$. The function*

$$z \longmapsto (y\nabla f(z)) \cdot \overline{(y\nabla g(z))}$$

is Γ-invariant.

Proof Let $\gamma = \begin{pmatrix} a & b \\ c & d \end{pmatrix} \in \Gamma$, and let $w = \gamma(z) = \frac{az+b}{cz+d}$. Then

$$\frac{\partial}{\partial z} = \frac{\partial w}{\partial z}\frac{\partial}{\partial w} + \frac{\partial \overline{w}}{\partial z}\frac{\partial}{\partial \overline{w}} = (cz+d)^{-2}\frac{\partial}{\partial w}.$$

One deduces from this that

$$(w - \overline{w})\frac{\partial}{\partial w} = \left(\frac{cz+d}{c\overline{z}+d}\right)(z - \overline{z})\frac{\partial}{\partial z}. \tag{3.36}$$

Denote by L the differential operator

$$L = -iy\frac{\partial}{\partial x} + y\frac{\partial}{\partial y} = -(z - \overline{z})\frac{\partial}{\partial \overline{z}}.$$

Equation (3.36) implies

$$\left(\frac{cz+d}{c\overline{z}+d}\right)^{-1}(Lf)\left(\frac{az+b}{cz+d}\right) = \left[(w - \overline{w})\frac{\partial}{\partial w}\right]f(w)$$

$$= \left[(z - \overline{z})\frac{\partial}{\partial z}\right]f(w)$$

$$= L\left(f\left(\frac{az+b}{cz+d}\right)\right).$$

One sees then that Lf and Lg – and hence also $y\nabla f$ and $y\nabla g$ – belong to the space of C^∞-functions ϕ on $\mathcal{H}$ such that

$$\phi(z) = \left(\frac{cz+d}{c\overline{z}+d}\right)\phi\left(\frac{az+b}{cz+d}\right), \quad \begin{pmatrix} a & b \\ c & d \end{pmatrix} \in \Gamma.$$

It follows that the function $(y\nabla f) \cdot \overline{(y\nabla g)}$ is Γ-invariant. $\quad\square$

Let

$$\mathcal{D}(\Gamma\backslash\mathcal{H}) = \{f \in C^\infty(\Gamma\backslash\mathcal{H}) \mid f \text{ bounded and } \Delta f \text{ bounded}\}.$$

It is clear that $\mathcal{D}(\Gamma\backslash\mathcal{H})$ is dense in $L^2(\Gamma\backslash\mathcal{H})$ and we have already remarked that an eigenfunction is necessarily C^∞. By the *Laplacian* on the surface $\Gamma\backslash\mathcal{H}$ we mean the (unbounded) operator $(\Delta, \mathcal{D}(\Gamma\backslash\mathcal{H}))$ on the Hilbert space $L^2(\Gamma\backslash\mathcal{H})$.

Proposition 3.16 *The Laplacian Δ is an (unbounded) symmetric positive operator on $L^2(\Gamma\backslash\mathcal{H})$.*

Proof Let f and g be two functions in $\mathcal{D}(\Gamma\backslash\mathcal{H})$. Consider the form

$$\omega = g\left(\frac{\partial f}{\partial x}dy - \frac{\partial f}{\partial y}dx\right).$$

The 1-form df is Γ-invariant; moreover, the 2-form

$$df \wedge \left(\frac{\partial f}{\partial x}dy - \frac{\partial f}{\partial y}dx\right) = |y\nabla f|^2 d\mu(z)$$

is Γ-invariant by Lemma 3.15. Thus the form ω is Γ-invariant and we can apply Lemma 3.14 to it. We then deduce from (3.4) that

$$\langle \Delta f, g\rangle = -\int_D \bar{g}\Delta^e f dx \wedge dy = \int_D (\nabla \bar{g})\cdot(\nabla f)\, dx \wedge dy,$$

where D is a fundamental domain for Γ. In particular

$$\langle \Delta f, g\rangle = \langle f, \Delta g\rangle$$

and

$$\langle \Delta f, f\rangle = \int_D |\nabla f|^2\, dx \wedge dy \geqslant 0. \qquad\qquad \square$$

We now write

$$\mathrm{grad} f = y\nabla f.$$

It follows from Lemma 3.15 that the function $\mathrm{grad} f \cdot \overline{\mathrm{grad} g}$ is Γ-invariant and the proof of Proposition 3.16 shows that

$$\langle \Delta f, g\rangle = \int_{\Gamma\backslash\mathcal{H}} \mathrm{grad} f \cdot \overline{\mathrm{grad} g}\, d\mu(z). \qquad\qquad (3.37)$$

Proposition 3.16 implies that the eigenvalue $\lambda = s(1-s)$ of an eigenfunction $f \in \mathcal{D}(\Gamma\backslash\mathcal{H})$ is real and non-negative. Thus, either $s = 1/2 + it$ with $t \in \mathbf{R}$, or $0 \leqslant s \leqslant 1$.

3.7 Integral Operators on $\Gamma\backslash\mathcal{H}$

We begin by consider the case where Γ is a cocompact Fuchsian group. Such a group operates properly discontinuously on the hyperbolic plane $\mathcal{H}$ with a compact fundamental domain D that we fix once and for all. We shall realize the spectral decomposition of the Laplacian Δ in $L^2(\Gamma\backslash\mathcal{H})$ with the help of the invariant integral operators. Recall that such an operator is given by a C^∞ point-pair invariant

$$k(z,w) = k(\rho(z,w)), \quad z, w \in \mathcal{H}, \tag{3.38}$$

which induces

$$(T_k f)(z) = \int_{\mathcal{H}} k(z,w) f(w) \, d\mu(w)$$

for $f : \mathcal{H} \to \mathbf{C}$. If we assume that f is Γ-invariant (in other words, that f descends to a function $f : \Gamma\backslash\mathcal{H} \to \mathbf{C}$), we can write

$$(T_k f)(z) = \int_D K(z,w) f(w) \, d\mu(w),$$

where the kernel K is given by the series

$$K(z,w) = K_\Gamma(z,w) = \sum_{\gamma \in \Gamma} k(z, \gamma w). \tag{3.39}$$

We call such a kernel an *automorphic kernel*. We shall want to first take care of all convergence issues.

We shall always assume that the C^∞ function k satisfies the decay condition (3.23): $|k(\rho)| = O\left(e^{-\rho(1+\delta)}\right)$, for a certain real number $\delta > 0$.

Lemma 3.17 *Let $z \in D$. For $m = 0, 1, \ldots$, we put*

$$\Gamma_m = \{\gamma \in \Gamma \mid m \leqslant \rho(z, \gamma z) < m + 1\}.$$

Then Γ_m is of size $|\Gamma_m| = O(e^m)$ as $m \to +\infty$.

Proof Let d be the diameter of D. The subsets $\gamma(D)$, $\gamma \in \Gamma_m$, are contained in the disk of center z and radius $r = m + 1 + d$. The area of this disk is equal to $2\pi(\cosh r - 1)$. Since the subsets are disjoint, one immediately deduces the lemma.

$\square$

Lemma 3.18 *Let k be a C^∞ point-pair invariant satisfying* (3.23). *For every compact subset $A \subset \mathcal{H}$ and for all $\varepsilon > 0$ there exists a finite subset $\Lambda \subset \Gamma$ such that*

$$\sum_{\gamma \in \Gamma - \Lambda} |k(z, \gamma w)| < \varepsilon \quad \text{for all } z, w \in A.$$

Proof We fix an integer $M \geqslant 1$ and take $\Lambda = \bigcup_{m=0}^{M} \Gamma_m$. Then

$$\sum_{\gamma \in \Gamma - \Lambda} |k(z, \gamma w)| \leqslant \sum_{m=M+1}^{+\infty} |\Gamma_m| k(m).$$

It then follows from (3.23) and from Lemma 3.17 that there exists a constant c, not depending on M, such that

$$\sum_{\gamma \in \Gamma - \Lambda} |k(z, \gamma w)| \leqslant c \sum_{m=M+1}^{+\infty} e^{-\delta m}.$$

The result follows by taking M sufficiently large.   □

In view of Lemmas 3.17 and 3.18 we can give a legitimate sense to the function K defined by the sum (3.39). Recall that a kernel K is said to be *symmetric* if $K(z, w) = K(w, z)$.

Lemma 3.19 *The automorphic kernel K is bi-Γ-invariant:*

$$K(\gamma z, \delta w) = K(z, w) \quad \text{for all } \gamma, \delta \in \Gamma.$$

It induces a symmetric kernel $K : \Gamma \backslash \mathcal{H} \times \Gamma \backslash \mathcal{H} \to \mathbf{C}$.

Proof Since $k(z, w)$ depends only on the distance, we have $k(\gamma z, g\delta w) = k(z, \gamma^{-1} g\delta w)$. When g runs through Γ, $\gamma^{-1} g\delta$ runs through Γ as well, and the bi-Γ-invariance of K follows. The symmetry comes from

$$\sum_{\gamma \in \Gamma} k(z, \gamma w) = \sum_{\gamma \in \Gamma} k(\gamma w, z) = \sum_{\gamma \in \Gamma} k(w, \gamma^{-1} z).$$ □

We now concern ourselves with the differentiability of the automorphic kernels above. In practice, point-pair invariants depend frequently on an additional real parameter – such is the case, for example, for the heat kernel. The following lemma allows one to study the differentiability with respect to this supplementary parameter.

We thus let T denote the space of parameters on which a given automorphic kernel will depend. We shall always take T to be an open interval of the real line.

Lemma 3.20 *Let $k = k(\rho, t) : (-\infty, +\infty) \times T \to \mathbf{C}$ be an even function in the first variable, C^∞ (as a function of two variables) and whose partial derivatives satisfy*

$$k^{(n,v)}(\rho, t) = O(e^{-\rho(1+\delta)})$$

on $[0, \infty) \times T$ for a certain strictly positive constant δ. Then K belongs to $C^\infty(\Gamma \backslash \mathcal{H} \times \Gamma \backslash \mathcal{H} \times T, \mathbf{C})$.

Idea of the proof Let $U \subset \mathcal{H}$ be an open disk and let $z \in \mathcal{H}$ be a point at non-zero distance from U. Then

$$\Gamma = \bigcup_{m=0}^{+\infty} \Gamma_m$$

where $\Gamma_m = \{\gamma \in \Gamma \mid m \leq \rho(z, \gamma z) < m + 1\}$. According to Lemma 3.17, $|\Gamma_m| = O(e^m)$. As in the proof of Lemma 3.18, it suffices to show that for each partial derivative $\mathcal{D}$, there exists constants depending only on U such that

$$\mathcal{D}[k(z, \gamma w)] = O(e^{-m(1+\delta)}), \tag{3.40}$$

whenever $z, w \in U$ and $\gamma \in \Gamma_m$. Here the notation $\mathcal{D}[.]$ means that the derivative is applied to the function $(z, w) \mapsto k(z, \gamma w)$. This derivative is of the form

$$\mathcal{D}[k(z, \gamma w)] = \sum_{j=1}^{v} (\mathcal{D}_T k^{(j)})(\rho(z, \gamma w)) D_j[\rho(z, \gamma w)],$$

where $\mathcal{D}_T$ is a partial derivative with respect to T, $k^{(j)} = \partial^j k / \partial \rho^j$, and D_j is a partial derivative of order j with respect to the variables in $\mathcal{H} \times \mathcal{H}$. The result follows by showing[12] that each $|D_j[\rho(z, \gamma w)]|$ is uniformly bounded for all $z, w \in U$ and $\gamma \in \Gamma$. □

3.7.1 The Heat Kernel

The above results apply notably to the family of examples that we have already considered: the heat kernel. In particular, we obtain the following theorem.

---

[12]This is a fundamental property of hyperbolic geometry.

Theorem 3.21 *Let $p(z, w, t)$ be the heat kernel of the hyperbolic plane and let*

$$P(z, w, t) = P_\Gamma(z, w, t) = \sum_{\gamma \in \Gamma} p(z, \gamma w, t).$$

The function P defines an element in $C^\infty(\Gamma \backslash \mathcal{H} \times \Gamma \backslash \mathcal{H} \times (0, +\infty))$ which, moreover, is a fundamental solution to the heat equation on $\Gamma \backslash \mathcal{H}$.

Proof Lemma 3.13, and its proof, implies that the function p_t satisfies the hypotheses of Lemma 3.20. From this it follows that P belongs to $C^\infty(\Gamma \backslash \mathcal{H} \times \Gamma \backslash \mathcal{H} \times (0, +\infty))$. It is immediate that the equation $\Delta_z p = -\partial p / \partial t$ implies that $\Delta_z P = -\partial P / \partial t$. The symmetry of P results from Lemma 3.19.

Let us now verify the initial condition. Let $f : \Gamma \backslash \mathcal{H} \to \mathbf{R}$ be a continuous function, viewed as a Γ-invariant function on $\mathcal{H}$. Suppose that the boundary of D – the fundamental domain of Γ – is of zero measure (this is the case, for example, when D is a Dirichlet domain), then

$$\int_D P(z, w, t) f(w) \, d\mu(w) = \sum_{\gamma \in \Gamma} \int_D p(z, \gamma w, t) f(w) \, d\mu(w)$$

$$= \sum_{\gamma \in \Gamma} \int_{\gamma(D)} p(z, \gamma w, t) f(w) \, d\mu(w)$$

$$= \int_{\mathcal{H}} p(z, w, t) f(w) \, d\mu(w),$$

and the initial condition follows from the corresponding property for p.   □

3.7.2   The Non-compact Case

Contrary to most of the rest of this chapter, we shall assume in this subsection that Γ is of the first kind but non-cocompact. The fundamental domain then contains cusps, which complicates the situation. We begin by introducing the space $C(\Gamma \backslash \mathcal{H})$ of *cuspidal* functions, that is, the space of bounded C^∞ functions $f : \Gamma \backslash \mathcal{H} \to \mathbf{C}$ whose constant term is zero at each cusp.

Let us explicate this last condition. Let $\mathfrak{a}$ be a cusp of the group Γ. According to Lemma 1.3, there exists an element $\sigma_\mathfrak{a} \in G$ such that the group $\sigma_\mathfrak{a}^{-1} \Gamma \sigma_\mathfrak{a}$ has a cusp at infinity and that

$$(\sigma_\mathfrak{a}^{-1} \Gamma \sigma_\mathfrak{a})_\infty = \left\{ \begin{pmatrix} 1 & m \\ 0 & 1 \end{pmatrix} \,\Big|\, m \in \mathbf{Z} \right\}.$$

In particular, any Γ-invariant function $f : \mathcal{H} \to \mathbf{C}$ satisfies

$$f\left(\sigma_\mathfrak{a} \left(\begin{smallmatrix} 1 & m \\ 0 & 1 \end{smallmatrix}\right) z\right) = f(\sigma_\mathfrak{a} z),$$

for all $m \in \mathbf{Z}$. In other words, $f \circ \sigma_\mathfrak{a}(z + m) = f \circ \sigma_\mathfrak{a}(z)$. One can therefore expand $f \circ \sigma_\mathfrak{a}$ in a Fourier series,

$$f(\sigma_\mathfrak{a} z) = \sum_n f_{\mathfrak{a}n}(y) e(nx) \quad (z = x + iy), \tag{3.41}$$

where the coefficients are given by

$$f_{\mathfrak{a}n}(y) = \int_0^1 f(\sigma_\mathfrak{a} z) e(-nx) \, dx.$$

We write simply $f_\mathfrak{a}$ for the function $f_{\mathfrak{a}0}$. If f is C^∞, then the series (3.41) converges absolutely and uniformly on compacta. One says that f has a vanishing constant term in each cusp if, for every cusp $\mathfrak{a}$ of Γ, the function $f_\mathfrak{a}$ is identically zero. We denote by $\Gamma_\mathfrak{a}$ the parabolic subgroup of Γ satisfying $\sigma_\mathfrak{a}^{-1} \Gamma_\mathfrak{a} \sigma_\mathfrak{a} = (\sigma_\mathfrak{a}^{-1} \Gamma \sigma_\mathfrak{a})_\infty$; it is the stabilizer of the cusp $\mathfrak{a}$ in Γ.

In this subsection we assume that the point-pair invariant k is C^∞ and of compact support.

Proposition 3.22 *The invariant integral operator T_k associated with k sends $C(\Gamma \backslash \mathcal{H})$ to itself.*

Proof It is clear that T_k sends a function bounded by M to a function bounded by the product $M \|k\|_\infty A_{D/2}$, where $\|k\|_\infty$ is the supremum norm of k, D is the diameter of the support of k, and A_R is the area of the hyperbolic disc of radius R.

Let $f \in C(\Gamma \backslash \mathcal{H})$. We compute the constant term of $T_k f$ at the cusp $\mathfrak{a}$ of Γ:

$$
\begin{aligned}
(T_k f)_\mathfrak{a}(y) &= \int_0^1 T_k f\left(\sigma_\mathfrak{a} \left(\begin{smallmatrix} 1 & t \\ 0 & 1 \end{smallmatrix}\right) z\right) dt \\
&= \int_0^1 \left(\int_\mathcal{H} k\left(\sigma_\mathfrak{a} \left(\begin{smallmatrix} 1 & t \\ 0 & 1 \end{smallmatrix}\right) z, w\right) f(w) \, d\mu(w) \right) dt \\
&= \int_\mathcal{H} k(z, w) \left(\int_0^1 f\left(\sigma_\mathfrak{a} \left(\begin{smallmatrix} 1 & t \\ 0 & 1 \end{smallmatrix}\right) w\right) dt \right) d\mu(w) \\
&= \int_\mathcal{H} k(z, w) f_\mathfrak{a}(\mathrm{Im}(w)) \, d\mu(w).
\end{aligned}
$$

Thus, if $f_\mathfrak{a}$ is identically zero, the same is true for $(T_k f)_\mathfrak{a}$. $\qquad\qquad\square$

Now consider the automorphic kernel $K(z, w)$. Again let D be a fundamental domain for Γ that we fix once and for all in this subsection. The complication relative to the compact case comes from the fact that the kernel K is not bounded on $D \times D$, and this regardless of the size of the support of k. The reason is that, as z and w tend towards the same cusp, the sum (3.39) tends towards infinity. To remedy this problem one subtracts from $K(z, w)$ the *principal parts*

$$H_\mathfrak{a}(z, w) = \sum_{\gamma \in \Gamma_\mathfrak{a} \backslash \Gamma} \int_{-\infty}^{\infty} k(z, \sigma_\mathfrak{a} \left(\begin{smallmatrix} 1 & t \\ 0 & 1 \end{smallmatrix} \right) \sigma_\mathfrak{a}^{-1} \gamma w) dt. \tag{3.42}$$

The function $H_\mathfrak{a}(z, w)$ is clearly Γ-invariant in the second variable. One defines the *total principal part* of the kernel $K(z, w)$ by summing all of the principal parts $H_\mathfrak{a}(z, w)$ corresponding to nonequivalent cusps:

$$H(z, w) = \sum_\mathfrak{a} H_\mathfrak{a}(z, w). \tag{3.43}$$

We write

$$\widehat{K}(z, w) = K(z, w) - H(z, w) \tag{3.44}$$

for the difference, called the *compact part* of the kernel $K(z, w)$. This defines a kernel on $D \times D$ and therefore an integral operator $\widehat{T}_k$ acting on functions $f : D \to \mathbf{C}$.

Proposition 3.23 *If $f \in \mathcal{C}(\Gamma \backslash \mathcal{H})$ we have $T_k f = \widehat{T}_k f$.*

Proof This is an immediate consequence of the following lemma. $\square$

Lemma 3.24 *For all $z \in \mathcal{H}$, the principal part $H_\mathfrak{a}(z, \cdot)$ is bounded, belongs to $L^2(\Gamma \backslash \mathcal{H}) \cap C^\infty(\Gamma \backslash \mathcal{H})$ and is orthogonal to the subspace $\mathcal{C}(\Gamma \backslash \mathcal{H})$.*

Proof We begin by showing that $H_\mathfrak{a}(z, \cdot)$ is bounded. Since $k(\sigma_\mathfrak{a} z, \sigma_\mathfrak{a} w) = k(z, w)$, we have

$$H_\mathfrak{a}(\sigma_\mathfrak{a} z, \sigma_\mathfrak{a} w) = \sum_{\tau \in B \backslash \sigma_\mathfrak{a}^{-1} \Gamma \sigma_\mathfrak{a}} \int_{-\infty}^{\infty} k(z, t + \tau w) dt. \tag{3.45}$$

Here the function k is of compact support. Thus the sum and the integral range over a domain where $|z - t - \tau w|^2 = O(\mathrm{Im}(z) \, \mathrm{Im}(\tau w))$, see (1.12). Restricting to this domain we thus have $\mathrm{Im}(\tau w) = O(\mathrm{Im}(z))$ (as well as $\mathrm{Im}(z) = O(\mathrm{Im}(\tau w))$), and the integral is an $O(\mathrm{Im}(z))$.

Fact Let $z \in \mathcal{H}$ and $Y > 0$. We have

$$\left|\{\tau \in (\sigma_\mathfrak{a}^{-1} \Gamma \sigma_\mathfrak{a})_\infty \backslash \sigma_\mathfrak{a}^{-1} \Gamma \sigma_\mathfrak{a} \mid \mathrm{Im}(\tau z) > Y\}\right| = O\left(1 + 1/Y\right), \qquad (3.46)$$

where the implied constant in the O does not depend on z.

Proof of the fact The volume of the part of the strip

$$\{z = x + iy \in \mathcal{H} \mid |x| < 1/2\}$$

above the horocycle of length $\log 1/Y$ is $1/Y$. The argument of the Lemma 3.18 can then be modified to prove the fact. In the case of $\Gamma = \mathrm{SL}(2, \mathbf{Z})$, we can just as well give an elementary proof:

The group Γ has only one cusp $\mathfrak{a} = \infty$. By Γ-invariance, we can assume that z belongs to the closure of the fundamental domain

$$D = \{z = x + iy \in \mathcal{H} \mid |z| \geqslant 1, \ |x| < 1/2\}.$$

We therefore have $y \geqslant \sqrt{3}/2$. Let

$$\tau = \begin{pmatrix} * & * \\ c & d \end{pmatrix} \in \Gamma.$$

Suppose that

$$\mathrm{Im}(\tau z) = \frac{y}{|cz + d|^2} > Y.$$

If $c \neq 0$, we have $|cz + d|^2 \geqslant 3c^2/4 \geqslant 3/4$, so that

$$y > \frac{3Y}{4}.$$

It follows from this that $|c| < (\sqrt{y}Y)^{-1}$ and $|cx + d| < \sqrt{4y/3Y}$. For a fixed $c \neq 0$, this last inequality has at most $O(\sqrt{4y/3Y})$ solutions d, whence a total number of pairs (c, d), with $c \neq 0$, bounded by

$$O\left(y^{-1/2} Y^{-1/2} y^{1/2} Y^{-1/2}\right) = O(Y^{-1}).$$

The result follows by adding 1 (the contribution of B), which corresponds to $c = 0$.
$\square$

Since the sum (3.45) ranges over a domain where $\text{Im}(\tau) = O(\text{Im}(z))$, it follows from (3.46) that there are at most $O(1+1/\text{Im}(z))$ terms in the sum. But each integral is an $O(\text{Im}(z))$, so one gets

$$H_a(\sigma_a z, \sigma_a w) = O\left(\left(1 + \frac{1}{\text{Im } z}\right)\text{Im } z\right) = O(1 + \text{Im } z). \tag{3.47}$$

The function $H_a(z, \cdot)$ is in particular bounded.

Given a function $f \in \mathcal{C}(\Gamma \backslash \mathcal{H})$, unfolding the integral yields

$$\langle H_a(\sigma_a z, .), f \rangle = \int_0^{+\infty} \int_0^1 \left(\int_{-\infty}^{+\infty} k(z, u + iv + t)dt\right) \overline{f(\sigma_a(u + iv))} \frac{dv\,du}{v^2}$$

$$= \int_0^{+\infty} \left(\int_{-\infty}^{+\infty} k(z, t + iv)dt\right) \left(\int_0^1 \overline{f(\sigma_a(u + iv))}du\right) v^{-2}dv$$

$$= 0,$$

since $0 = f_a(v) = \int_0^1 \overline{f(\sigma_a(u + iv))}du$. $\square$

Proposition 3.25 *Suppose that D is a polygonal and that its vertices at infinity are pairwise distinct modulo Γ. Then the kernel $\widehat{K}(z, w)$ is bounded on $D \times D$.*

Proof Since k is of compact support, the non-parabolic elements only contribute a finite number of terms to the sum $K(z, w)$: the set of non-parabolic $\gamma \in \Gamma$ for which some $z, w \in D$ satisfy $\text{dist}_{\mathcal{H}}(z, \gamma w) = O(1)$ is finite. We therefore have

$$K(z, w) = \sum_a \sum_{\gamma \in \Gamma_a} k(z, \gamma w) + O(1),$$

where the first sum ranges over the set of nonequivalent cusps. The upper bound (3.47) moreover implies that, in the sum (3.42), all of the terms, except for the class Γ_a, contributes a uniformly bounded amount. (These terms correspond to the $O(1) = O(\text{Im}(z)/\text{Im}(z))$ in (3.47).) We deduce that

$$H_a(z, w) = \int_{-\infty}^{+\infty} k\left(z, \sigma_a \begin{pmatrix} 1 & t \\ 0 & 1 \end{pmatrix} \sigma_a^{-1} w\right)dt + O(1)$$

so that

$$\widehat{K}(z, w) = \sum_a J_a(z, w) + O(1),$$

where

$$J_a(z, w) = \sum_{\gamma \in \Gamma_a} k(z, \gamma w) - \int_{-\infty}^{+\infty} k\left(z, \sigma_a \begin{pmatrix} 1 & t \\ 0 & 1 \end{pmatrix} \sigma_a^{-1} w\right)dt.$$

It remains to show that $J_a(z, w)$ is bounded on $D \times D$. This is a consequence of the Euler-MacLaurin formula

$$\sum_{b \in \mathbf{Z}} F(b) = \int F(t)dt + \int \psi(t)dF(t),$$

where $\psi(t) = t - [t] - 1/2$. Indeed, this implies that

$$J_a(\sigma_a z, \sigma_a w) = \sum_{b \in \mathbf{Z}} k(z, w + b) - \int_{-\infty}^{+\infty} k(z, w + t)dt$$

$$= \int_{-\infty}^{+\infty} \psi(t)dk(z, w + t) = O\left(\int_0^{+\infty} |dk(u)|\right) = O(1).$$

Since k is of compact support one therefore finds that the function $J_a(z, w)$ is in fact bounded on $\mathcal{H} \times \mathcal{H}$.   □

3.8 Review of Functional Analysis

In this chapter we aim to show that the spectrum of the Laplacian Δ is discrete whenever the surface $\Gamma \backslash \mathcal{H}$ is compact. Since the Laplacian is an unbounded operator, there is no general theorem implying that its spectrum is discrete. We shall get around this problem by deducing the spectral theorem for Δ from the spectral theorem for certain integral operators commuting with the Laplacian and which are themselves compact operators. In this section we recall the statements of general theorems concerning compact operators. These theorems are classical, see for example [18, Chap. VI], [103, §97].

Let $\mathbf{H}$ be a separable Hilbert space whose underlying norm we denote by $|\cdot|$. An operator $T : \mathbf{H} \to \mathbf{H}$ is *compact* if T sends bounded sets to relatively compact sets. Since $\mathbf{H}$ is separable, a subset of $\mathbf{H}$ is compact if and only if it is sequentially compact. The operator T is therefore compact if and only if, for any sequence of unitary vectors $x_n \in \mathbf{H}$, there exists a subsequence x_{n_k} such that the sequence $T(x_{n_k})$ converges. A compact operator is automatically bounded and is therefore continuous. Moreover, if λ is a non-zero eigenvalue then the λ-eigenspace is finite-dimensional.

Theorem 3.26 (Spectral theorem for compact operators) *Let T be a compact self-adjoint operator on a separable Hilbert space $\mathbf{H}$. Then the space $\mathbf{H}$ admits an orthonormal basis ϕ_i $(i = 1, 2, 3, \ldots)$ of eigenvectors of T, say $T\phi_i = \lambda_i \phi_i$. Moreover, the sequence of eigenvalues λ_i tends towards 0 as $i \to +\infty$.*

Let X be a locally compact space equipped with a positive Borel measure. Assume that $\mathbf{H} = L^2(X)$ is a separable Hilbert space. Let $K \in L^2(X \times X)$. One calls the operator

$$(Tf)(x) = \int_X K(x, y) f(y) \, dy \tag{3.48}$$

a *Hilbert-Schmidt operator*.

Theorem 3.27 (Hilbert-Schmidt) *A Hilbert-Schmidt operator is a compact operator.*

A Hilbert-Schmidt operator T whose kernel is real and symmetric is self-adjoint. One can thus apply the spectral theorem for self-adjoint compact operators. The image of T in $\mathbf{H}$ is generated by its eigenfunctions: there exists a maximal orthonormal system of eigenfunctions $\phi_1, \phi_2, \phi_3, \ldots$ of T in $\mathbf{H}$, i.e.,

$$\langle \phi_j, \phi_k \rangle = \delta_{jk}, \quad T\phi_j = \lambda_j \phi_j \text{ with } |\lambda_1| \geq |\lambda_2| \geq \cdots$$

Assume now that K is continuous, then every eigenfunction ϕ_j with $\lambda_j \neq 0$ is continuous and any function f in the image of T can be represented by an absolutely and uniformly convergent series

$$f(x) = \sum_{j \geq 1} \langle f, \phi_j \rangle \phi_j(x). \tag{3.49}$$

A compact self-adjoint operator is a *trace class operator* if the sum $\sum \lambda_j$ of its eigenvalues is absolutely convergent. We then call this sum the *trace* of T, written $\operatorname{tr}(T)$.

The following theorem implies, in particular, that a Hilbert-Schmidt operator T on $L^2(X)$ with X compact and whose kernel K is positive and continuous is trace class. One has then

$$\operatorname{tr}(T) = \int_X K(x, x) \, dx. \tag{3.50}$$

Theorem 3.28 (Mercer) *Let T be a Hilbert-Schmidt operator with positive kernel K. There exists a maximal orthonormal system $\phi_1, \phi_2, \phi_3, \ldots$ of T-eigenfunctions in $\mathbf{H}$ such that the sequence of corresponding eigenvalues (λ_j) is positive. Moreover, if the kernel K is continuous, it can be represented by an absolutely and uniformly convergent series*

$$K(x, y) = \sum_{j \geq 1} \lambda_j \phi_j(x) \phi_j(y). \tag{3.51}$$

3.9 Proof of the Spectral Theorem

We begin by assuming that the surface $\Gamma\backslash\mathcal{H}$ is compact.

Let $k : \mathcal{H} \times \mathcal{H} \to \mathbf{R}$ be a compactly supported C^∞ point-pair invariant. Denote by $K : \Gamma\backslash\mathcal{H} \times \Gamma\backslash\mathcal{H} \to \mathbf{R}$ the associated automorphic kernel. Let T_k be the invariant integral operator defined in § 3.7 by

$$(T_k\varphi)(z) = \int_{\Gamma\backslash\mathcal{H}} K(z, w)\varphi(w)\, d\mu(w).$$

Then the spectral theorem for self-adjoint compact operators applies to this Hilbert-Schmidt operator, and consequently there exists a maximal orthonormal system $\phi_1, \phi_2, \phi_3, \ldots$ of T_k-eigenfunctions in $L^2(\Gamma\backslash\mathcal{H})$ such that every function f in the closure of the image of T_k can be expanded as

$$f(z) = \sum_{j\geq 1} \langle f, \phi_j\rangle \phi_j(z). \tag{3.52}$$

Moreover, the eigenspace associated to a non-zero T_k-eigenvalue is of finite dimension. Since T_k and Δ commute and the Laplacian is a symmetric operator, every function $f \in C^\infty(\Gamma\backslash\mathcal{H}) \cap \mathrm{Im}(T_k)$ admits an expansion (3.52) where the functions $\phi_j \in \mathcal{D}(\Gamma\backslash\mathcal{H})$ are eigenfunctions of the Laplacian.

To prove the spectral theorem, it suffices then to choose K (or rather k) in such a way that the image $T_k(C^\infty(\Gamma\backslash\mathcal{H}))$ is dense in $C^\infty(\Gamma\backslash\mathcal{H})$. We begin with the following lemma.

Lemma 3.29 *Let $f \in L^2(\Gamma\backslash\mathcal{H})$ be non-zero. Fix a positive real number ε. Then there is an automorphic kernel K (associated to a compactly supported function k) such that $|T_k f - f| < \varepsilon$, where T_k is the integral operator associated with the kernel K and $|\cdot|$ denotes the L^2-norm. In particular, if $\varepsilon < |f|$ then $T_k f \neq 0$.*

Proof From the density of continuous functions in $L^2(\Gamma\backslash\mathcal{H})$, it suffices to prove the lemma for f continuous. Since $S = \Gamma\backslash\mathcal{H}$ is compact, the function f is uniformly continuous and there exists a constant $\eta > 0$ such that for all $z, w \in \mathcal{H}$ of distance $\rho(z, w) < \eta$, one has

$$|f(w) - f(z)| < \sqrt{\varepsilon/\mathrm{area}(S)}.$$

Let $k \in C^\infty(\mathbf{R})$ be even, compactly supported in $]-\eta, \eta[$ and such that $\int_0^{+\infty} k(\rho)\sinh\rho\, d\rho = \frac{1}{2\pi}$. For all $z \in \mathcal{H}$ one then has

$$\int_{\mathcal{H}} k(z, w)\, d\mu(w) = 1 \tag{3.53}$$

and if T_k is the invariant integral operator associated to the kernel k,

$$(T_k f - f)(z) = \int_{\mathcal{H}} k(z, w) f(w)\, d\mu(w) - f(z)$$

$$= \int_{\mathcal{H}} k(z, w)(f(w) - f(z))\, d\mu(w).$$

Using (3.53) again one finds that

$$\sup_{z \in \mathcal{H}} |(T_k f - f)(z)| \leq \sqrt{\frac{\varepsilon}{\mathrm{area}(S)}},$$

from which the lemma follows.   □

Proposition 3.30 *Let* **H** *be a non-zero Hilbert subspace of* $L^2(\Gamma \backslash \mathcal{H})$, *stable under all invariant integral operators. Then there exists a non-zero function* $f \in \mathbf{H} \cap C^\infty(\Gamma \backslash \mathcal{H})$ *which is an eigenfunction of the Laplacian.*

Proof The crucial property is that the invariant integral operators commute with the Laplacian. Let f_0 be a non-zero vector in **H**. According to Lemma 3.29, there exists an invariant integral operator T_k such that $T_k f_0 \neq 0$. According to Theorem 3.27, the operator T_k induces a non-zero compact self-adjoint operator on the closed subspace **H**. According to the spectral theorem for compact operators, the operator T_k admits a non-zero eigenvector with corresponding eigenvalue $\lambda \neq 0$. Moreover, the λ-eigenspace is finite-dimensional. Since Δ commutes with T_k, this eigenspace is invariant under the action of Δ, and since every linear transformation on a finite-dimensional vector space admits a non-zero eigenvector, a vector in the λ-eigenspace of T_k is also an eigenvector of Δ. Such an eigenvector (viewed as a function f on $\Gamma \backslash \mathcal{H}$) is necessarily C^∞: the kernel k is C^∞ and $f = \frac{1}{\lambda} T_k f$ (the operator T_k is regularizing).   □

We deduce now the first part of the spectral theorem, namely that the space $L^2(\Gamma \backslash \mathcal{H})$ decomposes into a Hilbert direct sum of Δ-eigenspaces.

Indeed, let Σ be the set of all the sets E of Δ-eigenspaces such that any two elements of E are mutually orthogonal. According to Zorn's lemma, Σ has a maximal element E. Let **H** be the orthogonal supplement of the closure of the direct sum of all the elements of E. If $\mathbf{H} \neq 0$, Proposition 3.30 implies that there exists a Δ-eigensubspace of **H**, contradicting the maximality of E.

Let us now show that the sequence of Δ-eigenvalues λ_i tends toward infinity. For this we let $\mathcal{P}_t$ denote the invariant integral operator associated with the heat kernel p_t; this is a (regularizing) compact self-adjoint operator. Moreover, we have

$$\mathcal{P}_s \circ \mathcal{P}_t = \mathcal{P}_{s+t}. \tag{3.54}$$

In particular, $\mathcal{P}_s \circ \mathcal{P}_t = \mathcal{P}_t \circ \mathcal{P}_s$ and $\mathcal{P}_t = \mathcal{P}_{t/2} \circ \mathcal{P}_{t/2}$. Since $\mathcal{P}_t$ is self-adjoint this last identity immediately implies that every operator $\mathcal{P}_t$ is positive.

Lemma 3.31 *For every* $f \in L^2(\Gamma\backslash\mathcal{H})$,

$$\lim_{t\downarrow 0} \mathcal{P}_t f = f$$

in $L^2(\Gamma\backslash\mathcal{H})$.

Proof Assume first that f is continuous. The function $\mathcal{P}_t f$ is a solution to the heat equation and thus

$$\frac{d}{dt}\int_{\Gamma\backslash\mathcal{H}}(\mathcal{P}_t f)^2 d\mu = -2\int_{\Gamma\backslash\mathcal{H}}(\mathcal{P}_t f)\Delta(\mathcal{P}_t f)\,d\mu \leq 0;$$

in other words[13] $\frac{d}{dt}|\mathcal{P}_t f| \leq 0$. Since f is continuous it follows from the definition of $\mathcal{P}_t$ that $\lim_{t\downarrow 0}\mathcal{P}_t f = f$ uniformly on $\Gamma\backslash\mathcal{H}$. Thus $|\mathcal{P}_t f| \leq |f|$. Since $C^0(\Gamma\backslash\mathcal{H})$ is dense in $L^2(\Gamma\backslash\mathcal{H})$ and the heat kernel $\mathcal{P}_t$ is continuous, we have

$$|\mathcal{P}_t f| \leq |f| \quad \text{for all } f \in L^2(\Gamma\backslash\mathcal{H}).$$

Invoking again the density of $C^0(\Gamma\backslash\mathcal{H})$ in $L^2(\Gamma\backslash\mathcal{H})$, the lemma follows. $\square$

We now apply the Hilbert-Schmidt theorem to each of the operators $\mathcal{P}_t$. We begin with $t = 1$ and let

$$\varphi_0, \varphi_1, \varphi_2, \ldots$$

be a complete orthonormal system in $L^2(\Gamma\backslash\mathcal{H})$ made up of eigenfunctions of $\mathcal{P}_1$ with corresponding eigenvalues

$$\eta_0, \eta_1, \eta_2, \cdots \geq 0; \quad \eta_j \longrightarrow 0 \text{ as } j \longrightarrow \infty.$$

The φ_j are eigenfunction of all of the operators $\mathcal{P}_t$. Indeed, consider first $t = 1/k$ where $k \in \mathbf{N}^*$. If φ is an eigenfunction of $\mathcal{P}_{1/k}$ of eigenvalue μ, then, since $\mathcal{P}_1 = (\mathcal{P}_{1/k})^k$, the function φ is an eigenfunction of $\mathcal{P}_1$ of eigenvalue μ^k. Since the system of eigenspaces for $\mathcal{P}_1$ is complete, the operator $\mathcal{P}_{1/k}$ admits the same complete orthonormal system of eigenfunctions. The corresponding eigenvalues are $\eta_0^{1/k}, \eta_1^{1/k}, \eta_2^{1/k}, \ldots$ According to (3.54) the functions φ_j are then eigenfunctions of $\mathcal{P}_t$ with corresponding eigenvalues η_j^t for all rationals $t > 0$. By continuity of the heat kernel we thus have

$$\mathcal{P}_t\varphi_j = \eta_j^t\varphi_j, \quad j = 0, 1, 2, \ldots$$

---

[13]Recall that $|\cdot|$ denotes the Hilbert norm of $L^2(\Gamma\backslash\mathcal{H})$.

for all $t > 0$. According to Lemma 3.31, $\lim_{t \downarrow 0} \mathcal{P}_t \varphi_j = \varphi_j$, so that $\eta_j^t \to 1$ as $t \downarrow 0$. We deduce that

$$\eta_j > 0 \quad \text{for all } j.$$

Since the kernel p_t is C^∞, the functions φ_j belong to the space $C^\infty(\Gamma \backslash \mathcal{H})$. By the compactness of the operator $\mathcal{P}_1$, all of the eigenspaces of finite dimension. We can thus assume that the eigenvalues are arranged in decreasing order. Let us show that

$$1 = \eta_0 > \eta_1 \geqslant \eta_2 \geqslant \cdots > 0.$$

It is clear that the function constantly equal to 1 is a solution to the heat equation. Hence it is an eigenfunction of the operators $\mathcal{P}_t$ for the eigenvalue 1. Let φ_j be an non-constant eigenfunction of $\mathcal{P}_1$. We have

$$\frac{d}{dt}|\mathcal{P}_t \varphi_j|^2 = -2 \int_{\Gamma \backslash \mathcal{H}} (\mathcal{P}_t \varphi_j) \Delta(\mathcal{P}_t \varphi_j) \, d\mu$$

$$= -2\eta_j^{2t} \int_{\Gamma \backslash \mathcal{H}} \varphi_j \Delta \varphi_j d\mu < 0.$$

Here we used $\eta_j > 0$ and the fact that if φ is non-constant then

$$\langle \varphi, \Delta \varphi \rangle = \int_{\Gamma \backslash \mathcal{H}} |\operatorname{grad} \varphi|^2 d\mu > 0.$$

We deduce that $|\mathcal{P}_t \varphi_j| < |\varphi_j|$ so that $\eta_j < 1$.

We can now prove the spectral theorem. More precisely, we demonstrate the following theorem.

Theorem 3.32 (Spectral theorem) *Let $S = \Gamma \backslash \mathcal{H}$ be a compact hyperbolic surface. The spectral problem*

$$\Delta \varphi = \lambda \varphi$$

admits a complete orthonormal system $\varphi_0, \varphi_1, \ldots$ of C^∞-eigenfunctions in $L^2(S)$. The corresponding eigenvalues $\lambda_0, \lambda_1, \ldots$ satisfy the following properties

1. $0 = \lambda_0 < \lambda_1 \leqslant \lambda_2 \leqslant \cdots$, $\lambda_n \to +\infty$ as $n \to +\infty$.
2. *The series*

$$\sum_{j=1}^{+\infty} \lambda_j^{-\sigma}$$

converges for all real $\sigma > 1$.

Proof Keeping the above notation we write

$$\lambda_j = -\log \eta_j, \quad j = 0, 1, 2, \ldots$$

From the equation

$$
\begin{aligned}
0 &= \Delta P_t \varphi_j + \frac{\partial}{\partial t} P_t \varphi_j \\
&= e^{-t\lambda_j}(\Delta \varphi_j - \lambda_j \varphi_j)
\end{aligned}
$$

one deduces that the φ_j are eigenfunctions of the Laplacian, and one obtains point 1 of the theorem. The Hilbert-Schmidt theorem then implies that

$$P_\Gamma(z, w, t) = \sum_{j=0}^{\infty} e^{-\lambda_j t} \varphi_j(z) \varphi_j(w), \qquad (3.55)$$

in the L^2-sense. Next Mercer's theorem implies that the series converges absolutely and uniformly for all $t > 0$. To prove point 2, we must pass from $e^{-\lambda_j t}$ to $\lambda_j^{-\sigma}$ and show that the series

$$\sum_{j=1}^{+\infty} \lambda_j^{-\sigma} \varphi_j(z) \varphi_j(w) \qquad (3.56)$$

converges absolutely and uniformly on $S \times S$.

The idea is to take $e^{-\lambda_j t}$, multiply it by $t^{\sigma-1}$ and then integrate. The integral expression for the Γ-function implies that for $\sigma > 1$,

$$\int_0^{+\infty} t^{\sigma-1} e^{-\lambda_j t} dt = \Gamma(\sigma) \lambda_j^{-\sigma}.$$

One can therefore consider the functions φ_j as eigenfunction of a new invariant integral operator whose kernel K_σ is given by

$$K_\sigma(z, w) = \int_0^{+\infty} \left(P_\Gamma(z, w, t) - \frac{1}{\text{area}(S)} \right) t^{\sigma-1} dt.$$

Let us show that the kernel K_σ is well-defined and continuous on $S \times S$. We begin by remarking that for all $\sigma > 1$, there exists an integer j_0 such that, for all $j \geq j_0$ and for all $\varepsilon \in (0, 1/2)$, one has

$$\int_\varepsilon^{+\infty} e^{-\lambda_j t} t^{\sigma-1} dt \leq e^{-\lambda_j t}.$$

It then follows from (3.55) that for all $\varepsilon > 0$ the function[14]

$$K_{\sigma,\varepsilon}(z, w) = \int_{\varepsilon}^{+\infty} \left(P_\Gamma(z, w, t) - \frac{1}{\text{area}(S)} \right) t^{\sigma-1} dt$$

is well-defined and continuous on $S \times S$. It satisfies

$$\lim_{\varepsilon \downarrow 0} \int_S K_{\sigma,\varepsilon}(z, w)\varphi_j(w)\, d\mu(w) = \Gamma(\sigma)\lambda_j^{-\sigma}\varphi_j(z)$$

for all $z \in S$ and for all $j \geqslant 1$.

Moreover, it follows from Lemmas 3.13 and 3.17 that $P_\Gamma(z, w, t) \leqslant (\text{const})t^{-1}$ for all $t < 1$. Thus for every $\delta \in (0, \varepsilon)$ we have

$$\int_{\delta}^{\varepsilon} P_\Gamma(z, w, t)t^{\sigma-1} dt = O\left(\frac{\varepsilon^{\sigma-1}}{\sigma - 1} \right).$$

We deduce that $K_\sigma(z, w)$ is well-defined on $S \times S$ and that $K_{\sigma,\varepsilon}$ converges uniformly to K_σ as ε tends toward 0. Furthermore, we have

$$\int_S K_\sigma(z, w)\varphi_j(w)\, d\mu(w) = \Gamma(\sigma)\lambda_j^{-\sigma}\varphi_j(z).$$

The functions $\varphi_1, \varphi_2, \ldots$ then form a complete system of eigenfunction for the compact operator with kernel K_σ. The associated eigenvalues are $\Gamma(\sigma)\lambda_j^{-\sigma}, j = 1, 2, \ldots$ In particular, they are positive and Mercer's theorem implies that the series (3.56) converges uniformly on $S \times S$ (and that the sum is equal to $K_\sigma(z, w)$). On integrating (3.56) along the diagonal, one obtains point 2 of the theorem. $\quad\square$

In the case where the surface $\Gamma \backslash \mathcal{H}$ is of finite area but non-compact, we will content ourselves with the following theorem.

Theorem 3.33 *Let $S = \Gamma \backslash \mathcal{H}$ be a hyperbolic surface of finite area. The spectral problem for cuspidal functions*

$$\Delta\varphi = \lambda\varphi \quad (\varphi \in \mathcal{C}(\Gamma \backslash \mathcal{H}))$$

admits a complete orthonormal system of eigenfunctions.

Proof The statement to be established is just one part of a more general spectral theorem for such surfaces. The proof is a straightforward generalization of the analogous one for the compact case (cf. the passage directly following the proof of Proposition 3.30): one simply replaces the space $L^2(\Gamma \backslash \mathcal{H})$ by the closure of $\mathcal{C}(\Gamma \backslash \mathcal{H})$ and the operators T_k by the operators $\widehat{T}_k$ associated to the kernel $\widehat{K}$ defined

---

[14]Note that φ_0 is the function constantly equal to $1/\sqrt{\text{area}(S)}$.

in Eq. (3.44). Then one uses Propositions 3.23 and 3.25; note that $\widehat{T}_k$ sends an L^2-function to a bounded C^∞-function.

Finally, we show that the first eigenvalue is strictly positive. This comes from the fact that the functions $f \in C(\Gamma \backslash \mathcal{H})$ are bounded. The equation $\Delta f = 0$ then implies that f on $\mathcal{H}$ is harmonic – in the Euclidean sense – and bounded, and thus constant. But a non-zero constant function cannot be cuspidal. $\square$

3.10 The Minimax Principle

A better geometric understanding of the spectrum of the Laplacian comes from the minimal principle that we now describe. Let $S = \Gamma \backslash \mathcal{H}$ be a compact hyperbolic surface. For $k = 0, 1, \ldots$, we write $\lambda_k(S)$ for the k-th eigenvalue of the Laplacian on S.

Theorem 3.34 (Minimal principle)

1. *Let $f_0, \ldots, f_k \in C^\infty(S)$ be $k + 1$ functions of L^2-norm equal to 1 whose supports have measure zero pairwise intersections. Then*

$$\lambda_k(S) \leq \max_{0 \leq j \leq k} \int_S |\operatorname{grad} f_j(z)|^2 d\mu(z).$$

2. *Let $M = N_1 \cup \cdots \cup N_k$ be a partition into relatively compact subsets of strictly positive measure and having measure zero pairwise intersections. Put*

$$\nu(N_j) = \inf \int_{N_j} |\operatorname{grad} f(z)|^2 d\mu(z),$$

where f runs over the set of C^∞-functions such that

$$\int_{N_j} |f(z)|^2 d\mu(z) = 1 \quad and \quad \int_{N_j} f(z) \, d\mu(z) = 0.$$

Then

$$\lambda_k(S) \geq \min_{1 \leq j \leq k} \nu(N_j).$$

Proof Let $\varphi_0, \varphi_1, \ldots$ be a complete orthonormal system of C^∞-eigenfunctions with corresponding eigenvalues $\lambda_0(S), \lambda_1(S), \ldots$ Since the f_j in the first part of the theorem are linearly independent, one can form a linear combination $f = \beta_0 f_0 + \cdots + \beta_k f_k$ of L^2-norm equal to 1 and orthogonal to the functions $\varphi_0, \ldots, \varphi_{k-1}$. Let α_j be the "Fourier coefficient"

$$\alpha_j = \int_S f \overline{\varphi_j} \, d\mu.$$

For all $m > k$, we have

$$\int_S |\operatorname{grad} f(z)|^2 d\mu(z) - \sum_{j=k}^m |\alpha_j|^2 \lambda_j$$

$$= \int_S f \overline{\Delta f} d\mu - \int_S \left(\sum_{j=k}^m \alpha_j \varphi_j\right)\overline{\left(\sum_{j=k}^m \alpha_j \lambda_j \varphi_j\right)} d\mu$$

$$= \int_S \left(f - \sum_{j=k}^m \alpha_j \varphi_j\right)\overline{\Delta\left(f - \sum_{j=k}^m \alpha_j \varphi_j\right)} d\mu$$

$$= \int_S |\operatorname{grad}(f - \sum_{j=k}^m \alpha_j \varphi_j)|^2 d\mu \geq 0.$$

Here $\lambda_j(S) \geq \lambda_k(S)$, and according to Parseval's theorem

$$\sum_{j=k}^{+\infty} |\alpha_j|^2 = 1.$$

Since the pairwise supports of the f_j meet each other in measure zero sets and the L^2-norm of f is 1, we have

$$\sum_{i=0}^k |\beta_i|^2 = 1,$$

from which one deduces that

$$\lambda_k(S) \leq \int_S |\operatorname{grad} f(z)|^2 d\mu(z)$$

$$\leq \sum_{j=0}^k |\beta_i|^2 \int_S |\operatorname{grad} f_j(z)|^2 d\mu(z)$$

$$\leq \max_{0 \leq j \leq k} \int_S |\operatorname{grad} f_j(z)|^2 d\mu(z).$$

This proves the first point of the theorem.

We proceed to the proof of the second point. Let χ_j be the characteristic function of the subset $N_j \subset S$. Consider a linear combination $\varphi = \alpha_0 \varphi_0 + \cdots + \alpha_k \varphi_k$ of L^2-norm equal to 1 and orthogonal to the functions $\chi_1, \ldots, \chi_k$. Then, for $j = 1, \ldots, k$,

$$\int_{N_j} \varphi \, d\mu = 0.$$

By definition of $\nu(N_j)$, we then have

$$\int_{N_j} |\operatorname{grad} \varphi|^2 \, d\mu \geq \nu(N_j) \int_{N_j} |\varphi(z)|^2 \, d\mu(z)$$

and so (since $\sum_j \int_{N_j} |\varphi(z)|^2 \, d\mu(z) = 1$)

$$\int_S |\operatorname{grad} \varphi|^2 d\mu \geq \min_{1 \leq j \leq k} \nu(N_j).$$

On the other hand, since

$$\int_S |\varphi(z)|^2 d\mu(z) = |\alpha_0|^2 + \cdots + |\alpha_k|^2 = 1,$$

we have

$$\int_S |\operatorname{grad} \varphi|^2 d\mu = \int_S \varphi \overline{\Delta \varphi} d\mu = \sum_{j=0}^k \lambda_j(S) |\alpha_j|^2 \leq \lambda_k(S). \qquad \square$$

3.10.1   Small Eigenvalues I: Geometric Existence Criterion

Proposition 3.35 *Let S be a compact hyperbolic surface which is the union of two compact connected subsets A and B which intersect along a finite union $\bigcup_{i=1}^k \gamma_i$ of simple closed geodesics. Let $\ell(\gamma_i)$ be the lengths of the γ_i and*

$$h = \frac{\sum_{i=1}^k \ell(\gamma_i)}{\min(\operatorname{area}(A), \operatorname{area}(B))}.$$

Let ε be a positive real number such that the ε-neighborhood of each geodesic γ_i is embedded in S. Then there exists a constant $C = C(\varepsilon)$ depending only on ε such that

$$\lambda_1(S) \leq C(h + h^2).$$

Proof According to the minimax principle, to majorize $\lambda_1(S)$ it suffices to majorize the quotient $\int_S |\operatorname{grad} f|^2 d\mu / |f|_{L^2(S)}^2$ for a well chosen test function f with support in A or B. Suppose, for example,

$$h = \frac{\sum_{i=1}^k \ell(\gamma_i)}{\operatorname{area}(A)}.$$

Let $X = \bigcup_{i=1}^{k} \gamma_i$ be the boundary of A. Let

$$A(t) = \{z \in A \mid \mathrm{dist}(z, X) \leqslant t\}$$

and for sufficiently small t put

$$f(z) = \begin{cases} \dfrac{1}{t}\,\mathrm{dist}(z, X) & \text{if } z \in A(t), \\ 1 & \text{if } z \in A - A(t). \end{cases}$$

The function f satisfies $|\mathrm{grad}f(z)|^2 \leqslant t^{-2}$ for all $z \in A(t)$, and $|\mathrm{grad}f(z)| = 0$ for all $z \in A - A(t)$. We now estimate the area of $A(t)$. In Fermi coordinates (see Appendix A) and for t sufficiently small ($t \leq \varepsilon$), we find

$$\mathrm{area}(A(t)) \leqslant \sum_{i=1}^{k} \ell(\gamma_i) \int_0^t \cosh \tau\, d\tau = \sum_{i=1}^{k} \ell(\gamma_i) \sinh t.$$

For $t = \min(\varepsilon, \mathrm{argsh}(1/2h))$, we thus have

$$\int_S |\mathrm{grad}f|^2 d\mu / |f|_{L^2(S)}^2 \leqslant \frac{\sum_{i=1}^{k} \ell(\gamma_i) \sinh t}{t^2(\mathrm{area}(A) - \sum_{i=1}^{k} \ell(\gamma_i) \sinh t)}$$

$$\leqslant \frac{h \sinh t}{t^2(1 - h \sinh t)}$$

$$\leqslant C(h + h^2),$$

for a constant C depending only on ε. $\square$

Corollary 3.36 *Let S be a compact hyperbolic surface. For every real $\varepsilon > 0$, the surface S admits a finite cover S' whose first non-zero eigenvalue satisfies*

$$\lambda_1(S') \leqslant \varepsilon.$$

Proof The Euler characteristic $\chi(S)$ of the compact hyperbolic surface S is negative. Since $\chi(S) = 2 - 2g$ where g is the genus of S, we have $g \geqslant 2$. In particular, $g \geqslant 1$ and there exists a simple closed curve γ in S such that $S - \gamma$ is connected. Upon replacing γ by another curve homotopic to it, if necessary, we can assume that γ is a closed geodesic. Corresponding to γ there is a surjective homomorphism from the fundamental group of S onto $\mathbf{Z}$: take the algebraic intersection of a loop with γ. We form S_N, the cyclic cover of degree $2N$ above S. This is the surface obtained by gluing $2N$ copies of $S - \gamma$ boundary to boundary in a circular arrangement. There then exists two lifts γ_1 and γ_2 of γ in S_N such that the union $\gamma_1 \cup \gamma_2$ cuts the surface S_N into two pieces A and B, each consisting of N fundamental domains for

the action of the transformation group of the cover of S_N over S. Proposition 3.35
then implies the existence of a constant C independent of N such that

$$\lambda_1(S_N) \leqslant C \left(\frac{2\ell(\gamma)}{N \text{ area}(S)} + \frac{4\ell(\gamma)^2}{N^2 \text{ area}(S)^2} \right).$$

We deduce the corollary by letting N tend towards infinity.   □

3.10.2 Small Eigenvalues II: The Selberg Conjecture

Conjecture 3.37 (The Selberg conjecture) *Let a, b, and N be positive integers.
Then the first non-zero eigenvalue of the Laplacian on $C(\Gamma_{a,b}(N)\backslash\mathcal{H})$ satisfies*

$$\lambda_1(\Gamma_{a,b}(N)\backslash\mathcal{H}) \geqslant 1/4.$$

In as much as there should exist *some* uniform lower bound on λ_1, Selberg's
conjecture states that congruence covers of arithmetic surfaces are, in a geometric
sense, opposite to the case of cyclic covers explored in the preceding subsection.
The precise value of $1/4$ in the above conjecture lies deeper; it is a reflection of the
underlying arithmetic defining these surfaces. We shall return to this point later.
    In the case of the modular surface, the conjecture has been proved for its
congruence subgroups of small level.

Theorem 3.38 *Let $\Gamma = \text{SL}(2, \mathbf{Z})$. Then*

$$\lambda_1(\Gamma\backslash\mathcal{H}) \geqslant \frac{3\pi^2}{2} > \frac{1}{4}.$$

Proof Let D denote the standard fundamental domain of the modular group. Let
$f \in C(\Gamma\backslash\mathcal{H})$ have L^2-norm 1 and satisfy

$$\lambda_1(\Gamma\backslash\mathcal{H}) = \lambda := \int_D |\operatorname{grad} f|^2 \, d\mu.$$

Write

$$F = \{z \mid |x| < 1/2, \ y > \sqrt{3}/2\}.$$

The subset F is contained in the union of D with its translate by $\left(\begin{smallmatrix} 0 & -1 \\ 1 & 0 \end{smallmatrix} \right) \in \Gamma$. We
thus have

$$2\lambda \geqslant \int_{\sqrt{3}/2}^{+\infty} \int_0^1 |\operatorname{grad} f(z)|^2 d\mu(z).$$

The function $f(z)$ being 1-periodic in x, we can expand it in its Fourier series

$$f(z) = \sum_{n \neq 0} c_n(y) e(nx).$$

Hence

$$\operatorname{grad} f(z) = \begin{pmatrix} y \sum_{n \neq 0} 2\pi n c_n(y) e(nx) \\ y \sum_{n \neq 0} c_n'(y) e(nx) \end{pmatrix},$$

and

$$\int_{\sqrt{3}/2}^{+\infty} \int_0^1 |\operatorname{grad} f(z)|^2 \, d\mu(z) \geq \int_{\sqrt{3}/2}^{+\infty} \sum_{n \neq 0} |2\pi n c_n(y)|^2 dy$$

$$\geq 3\pi^2 \int_{\sqrt{3}/2}^{+\infty} \sum_{n \neq 0} |c_n(y)|^2 y^{-2} dy$$

$$\geq 3\pi^2 \int_{\sqrt{3}/2}^{+\infty} \int_0^1 |f(z)|^2 \, d\mu(z)$$

$$\geq 3\pi^2 \int_D |f(z)|^2 \, d\mu(z).$$

From this one deduces that $2\lambda \geq 3\pi^2$.   □

Remark 3.39 In [15], the authors compute to more than 1000 decimal places the first non-zero eigenvalue $\lambda_1 = \lambda_1(\Gamma \backslash \mathcal{H})$:

$$\lambda_1 \approx 91, 14134533635527808180\ldots$$

In [14] the Selberg conjecture is verified for several congruence subgroups of small level.

3.11 Commentary and References

The main result of this chapter is Theorem 3.32. Two classical references concerning this chapter are the books by Iwaniec and Buser [24, 63]. We have borrowed a lot from their presentations, following in large part Iwaniec's book.

§ 3.1

Unbounded operators are frequently encountered in mathematical physics and have been widely studied; we refer the reader to the books of Rudin [107], Dunford and Schwartz [38], or Taylor [127] for the general theory.

It is a general fact that on every Riemannian manifold the Laplacian is a symmetric operator. This can be deduced from Green's theorem, see for example [41, p. 184]. In our exposition, with its focus on surfaces, we have opted to reproduced the "elementary" proof to be found in [20, p. 133]; this proof uses only the classical Stokes' theorem (in the plane).

§ 3.2

Lemma 3.3 is taken from [63, Lem. 1.7] and Proposition 3.11 is proved in [63, (1.64')].

§ 3.5

One can study the heat equation on any Riemannian manifold. The book [8] proves, for example, the existence and uniqueness of solutions on compact manifolds. The book by Davies [33] is a good reference for the case of the hyperbolic plane. The Riemannian manifold $\mathcal{H}$ being non-compact, a fundamental solution to the heat equation is not necessarily unique.

One can find in the literature [33] good upper bounds on the heat kernel for general spaces. Lemma 3.13 is taken from [24, Lem. 7.4.26].

§ 3.6

The Laplacian defines a symmetric positive operator on any compact Riemannian manifold, see for example [8, p. 53]. Here again we have opted for an "elementary" proof close to that which one finds in the Bump's book [20] and which uses only Stokes' theorem for the hyperbolic surface $S = \Gamma \backslash \mathcal{H}$.

A classical theorem of Friedrichs implies that every positive symmetric operator admits a self-adjoint extension. Since $\mathcal{D}(\Gamma \backslash \mathcal{H})$ is dense in $L^2(\Gamma \backslash \mathcal{H})$ and the area of the surface $\Gamma \backslash \mathcal{H}$ is finite, it follows from Proposition 3.16 that $(\Delta, \mathcal{D}(\Gamma \backslash \mathcal{H}))$ admits a unique self-adjoint extension to $L^2(\Gamma \backslash \mathcal{H})$. It is this extension that one classically refers to as the Laplacian.

§ 3.7

Lemma 3.20 is extracted from [24, §7.5] which gives more details.

In the study of integral operators on finite area surfaces we have followed Iwaniec [63, §4.2].

§ 3.9

The proof of Lemma 3.29 and the proof of the first part of the spectral theorem generalize easily to show that if G is a locally compact topological group and Γ a discrete cocompact subgroup of G, the space $L^2(\Gamma \backslash G)$ decomposes into a (Hilbert) direct sum of invariant and irreducible subspaces under the right-regular representation of G.

In the non-compact case, there is a priori no reason for the cuspidal space $\mathcal{C}(\Gamma \backslash \mathcal{H})$ to be different from $\{0\}$. It is in fact still an open question to decide in which cases $\mathcal{C}(\Gamma \backslash \mathcal{H}) \neq \{0\}$. With the help of the trace formula, Selberg showed that this is in fact the case whenever Γ is commensurable with $SL(2, \mathbf{Z})$ (see Chap. 5). In the next chapter we present a recent alternative proof of this fact due to Lindenstrauss and Venkatesh. On the basis of these arithmetic examples, it was for a long time believed that $\mathcal{C}(\Gamma \backslash \mathcal{H})$ is always non-zero and in fact of infinite dimension. The remarkable works of Phillips-Sarnak [98], and their continuations by Deshouillers-Iwaniec [36], Wolpert [141], and Luo [85] (among others), showed that under a widely believed conjecture on eigenvalue multiplicities one has $\mathcal{C}(\Gamma \backslash \mathcal{H}) = \{0\}$ for generic Γ. We refer the reader to the above cited works for the precise meaning of "generic". The aforementioned conjecture is that the multiplicity of an eigenvalue λ for the group $\Gamma_0(N)$ is $O_N(1)$, or at least $O_N(\log(\lambda + 1))$; it is considered out of reach.

§ 3.10

The minimax principle is reproduced from [24, §8.2]. Proposition 3.35 is an easy case of the Buser inequality [23]. A first fundamental link between an isoperimetric constant – closely related to the constant h of Proposition 3.35 – and the first non-zero eigenvalue λ_1 is a famous inequality due to Cheeger [26].

In Proposition 3.35, the constant C can be chosen uniformly, see [23].

Theorem 3.38 is due to Roelcke [105].

3.12   Exercises

Exercise 3.40

1. Show that $\Delta = -4y^2 \partial \overline{\partial}$, where $\partial = \frac{1}{2} \left(\frac{\partial}{\partial x} - i \frac{\partial}{\partial y} \right)$ and $\overline{\partial} = \frac{1}{2} \left(\frac{\partial}{\partial x} + i \frac{\partial}{\partial y} \right)$. Verify the equality (1.20).
2. Show that in polar coordinates (r, θ) of the hyperbolic plane (see Appendix A), we have

$$\Delta = -\frac{\partial^2}{\partial r^2} - \frac{1}{\tanh r} \frac{\partial}{\partial r} - \frac{1}{(2 \sinh r)^2} \frac{\partial^2}{\partial \theta^2}.$$

Exercise 3.41 Let k be a non-negative integer. The *Maaß operators* are the following differential operators on $C^\infty(\mathcal{H})$:

$$R_k = (z - \bar{z})\frac{\partial}{\partial z} + \frac{k}{2},$$

$$L_k = -(z - \bar{z})\frac{\partial}{\partial z} - \frac{k}{2}.$$

We define the *weight k Laplacian* by the formula

$$\Delta_k = -y^2\left(\frac{\partial^2}{\partial x^2} + \frac{\partial^2}{\partial y^2}\right) + iky\frac{\partial}{\partial x}.$$

1. Show that

$$\Delta_k = -L_{k+2}R_k - \frac{k}{2}\left(1 + \frac{k}{2}\right) = -R_{k-2}L_k + \frac{k}{2}\left(1 - \frac{k}{2}\right).$$

2. Show that $R_k \circ \Delta_k = \Delta_{k+2} \circ R_k$ and $L_k \circ \Delta_k = \Delta_{k-2} \circ L_k$.

Exercise 3.42 For all $k \in \mathbf{Z}$, one defines an action of G on $C^\infty(\mathcal{H})$ by

$$f|_k g = \left(\frac{c\bar{z} + d}{|cz + d|}\right)^k f\left(\frac{az + b}{cz + d}\right), \qquad g = \begin{pmatrix} a & b \\ c & d \end{pmatrix}.$$

1. Show that if $f \in C^\infty(\mathcal{H})$ and $g \in G$, then

$$(R_k f)|_{k+2} g = R_k(f|_k g) \quad \text{and} \quad (L_k f)|_{k-2} g = L_k(f|_k g).$$

2. Deduce from the preceding question and the previous exercise that

$$(\Delta_k f)|_k g = \Delta_k(f|_k g).$$

Exercise 3.43 Show that the weight k Laplacian is a symmetric operator on $L^2(\mathcal{H})$ with domain $C_c^\infty(\mathcal{H})$.

Exercise 3.44 Let Γ be a cocompact Fuchsian group. Assume that $-I \in \Gamma$. Let χ be a homomorphism from Γ to $\mathbf{C}^\times$. We denote by $C^\infty(\Gamma\backslash\mathcal{H}, \chi, k)$ the space of C^∞-functions on $\mathcal{H}$ satisfying

$$\chi(\gamma)f(z) = \left(\frac{c\bar{z} + d}{|cz + d|}\right)^k f\left(\frac{az + b}{cz + d}\right), \qquad \gamma = \begin{pmatrix} a & b \\ c & d \end{pmatrix} \in \Gamma.$$

1. Show that $C^\infty(\Gamma\backslash\mathcal{H}, \chi, k) = \{0\}$ if $\chi(-I) \neq (-1)^k$. In what follows we shall always assume that $\chi(-I) = (-1)^k$.
2. Let $f, g \in C^\infty(\Gamma\backslash\mathcal{H}, \chi, k)$. Show that $f\bar{g}$ is a Γ-invariant function and that the expression

$$\langle f, g \rangle = \int_{\Gamma\backslash\mathcal{H}} f(z)\overline{g(z)}\frac{dxdy}{y^2}$$

   defines a scalar product on $C^\infty(\Gamma\backslash\mathcal{H}, \chi, k)$. We shall denote by $L^2(\Gamma\backslash\mathcal{H}, \chi, k)$ the Hilbert space completion of $C^\infty(\Gamma\backslash\mathcal{H}, \chi, k)$ with respect to this scalar product.
3. Deduce from the preceding exercise that the operators R_k and L_k send $C^\infty(\Gamma\backslash\mathcal{H}, \chi, k)$ to $C^\infty(\Gamma\backslash\mathcal{H}, \chi, k + 2)$ and $C^\infty(\Gamma\backslash\mathcal{H}, \chi, k - 2)$, resp.
4. Let $f \in C^\infty(\Gamma\backslash\mathcal{H}, \chi, k)$ and $g \in C^\infty(\Gamma\backslash\mathcal{H}, \chi, k + 2)$. Show that

$$\langle R_k f, g \rangle = \langle f, -L_{k+2}g \rangle.$$

5. Deduce that Δ_k defines a symmetric operator on the Hilbert space $L^2(\Gamma\backslash\mathcal{H}, \chi, k)$.

Exercise 3.45 We keep the notation from the preceding exercise. A *holomorphic modular form of weight k with respect to* Γ is a holomorphic function $f : \mathcal{H} \to \mathbf{C}$ such that

$$f(\gamma z) = \chi(\gamma)(cz + d)^k f(z), \quad \gamma = \begin{pmatrix} a & b \\ c & d \end{pmatrix} \in \Gamma.$$

1. Show that if f is a holomorphic modular form of k with respect to Γ, then the function $z \mapsto y^{k/2}f(z)$ is an eigenfunction of the weight k Laplacian with eigenvalue $\frac{k}{2}(1 - \frac{k}{2})$, and that f lies in the kernel of L_k.
2. Conversely, show that if $g \in C^\infty(\Gamma\backslash\mathcal{H}, \chi, k)$ is an eigenfunction for Δ_k and lies in the kernel of L_k, then the function $z \mapsto y^{-k/2}g(z)$ is a holomorphic modular form of weight k with respect to Γ.

Exercise 3.46 Show that if $k \in \mathbf{N}$ and if λ is an eigenvalue of Δ_k, then either $\lambda = \frac{\ell}{2}(1 - \frac{\ell}{2})$, where ℓ is an integer between 1 and k having the same parity as k, or $\lambda \geq 0$, and in fact $\lambda \geq \frac{1}{4}$ if k is odd.

Chapter 4
Maaß Forms

In this chapter we consider the particular case of congruence covers $Y(N) = \Gamma(N)\backslash\mathcal{H}$ or $Y_0(N) = \Gamma_0(N)\backslash\mathcal{H}$ of the modular surface. In the first sections we pay particular attention to the case of the modular group $\mathrm{SL}(2,\mathbf{Z})$.

Let us recall that a *Maaß form* is an eigenfunction of the Laplacian on a hyperbolic surface $\Gamma\backslash\mathcal{H}$, where $\Gamma \subset \mathrm{SL}(2,\mathbf{Z})$ is a congruence subgroup, that grows at most polynomially in the cusps of Γ. As we have already remarked, it is not obvious that non-constant cuspidal – or even square integrable – Maaß forms exist. In this chapter we show that Maaß forms do indeed exist, and plentifully so. We then recall how Maaß came to explicitly construct such functions from quadratic extensions of $\mathbf{Q}$.

4.1 Eisenstein Series for $\mathrm{SL}(2,\mathbf{Z})$

In this section $\Gamma = \mathrm{SL}(2,\mathbf{Z})$. Fix $z = x + iy \in \mathcal{H}$ and consider the series in the s-variable given by the expression[1]:

$$E(z,s) = \frac{1}{2} \sum_{\substack{c,d\in\mathbf{Z} \\ (c,d)=1}} \frac{y^s}{|cz+d|^{2s}}$$

$$= \frac{1}{2} \sum_{\gamma\in\Gamma_\infty\backslash\Gamma} (\mathrm{Im}(\gamma z))^s. \tag{4.1}$$

---

[1] The coefficient $\frac{1}{2}$ comes from the fact that the element $-I \in \mathrm{SL}(2,\mathbf{Z})$ acts trivially on $\mathcal{H}$.

© Springer International Publishing Switzerland 2016
N. Bergeron, *The Spectrum of Hyperbolic Surfaces*, Universitext,
DOI 10.1007/978-3-319-27666-3_4

One passes from the first to the second equality by associating with each pair of relatively prime integers (c, d) the set of matrices in Γ whose second row is (c, d), this being a coset in $\Gamma_\infty \backslash \Gamma$.

Lemma 4.1 *The series* (4.1) *is absolutely convergent on* $\mathrm{Re}(s) > 1$. *The sum of this series,* $E(z, s)$, *is called the* Eisenstein series *of the modular group* Γ. *It satisfies*

$$E(\gamma(z), s) = E(z, s), \quad \text{for all } \gamma \in \Gamma.$$

Proof It suffices to show that the expression (4.1) is absolutely convergent. Fix

$$z_0 \in D = \{z \in \mathcal{H} \mid |x| < 1/2, \ |z| > 1\}$$

and let δ be a positive real number such that the (hyperbolic) ball $B(z_0, \delta)$ is contained in D. The domain D and the strip $S = \{z \in \mathcal{H} \mid |x| < 1/2\}$ are the respective fundamental domains for the groups Γ and Γ_∞. Thus there exists a set of coset representatives $E \subset \Gamma$ for $\Gamma_\infty \backslash \Gamma$ such that

$$\bigcup_{\gamma \in E} \gamma(D) = S.$$

Each ball $B(\gamma z_0, \delta) = \gamma(B(z_0, \delta))$ $(\gamma \in E)$ is contained in $\gamma(D)$. In particular, these balls are pairwise disjoint and, if T is a real number sufficiently large for

$$B(z_0, \delta) \subset \{z = x + iy \in \mathbf{C} \mid y \leqslant T\},$$

then all of the sets $B(\gamma z_0, \delta)$ $(\gamma \in E)$ are contained in $\{z \in \mathcal{H} \mid |x| < 1/2, \ y \leqslant T\}$. Besides, each γz_0, except possibly z_0 itself, has imaginary part less than 1 since it lies in $S - D$. If two points $z = x + iy$ and $z' = x' + iy'$ in $\mathcal{H}$ are at (hyperbolic) distance less than or equal to 2δ, we have $ye^{-2\delta} \leqslant y' \leqslant ye^{2\delta}$. Hence there exists a constant $C = C(\delta)$ such that, if $z_0' = x_0' + iy_0' \in \mathcal{H}$ satisfies $y_0' \leqslant 1$, then for every $z = x + iy \in B(z_0', \delta)$ we have $y_0' \leqslant Cy$. We deduce that for a real number s strictly greater than 1:

$$\sum_{\substack{\gamma \in \Gamma_\infty \backslash \Gamma}} y(\gamma z_0)^s \leqslant y_0^s + C^s \sum_{\substack{\gamma \in \Gamma_\infty \backslash \Gamma \\ \gamma \neq 1}} \frac{1}{\mathrm{vol}(B(z_0, \delta))} \int_{\gamma B(z_0, \delta)} y^s \, \frac{dx \, dy}{y^2},$$

$$\leqslant y_0^s + \frac{C^s}{\mathrm{vol}(B(z_0, \delta))} \int_{-1/2}^{1/2} \int_0^T y^s \, \frac{dx \, dy}{y^2},$$

$$\leqslant y_0^s + \frac{C^s T^{s-1}}{(s-1) \, \mathrm{vol}(B(z_0, \delta))}.$$

The series (4.1) is therefore absolutely convergent. The automorphy follows immediately from the definition.   □

The fundamental properties of the Eisenstein series are its analytic continuation and functional equation. As for the Riemann zeta function, it is more natural to complete the Eisenstein series, which we do by writing

$$E^*(z, s) = \pi^{-s} \Gamma(s) \frac{1}{2} \sum_{\substack{m,n \in \mathbf{Z} \\ (m,n) \neq 0}} \frac{y^s}{|mz + n|^{2s}}$$

$$= \pi^{-s} \Gamma(s) \zeta(2s) E(z, s). \tag{4.2}$$

The passage from the first to the second equality comes from the fact that summing over all non-zero pairs $(m, n) \in \mathbf{Z}^2$ is the same thing as summing over all positive integers N and all pairs of relatively prime integers (c, d) by taking $(m, n) = (Nc, Nd)$. Of course ζ is the Riemann zeta function.

Theorem 4.2 *The function $E^*(z, s)$, defined by (4.2) for $\mathrm{Re}(s) > 1$, admits a meromorphic continuation to all $s \in \mathbf{C}$; it is holomorphic everywhere except at $s = 0$ and $s = 1$, where it has simple poles. The completed Eisenstein series satisfies the functional equation*

$$E^*(z, s) = E^*(z, 1 - s). \tag{4.3}$$

Moreover, it is bounded in vertical strips $\sigma_1 \leqslant \mathrm{Re}(s) \leqslant \sigma_2$ ($\sigma_1, \sigma_2 \in \mathbf{R}$) for s uniformly away from the poles.

Remark 4.3 Since the Γ function and the ζ function both admit meromorphic continuations to all of $\mathbf{C}$ (see Exercise 1.16), it follows from Theorem 4.2 that the function $E(z, s)$ admits a meromorphic continuation to all $s \in \mathbf{C}$. We shall see in §4.2.3 that this continuation is holomorphic on the half-plane $\mathrm{Re}(s) \geqslant 1/2$.

Proof As in the case of the Riemann zeta function (see Exercise 1.16), we shall use the so-called "theta series" given by the formula

$$\Theta(t)(= \Theta(z, t)) = \sum_{(m,n) \in \mathbf{Z}^2} e^{-\pi |mz + n|^2 t / y},$$

for $t > 0$. It is immediate that

$$E^*(z, s) = \frac{1}{2} \int_0^{+\infty} (\Theta(t) - 1) t^s \frac{dt}{t}$$

$$= \frac{1}{2} \int_0^1 (\Theta(t) - 1) t^s \frac{dt}{t} + \frac{1}{2} \int_1^{+\infty} (\Theta(t) - 1) t^s \frac{dt}{t}$$

$$= E_0^*(z, s) + E_1^*(z, s).$$

Since y and t are positive, the function $\Theta(t) - 1$ is of rapid decay and the function $E_1^*(z, s)$ is both holomorphic on the entire s-plane and bounded in vertical strips.

Now consider the function $\mathbf{R}^2 \to \mathbf{C}$:

$$f_{z,t}(u_1, u_2) = e^{-\pi|u_1 z + u_2|^2 t/y} = e^{-\pi[(u_1 x + u_2)^2 + (u_1 y)^2]t/y}.$$

It follows from Lemma 1.1 that the function $u = (u_1, u_2) \mapsto e^{-\pi|u|^2} = e^{-\pi(u_1^2 + u_2^2)}$ is its own Fourier transform. One then has

$$\hat{f}_{z,t}(\xi_1, \xi_2) = \int_{\mathbf{R}^2} e^{-\pi[(u_1 x + u_2)^2 + (u_1 y)^2]t/y} e(-\langle \xi, u \rangle)\, du$$

$$= \int_{\mathbf{R}^2} e^{-\pi|v|^2} e(-\langle {}^t A^{-1}\xi, v \rangle)|\det(A)|^{-1} dv,$$

where

$$A = \sqrt{\frac{t}{y}} \begin{pmatrix} x & 1 \\ y & 0 \end{pmatrix}$$

and $v = Au$. From this one deduces that

$$\hat{f}_{z,t}(\xi_1, \xi_2) = t^{-1} e^{-\pi|{}^t A^{-1}\xi|^2} = t^{-1} e^{-\pi[(y\xi_2)^2 + (\xi_1 - x\xi_2)^2]/ty}.$$

Hence, the Poisson summation formula (see the introduction) applied to $f_{z,t}$ implies that $\Theta(t) = t^{-1}\Theta(t^{-1})$.

But the change of variables $t \mapsto t^{-1}$ then shows that if $\mathrm{Re}(s) < 0$,

$$E_1^*(z, s) = \frac{1}{2} \int_0^1 (t\Theta(t) - 1)t^{-s}\frac{dt}{t},$$

$$= \frac{1}{2} \int_0^1 \Theta(t)t^{1-s}\frac{dt}{t} + \frac{1}{2s},$$

$$= \frac{1}{2} \int_0^1 (t\Theta(t) - 1)t^{1-s}\frac{dt}{t} + \frac{1}{2 - 2s} + \frac{1}{2s},$$

$$= E_0^*(z, 1 - s) + \frac{1}{2 - 2s} + \frac{1}{2s}.$$

We find that the function $E_0^*(z, s)$ – and thus $E^*(z, s)$ as well – admits a meromorphic continuation to all $s \in \mathbf{C}$, and that

$$E^*(z, s) = E_1^*(z, s) + E_1^*(z, 1 - s) + \frac{1}{2s - 2} - \frac{1}{2s}. \qquad \Box$$

The following paragraph is a first aside on the links between the Eisenstein series and the Riemann zeta function.

4.1.1 Eisenstein Series and the Riemann Zeta Function I

Let s be a complex number with real part strictly larger than 1. The Eisenstein series $E(z, s)$ is invariant under the transformation $z \mapsto z + 1$. As a result, we can expand $E(z, s)$ in a Fourier series,

$$E(z, s) = \sum_{r \in \mathbf{Z}} a_r(y, s)e(rx). \tag{4.4}$$

We calculate the Fourier coefficients

$$a_r(y, s) = \int_0^1 E(x + iy, s)e(-rx)\, dx$$

or rather the coefficients

$$a_r^*(y, s) = \int_0^1 E^*(x + iy, s)e(-rx)\, dx,$$

via the first expression for $E^*(z, s)$ in (4.2). We begin by considering the contribution of the terms with $m = 0$ in the completed Eisenstein series. Such a term is independent of x and therefore doesn't contribute to a_0^*. Since n and $-n$ contribute equally, their joint contribution is

$$\pi^{-s}\Gamma(s)y^s \sum_{n=1}^{+\infty} n^{-2s} = \pi^{-s}\Gamma(s)\zeta(2s)y^s. \tag{4.5}$$

This is one part of a_0^*. Let us now calculate the contribution of the terms with $m \neq 0$ to each of the a_r^*. Since (m, n) and $(-m, -n)$ contribute equally, we can group them together and assume that $m > 0$. The contribution to the coefficient a_r^* is therefore equal to

$$\pi^{-s}\Gamma(s)y^s \sum_{m=1}^{+\infty} \sum_{n=-\infty}^{+\infty} \int_0^1 [(mx + n)^2 + m^2y^2]^{-s}e(-rx)\, dx$$

$$= \pi^{-s}\Gamma(s)y^s \sum_{m=1}^{+\infty} \sum_{n \in \mathbf{Z}/m\mathbf{Z}} \int_{-\infty}^{+\infty} [(mx + n)^2 + m^2y^2]^{-s}e(-rx)\, dx.$$

The change of variables $x \mapsto x - n/m$ transforms this last expression into

$$\pi^{-s}\Gamma(s)y^s \sum_{m=1}^{+\infty} m^{-2s} \sum_{n \in \mathbf{Z}/m\mathbf{Z}} e(-rn/m) \int_{-\infty}^{+\infty} [x^2 + y^2]^{-s} e(-rx)\, dx,$$

which, since

$$\sum_{n \in \mathbf{Z}/m\mathbf{Z}} e(-rn/m) = \begin{cases} m & \text{if } m|r, \\ 0 & \text{else,} \end{cases}$$

is equal to

$$\pi^{-s}\Gamma(s)y^s \sum_{m|r} m^{1-2s} \int_{-\infty}^{+\infty} [x^2 + y^2]^{-s} e(-rx)\, dx. \tag{4.6}$$

We distinguish two cases according to whether r is zero or not. Assume first that $r = 0$. The condition $m|r$ in (4.6) is then empty and, using the integral expression (1.26) of the Γ function, we find that (4.6) is equal to

$$\left(\frac{y}{\pi}\right)^s \Gamma(s) \int_{-\infty}^{+\infty} [x^2 + y^2]^{-s}\, dx = \int_{-\infty}^{+\infty} \int_0^{+\infty} e^{-t} \left(\frac{ty}{\pi(x^2 + y^2)}\right)^s \frac{dt}{t}\, dx$$

$$= \int_0^{+\infty} \int_{-\infty}^{+\infty} e^{-\pi t(x^2 + y^2)/y} t^s\, dx\, \frac{dt}{t},$$

where we have interchanged the order of integration (the integral is absolutely convergent for $\mathrm{Re}(s) > 1/2$) and applied the change of variable $t \mapsto \pi t(x^2 + y^2)/y$. But it is well known that

$$\int_{-\infty}^{+\infty} e^{-t\pi x^2/y}\, dx = \sqrt{\frac{y}{t}}.$$

Inserting this expression into the last integral and combining the result with Eq. (4.5), we obtain the following expression for the constant term:

$$a_0^*(y, s) = \pi^{-s}\Gamma(s)\zeta(2s)y^s + \pi^{-s+1/2}\Gamma(s - 1/2)\zeta(2s - 1)y^{1-s}. \tag{4.7}$$

Now assume that $r \neq 0$. Then

$$a_r^*(y, s) = \sigma_{1-2s}(|r|) \left(\frac{y}{\pi}\right)^s \Gamma(s) \int_{-\infty}^{+\infty} [x^2 + y^2]^{-s} e(-rx)\, dx,$$

where

$$\sigma_{1-2s}(r) = \sum_{m|r} m^{1-2s}.$$

But analogously to the case $r = 0$,

$$\left(\frac{y}{\pi}\right)^s \Gamma(s) \int_{-\infty}^{+\infty} [x^2 + y^2]^{-s} e(-rx)\, dx = \int_0^{+\infty} \int_{-\infty}^{+\infty} e^{-\pi t(x^2+y^2)/y} t^s e^{-2i\pi rx} dx\, \frac{dt}{t}.$$

According to Lemma 1.1,

$$\int_{-\infty}^{+\infty} e^{-t\pi x^2/y} e(-rx)\, dx = \sqrt{\frac{y}{t}}\, e^{-y\pi r^2/t}.$$

One thus obtains

$$\int_0^{+\infty} \int_{-\infty}^{+\infty} e^{-\pi t(x^2+y^2)/y} t^s e^{-2i\pi rx} dx\, \frac{dt}{t}$$

$$= \sqrt{y} \int_0^{+\infty} e^{-2\pi |r| y(|r|^{-1} t + (|r|^{-1} t)^{-1})/2} t^{s-1/2}\, \frac{dt}{t}.$$

We recognize this last expression – see Appendix B – as

$$2|r|^{s-1/2} \sqrt{y}\, K_{s-1/2}(2\pi |r| y),$$

where K_ν denotes the Bessel K function,

$$K_\nu(y) = \int_0^{+\infty} e^{-y \cosh \tau} \cosh(\nu\tau) d\tau = \frac{1}{2} \int_0^{+\infty} e^{-y(t+t^{-1})/2} t^\nu\, \frac{dt}{t}.$$

It follows from all of this that if $r \neq 0$,

$$a_r(y, s)^* = 2|r|^{s-1/2} \sigma_{1-2s}(|r|) \sqrt{y}\, K_{s-1/2}(2\pi |r| y). \tag{4.8}$$

We summarize these computations in the following proposition.

Proposition 4.4 *Assume that* $\text{Re}(s) > 1$. *Then the Eisenstein series* $E^*(z, s)$ *admits a Fourier expansion of the form*

$$E^*(z, s) = \xi(2s)y^s + \xi(2s - 1)y^{1-s}$$

$$+ 2 \sum_{r \in \mathbf{Z}^*} |r|^{s-1/2} \sigma_{1-2s}(|r|) \sqrt{y}\, K_{s-1/2}(2\pi |r| y) e(rx),$$

where $\xi(s) = \pi^{-s/2}\Gamma\left(\frac{s}{2}\right)\zeta(s)$ and the right-hand sum converges uniformly on compacta. In particular,

$$E(z, s) = y^s + \varphi(s)y^{1-s} + 2\sum_{r \in \mathbf{Z}^*} \frac{|r|^{s-1/2}\sigma_{1-2s}(|r|)}{\xi(2s)}\sqrt{y}\,K_{s-1/2}(2\pi|r|y)e(rx),$$

where $\varphi(s) = \xi(2s - 1)/\xi(2s)$.

We immediate remark that the meromorphic continuation and functional equation of the Riemann zeta function follow from the analogous results for the Eisenstein series (Theorem 4.2) and from the computation of the constant term (4.7) that we can rewrite as

$$a_0^*(y, s) = \xi(2s)y^s + \xi(2 - 2s)y^{1-s}. \tag{4.9}$$

From this discussion we shall principally retain the fact that the a_0 Fourier coefficient of the Eisenstein series (see (4.4)) satisfies

$$a_0(y, s) = y^s + \varphi(s)y^{1-s}, \tag{4.10}$$

where $\varphi(s) = \xi(2 - 2s)/\xi(2s)$. The function φ admits a meromorphic continuation to the entire complex plane, holomorphic for $\text{Re}(s) > 1/2$. It follows from the functional equation of the Riemann zeta function that $\varphi(s)\varphi(1 - s) = 1$. Finally, $\overline{\varphi(s)} = \varphi(\bar{s})$. Since $\bar{s} = 1 - s$ on the line $\text{Re}(s) = 1/2$, one finds

$$\left|\varphi\left(\frac{1}{2} + it\right)\right| = 1 \text{ for all } t \in \mathbf{R}. \tag{4.11}$$

4.2   Eisenstein Series and the Spectrum of the Laplacian

In this section we shall always take $\Gamma = \text{SL}(2, \mathbf{Z})$. Recall then that

$$D = \{z = x + iy \in \mathcal{H} \mid |x| < 1/2, \ |z| > 1\}$$

is a fundamental domain for Γ.

Since the function $z \mapsto \text{Im}(z)^s$ is a Laplacian eigenfunction with eigenvalue $\lambda = s(1 - s)$, and Δ commutes with the Γ action, the Eisenstein series is also a Laplacian eigenfunction with eigenvalue λ. In other words, if $\text{Re}(s) > 1$,

$$\Delta E(z, s) = s(1 - s)E(z, s).$$

If follows from the Fourier expansion that the function $z \mapsto E(z, s)$ is never square integrable[2] on D. Nevertheless, for $s \in \frac{1}{2} + i\mathbf{R}$, the expression (4.10) shows (at least formally – the function φ could have a singularity) that each series $E(z, s)$ is "almost L^2-integrable" on D. We now show that these series allow one to describe the orthocomplement of the cuspidal spectrum: the continuous spectrum.

Let us begin by truncating the series $E(z, s)$ in such a way as to render L^2-integrable. Let ψ be a C^∞ function with support in $(0, +\infty)$. The series

$$E(z|\psi) = \frac{1}{2} \sum_{\gamma \in \Gamma_\infty \backslash \Gamma} \psi(\text{Im}(\gamma z)) \tag{4.12}$$

is called an *incomplete Eisenstein series*. The series (4.12) defines a Γ-invariant function $z \mapsto E(z|\psi)$ from $\mathcal{H}$ to $\mathbf{C}$, bounded and in fact lying in $C_c^\infty(\Gamma \backslash \mathcal{H})$; in particular it is L^2-integrable on D. On the other hand, the incomplete Eisenstein series are not eigenfunctions of the Laplacian. Via Mellin inversion, we can nevertheless write them as weighted integrals of Eisenstein series. Let us make a brief detour to recall the *Mellin transform*, the multiplicative analog of the Fourier transform.

4.2.1 Mellin Inversion Formula, Phragmén-Lindelöf Principle

Lemma 4.5 *Let ψ be a continuous function on $(0, +\infty)$ and Ψ its Mellin transform:*

$$\Psi(s) := \int_0^{+\infty} \psi(y) y^s \frac{dy}{y}.$$

If non-empty, the domain of absolute convergence of $\Psi(s)$ is a vertical strip $\sigma_1 < \text{Re}(s) < \sigma_2$ ($\sigma_1, \sigma_2 \in [-\infty, +\infty]$). For every $\sigma \in (\sigma_1, \sigma_2)$ and $0 < y < +\infty$,

$$\psi(y) = \frac{1}{2i\pi} \int_{-\infty}^{+\infty} \Psi(\sigma + ir) y^{-\sigma - ir} dr.$$

Proof If the integral

$$\int_0^1 \psi(y) y^s \frac{dy}{y} \quad \left(\text{resp.} \int_1^{+\infty} \psi(y) y^s \frac{dy}{y} \right)$$

---

[2] When $\text{Re}(s) > 1$, this is obvious since $s(1 - s)$ doesn't lie in $\mathbf{R}_+$.

converges absolutely for a certain value σ of s, it remains absolutely convergent in the half-plane $\mathrm{Re}(s) > \mathrm{Re}(\sigma)$ (resp. $\mathrm{Re}(s) < \mathrm{Re}(\sigma)$). The domain of absolute convergence of $\Psi(s)$ is thus a vertical strip.

To continue we relate Mellin inversion to Fourier inversion. Let $\psi_\sigma(v) = \psi(e^v)e^{\sigma v}$. For $\sigma \in (\sigma_1, \sigma_2)$, the function ψ_σ is continuous and belongs to the space $L^1(\mathbf{R}) \cap L^2(\mathbf{R})$. The change of variables $y = e^v$ in the definition of Ψ gives the expression

$$\Psi(\sigma + ir) = \int_{-\infty}^{+\infty} \psi_\sigma(v)e^{ivr}dv,$$

which, by Fourier inversion, is equivalent to

$$\psi_\sigma(v) = \frac{1}{2i\pi} \int_{-\infty}^{+\infty} \Psi(\sigma + ir)e^{-ivr}dr.$$

We obtain the Mellin inversion formula by multiplying this last expression by $e^{-\sigma v}$ and putting $y = e^v$. □

Lemma 4.6 (Phragmén-Lindelöf principle) *Let $f(s)$ be a holomorphic function in the upper part of a vertical strip*

$$\sigma_1 \leqslant \mathrm{Re}(s) \leqslant \sigma_2, \quad \mathrm{Im}(s) > c,$$

such that $f(\sigma + it) = O\left(e^{t^\alpha}\right)$ for a real number $\alpha > 0$ when $\sigma_1 \leqslant \sigma \leqslant \sigma_2$. Suppose that $f(\sigma + it) = O(t^M)$ for $\sigma = \sigma_1$ and $\sigma = \sigma_2$. Then $f(\sigma + it) = O(t^M)$ uniformly in $\sigma \in [\sigma_1, \sigma_2]$.

Proof We write $s = \sigma + it = re^{i\theta}$. Replacing f by the function $s \mapsto f(s)/s^M$ if necessary, we can assume that $M = 0$. Moreover, it is enough to consider the function f in the half-plane $t > t_1$ for large t_1, in such a way that the argument θ of s stays close to $\pi/2$ as σ runs through the interval $[\sigma_1, \sigma_2]$. Let m be an integer congruent to 2 mod 4 and strictly greater than α. Then

$$s^m = r^m(\cos m\theta + i\sin m\theta),$$

and $m\theta$ is close to π mod 2π so that $\cos(m\theta)$ is negative. We fix a positive real number ε and consider the function

$$g_\varepsilon(s) = g(s) = f(s)e^{\varepsilon s^m}.$$

Then there exists a constant B such that for t sufficiently large one has

$$|g(s)| \leqslant Be^{t^\alpha}e^{\varepsilon r^m \cos m\theta}.$$

There is a real number T_ε (we may take T_ε greater than t_1) such that for all σ in the interval $[\sigma_1, \sigma_2]$ and for all t greater than T_ε, $|g(\sigma + it)| \leqslant B$. We fix a real number t_2 greater than T_ε. By hypothesis on f the function f is bounded on the vertical lines $\mathrm{Re}(s) = \sigma_1$ and $\mathrm{Re}(s) = \sigma_2$. And by definition of t_1 and m, if $\mathrm{Im}(s) \geqslant t_1$ and $\sigma \in [\sigma_1, \sigma_2]$ then $\cos m\theta$ is negative so that we have $e^{\varepsilon r^m \cos m\theta} \leqslant 1$. Changing B if necessary we can thus assume that $|g|$ is bounded above by B on the boundary of the rectangle defined by the conditions $t_1 \leqslant \mathrm{Im}(s) \leqslant t_2$ and $\sigma_1 \leqslant \sigma \leqslant \sigma_2$. According to the maximum modulus principle $|g(s)|$ is then bounded above by B in all of the rectangle. One deduces that

$$|f(s)| \leqslant B e^{-\varepsilon r^m \cos m\theta}$$

in this rectangle and thus (letting t_2 tend to ∞) for all s satisfying the conditions $\sigma \in [\sigma_1, \sigma_2]$ and $\mathrm{Im}(s) \geqslant t_1$. By letting ε tend towards 0, we find

$$|f(s)| \leqslant B$$

for all s such that $\sigma \in [\sigma_1, \sigma_2]$ and $t \geqslant t_1$ and the lemma is proved. $\qquad\square$

In order to apply the Phragmén-Lindelöf principle we shall need the following proposition. A holomorphic function $f(s)$ is said to be of *order* ρ if

$$f(s) = O(e^{|s|^{\rho+\varepsilon}})$$

for all $\varepsilon > 0$.

Proposition 4.7 *The function $s \mapsto s(s-1)\xi(s)$ is (holomorphic and) of order* 1.

Proof By virtue of the functional equation (1.27), it suffices to consider the half-plane $\mathrm{Re}(s) \geqslant 1/2$. We already know that this function is holomorphic. Moreover, it is well known – this is *Stirling's formula* – that when $\mathrm{Re}(s) \geqslant 1/2$ and $|s| \to +\infty$,

$$\Gamma(s) \sim \sqrt{2\pi}\, e^{-s} s^{s-1/2} = \sqrt{2\pi}\, e^{-s+(s-1/2)\log s}. \qquad (4.13)$$

In particular, there exists a constant $M > 0$ such that $\Gamma(s) = O(e^{M|s|\log|s|})$. The following lemma then allows us to conclude that when $\mathrm{Re}(s) \geqslant 1/2$ and $|s| \to +\infty$,

$$s(s-1)\pi^{-s/2}\Gamma(s/2)\zeta(s) = O(|s|^3 e^{\frac{M}{2}|s|\log|s|}) = O(e^{|s|^{1+\varepsilon}}),$$

for all $\varepsilon > 0$. $\qquad\square$

Lemma 4.8 *We have*

$$\zeta(s) - \frac{1}{s-1} = O(|s|)$$

for $\mathrm{Re}(s) \geqslant 1/2$.

Proof For $\mathrm{Re}(s) > 1$,

$$\zeta(s) - \frac{1}{s-1} = \sum_{n=1}^{+\infty} n^{-s} - \int_1^{+\infty} x^{-s} dx$$

$$= \sum_{n=1}^{+\infty} \int_n^{n+1} (n^{-s} - x^{-s}) \, dx. \tag{4.14}$$

The expression under the integral in (4.14) is bounded above in absolute value by

$$|n^{-s} - x^{-s}| = \left| \int_n^x s t^{-s-1} dt \right| \le |s| n^{-\mathrm{Re}(s)-1}.$$

We deduce that

$$\left| \zeta(s) - \frac{1}{s-1} \right| \le |s| \zeta(\mathrm{Re}(s) + 1).$$

The expression (4.14) thus furnishes an analytic continuation of $\zeta(s) - \frac{1}{s-1}$ to all of the half-plane $\mathrm{Re}(s) > 0$. In particular, $|\zeta(s) - \frac{1}{s-1}| \le |s| \zeta(3/2)$ for $\mathrm{Re}(s) \ge 1/2$. $\qquad\square$

It follows from Proposition 4.7 that the function $s \mapsto (s-1)\zeta(s)$ is entire and of order 1. Indeed, $s(s-1)\xi(s)$ and $1/s\Gamma(s/2)$ are both entire and of order 1.

4.2.2 The Space of Incomplete Eisenstein Series

We now apply the Mellin inversion formula to a function ψ with compact support in $(0, +\infty)$. We begin by observing that the integral

$$\widehat{\psi}(s) := \int_0^{+\infty} \psi(y) y^{-s-1} dy \tag{4.15}$$

converges absolutely for all s. The Mellin inversion formula then implies that

$$\psi(y) = \frac{1}{2i\pi} \int_{(\sigma)} y^s \widehat{\psi}(s) ds \tag{4.16}$$

where (σ) denotes the vertical line $\mathrm{Re}(s) = \sigma$. The integral (4.16) is independent of the real number σ: this of course follows from the Mellin inversion formula but one can also see this more directly. Given two real numbers $\sigma_1 < \sigma_2$, an iterated integration by parts shows that on the vertical strip $\sigma_1 \leqslant \mathrm{Re}(s) \leqslant \sigma_2$ we have

$$\widehat{\psi}(s) = O((|s|+1)^{-A}), \quad \text{for every constant } A > 0. \tag{4.17}$$

Let T be real and arbitrary and consider the line integral

$$\frac{1}{2i\pi} \int_C y^s \widehat{\psi}(s) ds,$$

where C is the rectangle with vertices $\sigma_1 - iT$, $\sigma_1 + iT$, $\sigma_2 - iT$ and $\sigma_2 + iT$, taken in the direct orientation. According to the Cauchy integration formula, since $s \mapsto y^s \widehat{\psi}(s)$ is holomorphic, we have

$$\frac{1}{2i\pi} \int_C y^s \widehat{\psi}(s) ds = 0.$$

The rapid decay (4.17) of $\widehat{\psi}$ in vertical strips (take $A = 2$) implies that the contributions to the integral of each of the two horizontal sides H_1 and H_2 are majorized by

$$O\left(\frac{1}{2\pi} \int_{\sigma_1}^{\sigma_2} y^t (1 + |t + iT|)^{-2} dt \right) = O(y^{\sigma_2} T^{-1}).$$

By letting T tend towards $+\infty$ (with the other parameters being fixed), we deduce that these two contributions disappear and it remains

$$\frac{1}{2i\pi} \int_{(\sigma_1)} y^s \widehat{\psi}(s) ds = \frac{1}{2i\pi} \int_{(\sigma_2)} y^s \widehat{\psi}(s) ds.$$

We shall make regular use of this method of *contour shift* along a vertical line. The key point is that the integrand decays rapidly at infinity in vertical strips, see (4.17). We shall encounter cases where the contour C encircles a finite number – bounded for all T – of poles of the integrand. One must then take into account the residues of the integrand of these poles.

It follows from this expression that for $\sigma > 1$,

$$E(z|\psi) = \frac{1}{2i\pi} \int_{(v)} E(z, s) \widehat{\psi}(s) ds. \tag{4.18}$$

This is formally obvious by summing over $\Gamma_\infty \backslash \Gamma$ the identity

$$\psi(\mathrm{Im}(\gamma z)) = \frac{1}{2i\pi} \int_{(\sigma)} \mathrm{Im}(\gamma z)^s \widehat{\psi}(s) ds.$$

This identity is valid for every real σ. The question is to justify the convergence of the sum and the interchange of the sum and integral. But according to Lemma 4.1 the series $E(z, s)$ is absolutely convergent for $\sigma = \mathrm{Re}(s) > 1$ and $|E(z, s)| \leq E(z, \sigma)$. It then follows from (4.17) that the integral (4.18) converges absolutely as soon as $\sigma > 1$.

Again we can show directly – by shifting contours – that (4.18) is independent of $\sigma > 1$. We shall in fact shift the contour to the vertical line $\sigma = 1/2$ so that we integrate $\widehat{\psi}$ against the "unitary Eisenstein series". Along the way we encounter a pole of $E(z, s)$.[3]

Let us from now on write $\mathcal{E}(\Gamma \backslash \mathcal{H})$ for the space of incomplete Eisenstein series.

Proposition 4.9 *The Hilbert space $L^2(\Gamma \backslash \mathcal{H})$ is the orthogonal sum*

$$L^2(\Gamma \backslash \mathcal{H}) = \overline{\mathcal{C}(\Gamma \backslash \mathcal{H})} \oplus \overline{\mathcal{E}(\Gamma \backslash \mathcal{H})}$$

of the closures of $\mathcal{C}(\Gamma \backslash \mathcal{H})$ and $\mathcal{E}(\Gamma \backslash \mathcal{H})$.

Proof We conserve the notation from the proof of Lemma 4.1; recall in particular that the strip $S = \{z \in \mathcal{H} \mid |x| < 1/2\}$ is a fundamental domain for Γ_∞. Consider a function $f \in L^2(\Gamma \backslash \mathcal{H})$. It follows from the Cauchy-Schwarz inequality and the finiteness of the area of D that $|f|$ is integrable on D. Thus

$$\langle f, E(\cdot | \psi) \rangle = \frac{1}{2} \int_D f(z) \sum_{\gamma \in \Gamma_\infty \backslash \Gamma} \overline{\psi}(\mathrm{Im}(\gamma z)) \, d\mu(z)$$

$$= \frac{1}{2} \sum_{\gamma \in E} \int_{\gamma D} f(z) \overline{\psi}(y) \, d\mu(z) = \int_S f(z) \overline{\psi}(y) \, d\mu(z)$$

$$= \int_0^{+\infty} \left(\int_{-1/2}^{1/2} f(z) \, dx \right) \overline{\psi}(y) y^{-2} dy.$$

The inner integral is the constant term $f_0(y)$ of the Fourier expansion of f.[4] We thus obtain[5]

$$\langle f, E(\cdot | \psi) \rangle = \int_0^{+\infty} f_0(y) \overline{\psi}(y) y^{-2} dy. \qquad (4.19)$$

---

[3]We shall see that this explains why the constant functions belong to the subspace $\overline{\mathcal{E}(\Gamma \backslash \mathcal{H})} \subset L^2(\Gamma \backslash \mathcal{H})$ in the decomposition of Proposition 4.9 below.

[4]Note that according to Fubini's theorem we can speak of the constant term (as a measurable function) f_0 of any integrable function f on D. The subspace $\overline{\mathcal{C}(\Gamma \backslash \mathcal{H})}$ then consists of functions whose constant term is almost everywhere equal to 0.

[5]This is no problem at zero since ψ is of compact support.

Assume now that f is orthogonal to the subspace $\mathcal{E}(\Gamma\backslash\mathcal{H})$. Then the integral (4.19) is zero for all functions ψ of compact support in $(0, +\infty)$. This last property is equivalent to the condition that $f \in \mathcal{C}(\Gamma\backslash\mathcal{H})$. □

It is immediate that the Laplacian Δ preserves the subspace $\mathcal{C}(\Gamma\backslash\mathcal{H})$ and $\mathcal{E}(\Gamma\backslash\mathcal{H})$. We have already verified that Δ has discrete spectrum in $\mathcal{C}(\Gamma\backslash\mathcal{H})$. In the rest of this section we show that, in $\mathcal{E}(\Gamma\backslash\mathcal{H})$, the spectrum of Δ is continuous, with the exception of the zero eigenvalue corresponding to the constant functions. We thus have to implement the meromorphic continuation of the Eisenstein series in (4.18) in such a way as to reduce to the case of $\sigma = 1/2$. We begin by showing that the Eisenstein series $E(z, s)$ is analytic on the vertical line $\operatorname{Re}(s) = 1/2$.

4.2.3 Regularity of $E(z, s)$ on the Vertical Line $\operatorname{Re}(s) = 1/2$

If follows from Theorem 4.2 (and from the meromorphic continuation of the Riemann zeta function) that the Eisenstein series $E(z, s)$ admits a meromorphic continuation to the entire complex plane, with only one pole on the half-plane $\operatorname{Re}(s) > 1/2$ at $s = 1$. In this subsection we show that the Eisenstein series remains analytic on the line $\operatorname{Re}(s) = 1/2$. We deduce this from the following lemma which expresses what is known as the *Maaß-Selberg relations*.

Given a real number $Y > 0$, write $E^Y(z, s)$ for the truncated Eisenstein series. This is a Γ-invariant function equal to $E(z, s)$ when $z \in D$ has imaginary part $\leqslant Y$ and equal to $E(z, s) - y^s - \varphi(s)y^{1-s}$ if $z \in D$ has imaginary part $> Y$.

Lemma 4.10 (Maaß-Selberg) *Let $\sigma > 1/2$ be a real number and $s = \sigma + ir$ and $s' = \sigma + ir'$ two complex numbers on the line (σ) whose imaginary parts r and r' are not opposite each other. Then*

$$\langle E^Y(\cdot, s), E^Y(\cdot, s') \rangle = \frac{1}{i(r+r')}\varphi(\sigma - ir')Y^{i(r+r')}$$

$$- \frac{1}{i(r+r')}\varphi(\sigma + ir)Y^{-i(r+r')} + \frac{1}{2\sigma - 1 + i(r-r')}Y^{2\sigma-1+i(r-r')}$$

$$- \frac{1}{2\sigma - 1 + i(r-r')}\varphi(\sigma + ir)\varphi(\sigma - ir')Y^{1-2\sigma-i(r-r')}.$$

Proof Note first that

$$\langle E^Y(\cdot, s), E^Y(\cdot, s') \rangle = \int_D E^Y(z, s)\overline{E^Y(z, s')}\, d\mu(z)$$

$$= \int_D E^Y(z, s)E^Y(z, \bar{s}')\, d\mu(z)$$

and that $E^Y = E$ for $z \in D^Y$, the subset of D consisting of complex numbers of imaginary part $\leqslant Y$.

Write $\lambda = s(1-s)$ and $\lambda' = \overline{s'}(1-\overline{s'})$ for the corresponding eigenvalues of the Eisenstein series $E(z,s)$ and $E(z,\overline{s'})$. According to Green's formula (3.5) and the Fourier expansions of the Eisenstein series,

$$(\lambda - \lambda') \int_{z \in D^Y} E(z,s) E(z,\overline{s'}) \, d\mu(z)$$

$$= \int_{z \in D^Y} (E(z,\overline{s'}) \Delta E(z,s) - E(z,s) \Delta E(z,\overline{s'})) \, d\mu(z)$$

$$= \int_0^1 \left(E(x+iY,\overline{s'}) \left(\frac{\partial}{\partial y} E \right) (x+iY,s) \right.$$

$$\left. - E(x+iY,s) \left(\frac{\partial}{\partial y} E \right) (x+iY,\overline{s'}) \right) dx$$

$$= \sum_{n \in \mathbf{Z}} (a_{-n}(Y,\overline{s'}) a_n'(Y,s) - a_{-n}'(Y,\overline{s'}) a_n(Y,s)).$$

But according to the differential equation of the Whittaker functions (see (3.10)),

$$\frac{d}{dy} \left(a_{-n}(y,\overline{s'}) a_n'(y,s) - a_{-n}'(y,\overline{s'}) a_n(y,s) \right)$$

$$= a_{-n}(y,\overline{s'}) a_n''(y,s) - a_{-n}''(y,\overline{s'}) a_n(y,s)$$

$$= (1 - \lambda y^{-2}) a_{-n}(y,\overline{s'}) a_n(y,s) - (1 - \lambda' y^{-2}) a_{-n}(y,\overline{s'}) a_n(y,s)$$

$$= (\lambda' - \lambda) y^{-2} a_{-n}(y,\overline{s'}) a_n(y,s).$$

If $n \neq 0$, this last expression is of exponential decay as $y \to +\infty$, so one can then integrate it over the set of $z \in D$ of imaginary part $\geqslant Y$. One obtains in this way

$$a_{-n}(Y,\overline{s'}) a_n'(Y,s) - a_{-n}'(Y,\overline{s'}) a_n(Y,s)$$

$$= (\lambda' - \lambda) \int_Y^{+\infty} a_{-n}(y,\overline{s'}) a_n(y,s) y^{-2} \, dy.$$

By summing over all $n \neq 0$, this gives

$$(\lambda - \lambda') \int_{z \in D^Y} E(z,s) E(z,\overline{s'}) \, d\mu(z)$$

$$= a_0(Y,\overline{s'}) a_0'(Y,s) - a_0'(Y,\overline{s'}) a_0(Y,s)$$

$$+ (\lambda' - \lambda) \int_Y^{+\infty} \int_0^1 E^Y(z,s) E^Y(z,\overline{s'}) \, d\mu(z).$$

We can rewrite this as

$$\langle E^Y(\cdot, s), E^Y(\cdot, s')\rangle = \frac{1}{\lambda - \lambda'}\left(a_0(Y, \bar{s'})a_0'(Y, s) - a_0'(Y, \bar{s'})a_0(Y, s)\right).$$

We deduce that

$$\langle E^Y(\cdot, s), E^Y(\cdot, s')\rangle$$

$$= \frac{1}{i(r + r')(1 - 2\sigma - i(r - r'))}\left(a_0(Y, \sigma - ir')a_0'(Y, \sigma + ir)\right.$$

$$- a_0'(Y, \sigma - ir')a_0(Y, \sigma + ir))$$

$$= \frac{1}{i(r + r')(1 - 2\sigma - i(r - r'))}\left\{\left(Y^{\sigma-ir'} + \varphi(\sigma - ir')Y^{1-\sigma+ir'}\right)\right.$$

$$\times \left((\sigma + ir)Y^{\sigma-1+ir} + (1 - \sigma - ir)\varphi(\sigma + ir)Y^{-\sigma-ir}\right)$$

$$- \left(Y^{\sigma+ir} + \varphi(\sigma + ir)Y^{1-\sigma-ir}\right)$$

$$\times \left((\sigma - ir')Y^{\sigma-1-ir'} + (1 - \sigma + ir')\varphi(\sigma - ir')Y^{-\sigma+ir'}\right)\right\}$$

$$\times \left\{(1 - 2\sigma - i(r - r'))\left(\varphi(\sigma + ir)Y^{-i(r+r')} - \varphi(\sigma - ir')Y^{i(r+r')}\right)\right.$$

$$+ i(r + r')\left(Y^{2\sigma-1+i(r-r')} - \varphi(\sigma + ir)\varphi(\sigma - ir')Y^{1-2\sigma-1-i(r-r')}\right)\right\}$$

$$= \frac{1}{i(r + r')}\varphi(\sigma - ir')Y^{i(r+r')} - \frac{1}{i(r + r')}\varphi(\sigma + ir)Y^{-i(r+r')}$$

$$+ \frac{1}{2\sigma - 1 + i(r - r')}Y^{2\sigma-1+i(r-r')}$$

$$- \frac{1}{2\sigma - 1 + i(r - r')}\varphi(\sigma + ir)\varphi(\sigma - ir')Y^{1-2\sigma-i(r-r')}. \qquad \Box$$

Given $\sigma > 1/2$ and $r \neq 0$, we in particular find that if $s_0 = \sigma + ir$,

$$|E^Y(\cdot, s_0)|^2 = (2\sigma - 1)^{-1}Y^{2\sigma-1} - r^{-1}\,\mathrm{Im}\left(\varphi(\sigma + ir)Y^{-2ir}\right)$$

$$- (2\sigma - 1)^{-1}Y^{1-2\sigma}|\varphi(\sigma + ir)|^2. \qquad (4.20)$$

We can now prove the following theorem.

Theorem 4.11 *The function $s \mapsto E(z, s)$ has no pole on the line* $\mathrm{Re}(s) = 1/2$.

Proof According to (4.11),

$$|\varphi(\sigma + ir)|^2 \longrightarrow 1$$

as σ tends to $1/2$. It then follows from (4.20) that $|E^Y(\cdot, \sigma + ir)|$ remains bounded as $\sigma \to 1/2$. But

$$E(z, s) = \begin{cases} E^Y(z, s) & \text{if } y \leqslant Y, \\ E^Y(z, s) + y^s + \varphi(s)y^{1-s} & \text{if } y > Y. \end{cases}$$

The function $s \mapsto E(\cdot, s)$ therefore has no pole on the line $\mathrm{Re}(s) = 1/2$. $\square$

4.2.4 Eisenstein Series and the Riemann Zeta Function II

The r-th Fourier coefficient of $E(z, s)$ takes the form

$$a_r(y, s) = 2 \frac{|r|^{s-1/2}\sigma_{1-2s}(|r|)}{\xi(2s)} \sqrt{y}\, K_{s-1/2}(2\pi|r|y).$$

We may therefore deduce from Theorem 4.11 the following well-known result.

Corollary 4.12 (Hadamard-De la Vallée Poussin) *The Riemann zeta function* $\zeta(s)$ *does not vanish on the line* $\mathrm{Re}(s) = 1$.

The prime number theorem is a corollary of the theorem of Hadamard-De la Vallée Poussin.

Theorem 4.13 (Prime Number Theorem) *One has the asymptotic*

$$\pi(x) := |\{p \in \mathbf{N} \text{ prime } \leqslant x\}| \sim \frac{x}{\log x}$$

as $x \in \mathbf{R}_+^*$ *tends toward* $+\infty$.

Proof For $\mathrm{Re}(s) > 1$, we have

$$-\frac{\zeta'(s)}{\zeta(s)} = \sum_{p \text{ prime}} \frac{\log p}{p^s - 1}$$

$$= \underbrace{\sum_p \frac{\log p}{p^s}}_{\Phi(s)} + \underbrace{\sum_p \frac{\log p}{p^s(p^s - 1)}}_{\text{converges for } \mathrm{Re}(s) > 1/2} .$$

The theorem of Hadamard-De la Vallée Poussin then implies that the function $s \mapsto \Phi(s)$ extends to the full half-plane $\mathrm{Re}(s) \geqslant 1$ with a simple pole at $s = 1$

of residue 1. But

$$\Phi(s) = \sum_{p} \frac{\log p}{p^s} = \int_{1}^{+\infty} \frac{d\theta(x)}{x^s}$$

$$= s \int_{1}^{+\infty} \frac{\theta(x)\, dx}{x^{s+1}} = s \int_{0}^{+\infty} e^{-st}\theta(e^t)dt, \qquad \text{where} \quad \theta(x) = \sum_{p \leqslant x} \log p.$$

We shall admit the following theorem due to Ikehara.

Theorem 4.14 *Let φ be a positive increasing function on $\mathbf{R}_+$ such that the integral*

$$f(s) = \int_{0}^{+\infty} e^{-st}\varphi(t)dt \ (s = \sigma + i\tau)$$

converges for $\sigma > 1$. Assume furthermore that there exists a constant A and a function $\tau \mapsto g(\tau)$ such that

$$\lim_{\sigma \to 1+} f(s) - \frac{A}{s-1} = g(\tau)$$

uniformly on compacta $\tau \in [-a, a]$. Then

$$\varphi(t) \sim A e^t.$$

Applied to the function $\varphi(t) = \theta(e^t)$, the Theorem 4.14 implies

$$\theta(x) \sim x.$$

Finally, for $\varepsilon > 0$, we have

$$\pi(x) \log x = \sum_{p \leqslant x} \log x \geqslant \sum_{p \leqslant x} \log p = \theta(x)$$

$$\geqslant \sum_{x^{1-\varepsilon} \leqslant p \leqslant x} \log p$$

$$\geqslant (1-\varepsilon) \sum_{x^{1-\varepsilon} \leqslant p \leqslant x} \log x$$

$$\geqslant (1-\varepsilon) \log x \left[\pi(x) + O(x^{1-\varepsilon}) \right].$$

From this we deduce the claim that $\pi(x) \sim x/\log x$. □

Let us now return to the problem of the spectral decomposition on $\Gamma \backslash \mathcal{H}$.

4.2.5   The Eisenstein Transform

We denote by $C_c^\infty((0, +\infty))$ the subspace of compactly supported C^∞ functions in the Hilbert space $L^2((0, +\infty))$ endowed with the scalar product

$$\langle f, g \rangle = \frac{1}{2\pi} \int_0^{+\infty} f(r)\overline{g}(r)\, dr.$$

The *Eisenstein transform* is the map

$$E : C_c^\infty((0, +\infty)) \longrightarrow C^\infty(\Gamma \backslash \mathcal{H})$$

defined by

$$(Ef)(z) = \frac{1}{2\pi} \int_0^{+\infty} f(r)E(z, 1/2 + ir)\, dr. \qquad (4.21)$$

The computation (4.10) of the constant term of $E(z, s)$ shows that the Eisenstein series $E(z, 1/2 + ir)$ are almost L^2 on D. Upon integrating by parts in the r variable we obtain a slightly better upper bound on the Eisenstein transform, namely

$$(Ef)(z) = O(y^{1/2}(\log y)^{-1}) \qquad (4.22)$$

as y tends toward infinity in the cusp. The gain by a logarithmic factor implies that the image of the Eisenstein transform is in $L^2(\Gamma \backslash \mathcal{H})$, i.e.,

$$E : C_c^\infty((0, +\infty)) \longrightarrow L^2(\Gamma \backslash \mathcal{H}).$$

Proposition 4.15 *The Eisenstein transform E is an isometry from $C_c^\infty((0, +\infty))$ into $L^2(\Gamma \backslash \mathcal{H})$.*

Proof We again denote by $E^Y(z, s)$ the truncated Eisenstein series. This is the Γ-invariant function equal to $E(z, s)$ if $z \in D$ has imaginary part $\leqslant Y$ and to $E(z, s) - y^s - \varphi(s)y^{1-s}$ if $z \in D$ has imaginary part $> Y$. Consider the truncated Eisenstein transform

$$(E^Yf)(z) = \frac{1}{2\pi} \int_0^{+\infty} f(r)E^Y(z, 1/2 + ir)\, dr.$$

For all $z \in D$ we have $(E^Yf)(z) = (Ef)(z)$ if $y(z) \leqslant Y$ and, according to (4.22),

$$(E^Yf)(z) = (Ef)(z) + O\left(\frac{y(z)^{1/2}}{\log y(z)}\right)$$

in general. In particular,

$$|(E - E^Y)f| = O((\log Y)^{-1/2}).$$

Moreover,

$$|E^Y f|^2 = \frac{1}{(2\pi)^2} \int_0^{+\infty} \int_0^{+\infty} f(r') \overline{f}(r) \langle E^Y(\cdot, 1/2 + ir'), E^Y(\cdot, 1/2 + ir) \rangle dr dr',$$

and according to Lemma 4.10,

$$\langle E^Y(\cdot, \tfrac{1}{2} + ir'), E^Y(\cdot, \tfrac{1}{2} + ir) \rangle = \frac{1}{i(r' - r)} \left(1 - \varphi(\tfrac{1}{2} + ir')\varphi(\tfrac{1}{2} - ir)\right) Y^{i(r-r')}$$

$$+ \frac{1}{i(r' - r)} \left(Y^{i(r'-r)} - Y^{i(r-r')}\right)$$

$$- \frac{1}{i(r + r')} \varphi(\tfrac{1}{2} + ir') Y^{-i(r+r')}$$

$$+ \frac{1}{i(r + r')} \varphi(\tfrac{1}{2} - ir) Y^{i(r+r')}.$$

The integral expression of $|E^Y f|^2$ then decomposes as the sum of four integrals corresponding to the four terms on the right-hand side of the above equality. We can again integrate by parts (in the r variable) in three of the four integrals to gain a factor of $\log Y$. We leave only the second integral untouched for the moment, yielding

$$|E^Y f|^2 = \frac{1}{(2\pi)^2} \int_0^{+\infty} \int_0^{+\infty} f(r') \overline{f}(r) \frac{Y^{i(r'-r)} - Y^{i(r-r')}}{i(r' - r)} dr dr' + O((\log Y)^{-1}).$$

Now we change variables from r' to $u = r' - r$ and then from u to $v = u \log Y$. We get

$$|E^Y f|^2 = \frac{1}{(2\pi)^2} \int_0^{+\infty} \overline{f}(r) \int_{-r}^{+\infty} f(r + u) 2 \sin(u \log Y) \frac{du}{u} dr + O((\log Y)^{-1})$$

$$- \frac{1}{(2\pi)^2} \int_0^{+\infty} \overline{f}(r) \int_{-r \log Y}^{+\infty} f(r + v/\log Y) 2 \sin(v) \frac{dv}{v} dr + O((\log Y)^{-1}).$$

Since f is C^∞ and compactly supported in $(0, +\infty)$, we have

$$\int_{-r\log Y}^{+\infty} f(r + v/\log Y)2\sin(v)\frac{dv}{v} = \int_{-\infty}^{+\infty} f(r + v/\log Y)2\sin(v)\frac{dv}{v}$$

$$= f(r)\int_{-\infty}^{+\infty} 2\sin(v)\frac{dv}{v} + O((\log Y)^{-1})$$

$$= 2\pi f(r) + O((\log Y)^{-1}),$$

where the constant in the O depends on f but not on r. Finally,

$$|E^Y f|^2 = \frac{1}{2\pi}\int_0^{+\infty} f(r)\bar{f}(r)\,dr + O((\log Y)^{-1}).$$

We conclude the proof of Proposition 4.15 by letting Y tend towards $+\infty$. $\qquad\square$

Proposition 4.15 allows us to extend E to an isometry from $L^2(\mathbf{R}^+)$ onto (the closure of) its image $\mathrm{Im}(E) \subset L^2(\Gamma\backslash\mathcal{H})$. This subspace is invariant under the action of the Laplacian Δ. Indeed, we have

$$\Delta E = EM,$$

where

$$(Mf)(r) = (r^2 + 1/4)f(r).$$

The proof of Lemma 4.10 – more precisely, the computation of $\langle E^Y(\cdot, s), E^Y(\cdot, s)\rangle$ – shows that, if $f \in \mathcal{C}(\Gamma\backslash\mathcal{H})$ is a Laplacian eigenfunction[6] with eigenvalue $((r')^2 + 1/4)$, then

$$\langle f, E^Y(\cdot, 1/2 + ir)\rangle = 0$$

(the constant term of f is trivial). The image of the Eisenstein transform is thus orthogonal to the space $\mathcal{C}(\Gamma\backslash\mathcal{H})$ of cusp forms. It is moreover orthogonal to the constant functions since the eigenvalue 0 is strictly less than $1/4 \leqslant 1/4 + r^2$ for $r \in \mathbf{R}$. We deduce that

$$\mathbf{C}\cdot 1 \oplus \mathrm{Im}(E) \subset \overline{\mathcal{E}(\Gamma\backslash\mathcal{H})}.$$

We are going to show that the space of constant functions $\mathbf{C}\cdot 1$ and the image of the Eisenstein transform generate a dense subspace of $\mathcal{E}(\Gamma\backslash\mathcal{H})$.

---

[6]Since f is cuspidal, $f^Y = f$.

4.2.6 The Spectral Theorem for the Modular Surface

Since the incomplete Eisenstein series are of compact support in $\Gamma\backslash\mathcal{H}$, we can in particular form the "scalar product"[7]

$$\langle E(\cdot|\psi), E(\cdot, 1/2 + ir)\rangle.$$

We may use the analog of the formula (4.19) and the Fourier expansion of the Eisenstein series to see that

$$\langle E(\cdot|\psi), E(\cdot, 1/2 + ir)\rangle$$
$$= \int_0^{+\infty} \left(y^{1/2-ir} + \varphi(1/2 - ir)y^{1/2+ir}\right)\psi(y)y^{-2}dy$$
$$= \widehat{\psi}(1/2 + ir) + \varphi(1/2 - ir)\widehat{\psi}(1/2 - ir). \qquad (4.23)$$

We put $f(r) = \widehat{\psi}(1/2 + ir) + \varphi(1/2 - ir)\widehat{\psi}(1/2 - ir)$. Thus we have

$$\frac{1}{4\pi} \int_{\mathbf{R}} \langle E(\cdot|\psi), E(\cdot, 1/2 + ir)\rangle E(z, 1/2 + ir)\, dr = E(f). \qquad (4.24)$$

The modular surface is of area $\pi/3$. We then denote by u_0 the constant function equal to $\sqrt{3/\pi}$ and by (u_j), for $j \geqslant 1$, an orthonormal basis of the cuspidal space $\mathcal{C}(\Gamma\backslash\mathcal{H})$, ordered by increasing eigenvalues λ_j. Thus $\Delta u_j = \lambda_j u_j$ and

$$0 = \lambda_0 < \lambda_1 \leqslant \lambda_2 \leqslant \cdots \leqslant \lambda_n \cdots$$

Theorem 4.16 Let $\Gamma = \mathrm{SL}(2, \mathbf{Z})$. For all $u \in \mathcal{C}(\Gamma\backslash\mathcal{H}) \oplus \mathcal{E}(\Gamma\backslash\mathcal{H})$, we have

$$u(z) = \sum_{j \geqslant 0} \langle u, u_j\rangle u_j(z) + \frac{1}{4\pi} \int_{\mathbf{R}} \langle u, E(\cdot, 1/2 + ir)\rangle E(z, 1/2 + ir)\, dr \qquad (4.25)$$

in the L^2 sense. If, moreover, u belongs to the domain $\mathcal{D}(\Gamma\backslash\mathcal{H})$ of Δ, the series (4.25) converges absolutely and uniformly on compacta. Finally, we have the Plancherel formula

$$|u|_{L^2(\Gamma\backslash\mathcal{H})} = \sum_{j \geqslant 0} |\langle u, u_j\rangle|^2 + \frac{1}{4\pi} \int_{\mathbf{R}} |\langle u, E(\cdot, 1/2 + ir)\rangle|^2 dr. \qquad (4.26)$$

---

[7]This is an abuse of notation: the Eisenstein series $E(\cdot, 1/2 + ir)$ is not square integrable.

Remark 4.17 As for the Fourier transform, the Plancherel formula allows one to extend the spectral decomposition to all of $L^2(\Gamma\backslash\mathcal{H})$. The notation $\langle u, E(\cdot, 1/2+ir)\rangle$ then becomes dangerous: the integral no longer converges!

Proof We can assume $u \in \mathcal{E}(\Gamma\backslash\mathcal{H})$, so that u is equal to an incomplete Eisenstein series $E(z|\psi)$, where $\psi \in C_c^\infty((0, +\infty))$.

The integral (4.18) converges absolutely for $\sigma > 1$. We then shift the integration contour to the left, stopping at $\sigma = 1/2$. To do this we need to control the growth of $E(z, s)$ in vertical strips in s. We shall apply the Phragmén-Lindelöf principle.

Let ε be a positive real number. We have already remarked that $|E(z, 1+\varepsilon+it)| \leq E(z, 1+\varepsilon)$, which is a strictly positive constant. The functional equation of the zeta function along with Stirling's formula (4.13) together imply that

$$|\zeta(-2\varepsilon + 2it)| \sim |\zeta(1 + 2\varepsilon - 2it)| \left|\frac{t}{\pi}\right|^{1/2+2\varepsilon}$$

as $|t|$ tends toward infinity. It then follows from Theorem 4.2 – and again from Stirling's formula – that

$$|E(z, -\varepsilon - it)| = O_{z,\varepsilon}(|t|^{1/2+\varepsilon})$$

as $|t|$ tends toward infinity.

As for the Riemann zeta function, the Eisenstein series $E(z, s)$ is of order 1. The Phragmén-Lindelöf principle then applies. We deduce that the Eisenstein series $E(z, s)$ remains of polynomial growth (away from the poles) in the vertical strip $-\varepsilon \leq \text{Re}(s) \leq 1 + \varepsilon$. Since, moreover, the function $\widehat{\psi}$ is of rapid decay in vertical strips (see (4.17)), we may shift the contour to $\sigma = 1/2$. We pass the unique pole at $s = 1$ of $E(z, s)$. The residue of $E(z, s)$ at $s = 1$ is $\pi/2\Gamma(1)\zeta(2) = 3/\pi$, and one finds

$$E(z|\psi) = \sqrt{\frac{3}{\pi}}\, \widehat{\psi}(1)u_0 + \frac{1}{2i\pi} \int_{(1/2)} \widehat{\psi}(s)E(z, s)ds. \tag{4.27}$$

Finally, since the Eisenstein series $E(z, s)$, on the line $\sigma = 1/2$, are orthogonal to the constant functions and since the area of the modular surface is $\pi/3$, we may rewrite the first term above as

$$\sqrt{\frac{3}{\pi}}\, \widehat{\psi}(1) = \langle E(\cdot|\psi), u_0\rangle. \tag{4.28}$$

The expansion (4.27) is not sufficient to prove Theorem 4.16: the coefficient $\widehat{\psi}(s)$ is not the projection of $E(\cdot|\psi)$ onto $E(\cdot, s)$. To obtain the desired decomposition we use (4.23) and the functional equation (4.3), written as

$$E(z, 1 - s) = \varphi(1 - s)E(z, s).$$

These two expressions imply

$$\langle E(\cdot|\psi), E(\cdot, 1/2 + ir)\rangle E(z, 1/2 + ir)$$
$$= \widehat{\psi}(1/2 + ir)E(z, 1/2 + ir) + \widehat{\psi}(1/2 - ir)E(z, 1/2 - ir).$$

Upon integrating this last expression over $\mathbf{R}$, we finally obtain

$$\frac{1}{2i\pi} \int_{(1/2)} \widehat{\psi}(s)E(z, s)ds = \frac{1}{4\pi} \int_{\mathbf{R}} \langle E(\cdot|\psi), E(\cdot, 1/2 + ir)\rangle E(z, 1/2 + ir)\, dr,$$

which is nothing other than the projection onto the Eisenstein series.

It remains then to establish the Plancherel formula. This follows from (4.23), (4.24) and Proposition 4.15:

$$\left| \frac{1}{4\pi} \int_{\mathbf{R}} \langle E(\cdot|\psi), E(\cdot, 1/2 + ir)\rangle E(z, 1/2 + ir)\, dr \right|^2_{L^2(\Gamma\backslash\mathcal{H})} = |E(f)|^2_{L^2(\Gamma\backslash\mathcal{H})}$$

$$= \frac{1}{2\pi} \int_0^{+\infty} |f|^2$$

$$= \frac{1}{2\pi} \int_0^{+\infty} |\langle E(\cdot|\psi), E(\cdot, 1/2 + ir)\rangle|^2 dr$$

$$= \frac{1}{4\pi} \int_{\mathbf{R}} |\langle E(\cdot|\psi), E(\cdot, 1/2 + ir)\rangle|^2 dr.$$

(The last equality is obtained by using the functional equation $E(z, 1 - s) = \varphi(1 - s)E(z, s)$ and the unitarity of φ along the line $\sigma = 1/2$.) □

4.3 Existence of Cusp Forms

We have already remarked that it is not at all obvious that $\mathcal{C}(\Gamma\backslash\mathcal{H})$ should not in general be reduced to $\{0\}$. A profound theorem of Selberg affirms that $\dim \mathcal{C}(\Gamma\backslash\mathcal{H}) = +\infty$ as soon as Γ is a finite index subgroup of $SL(2, \mathbf{Z})$. It suffices of course to show this for $\Gamma = SL(2, \mathbf{Z})$. This is the goal of this section.

4.3.1 Automorphic Wave Equation

Just as for the heat equation considered in Chap. 3, Theorem 4.16 allows one to construct solutions $u = u(z, t) \in C^\infty(\Gamma\backslash\mathcal{H} \times \mathbf{R})$ to the automorphic *wave equation*

$$\frac{\partial^2 u}{\partial t^2} + \Delta u = \frac{u}{4}, \tag{4.29}$$

with initial conditions

$$u(z,0) = f(z), \quad \frac{\partial u}{\partial t}(z,0) = 0, \quad f \in C_c^\infty(\Gamma \backslash \mathcal{H}). \tag{4.30}$$

We can think of a solution to (4.29) as the amplitude of a wave which propagates on the hyperbolic surface $\Gamma \backslash \mathcal{H}$.

For all $t \in \mathbf{R}$ we can define a linear operator U_t of $C_c^\infty(\Gamma \backslash \mathcal{H})$ in $L^2(\Gamma \backslash \mathcal{H})$ which, with a function $f \in C_c^\infty(\Gamma \backslash \mathcal{H})$, associates the unique solution $u = u(z,t)$ of (4.29) such that $u(\cdot, 0) = f$ and $u_t(\cdot, 0) = 0$. Formally, we want

$$U_t = \cos(t\sqrt{\Delta - 1/4}).$$

One can define U_t rigorously by

$$U_t f = \langle f, u_0 \rangle \cosh(t/2)u_0 + \sum_{j \geq 1} \langle f, u_j \rangle \cos(r_j t)u_j$$

$$+ \frac{1}{4\pi} \int_{\mathbf{R}} \langle f, E(\cdot, 1/2 + ir) \rangle \cos(rt)E(\cdot, 1/2 + ir) \, dr.$$

Here we have retained the notation of Theorem 4.16 and have written $\{1/4 + r_j^2\}$ for the spectrum of Δ in the cuspidal space $C(\Gamma \backslash \mathcal{H})$. The operator U_t is self-adjoint.

Similarly to the case of the heat kernel, we now seek to write U_t in the form of an invariant integral operator

$$T_k : f \longmapsto \int_{\mathcal{H}} k(\cdot, w)f(w) \, d\mu(w)$$

such that

$$\lim_{t \downarrow 0} T_k f = f, \quad \lim_{t \downarrow 0}(T_k f)_t = 0,$$

locally uniformly and for all $f \in C_c^\infty(\Gamma \backslash \mathcal{H})$. The kernel k, which depends on t, should in particular satisfy the equation

$$k_{tt} = -\Delta_z k + \frac{1}{4}k. \tag{4.31}$$

By integrating (4.31) against an eigenfunction of the Laplacian with eigenvalue $(1/4 + r^2)$, while using the fact that the Laplacian is a symmetric operator, we deduce from Theorem 3.7 that the Selberg transform h of k satisfies the ordinary differential equation

$$h_{tt}(r) = -\left(\frac{1}{4} + r^2\right)h(r) + \frac{1}{4}h(r) = -r^2h(r). \tag{4.32}$$

We thus expect (at least formally) to have

$$h(r) = \cos(rt).$$

The function h lies in the space $PW(\mathbf{C})$ (cf. § 3.4) and is of type t. It follows from Proposition 3.11 that

$$
\begin{aligned}
k(i, z) &= (\mathcal{S}^{-1}h)(z) \\
&= \frac{1}{4\pi} \int_0^{+\infty} (e^{irt} + e^{-irt}) \omega_{r^2+1/4}(i, z) r \tanh(\pi r) \, dr.
\end{aligned}
\tag{4.33}
$$

(Recall that $z \mapsto \omega_{r^2+1/4}(i, z)$ is the unique radial Laplacian eigenfunction of eigenvalue $(r^2 + 1/4)$ whose value at i is 1.)

Theorem 4.18 *Given $f \in C_c^\infty(\Gamma \backslash \mathcal{H})$, the solution $u(z, t)$ to Eq. (4.29) satisfying the initial conditions (4.30) can be written*

$$u(z, t) = \frac{1}{4\pi} \int_{\mathcal{H}} \int_0^{+\infty} (e^{irt} + e^{-irt}) \omega_{r^2+1/4}(z, w) r \tanh(\pi r) f(w) \, dr d\mu(w).$$

Given z and t, the value of $u(z, t)$ is independent of the values of f at points w such that $\rho(z, w) \leqslant |t|$.

Proof We leave as an exercise the verification of the first part of the theorem.

For the second part, note that the above integral has a kernel k given by (4.33). But, since h is of type t, Proposition 3.11 implies that $\rho \mapsto k(\rho)$ has support contained in the ball of radius $\leqslant |t|$. This establishes the second claim. □

4.3.2 Construction of Cusp Forms

We begin by noting that – at least formally – we have

$$2U_t E_{1/2+ir} = (e^{irt} + e^{-irt}) E_{1/2+ir}.$$

A lovely observation of Lindenstrauss and Venkatesh relates this identity to the existence, for every prime number p, of *Hecke operators* T_p, acting on functions on $\Gamma \backslash \mathcal{H}$. By definition, one has

$$T_p f(z) = \frac{1}{\sqrt{p}} \left(f(pz) + \sum_{k=0}^{p-1} f\left(\frac{z+k}{p}\right) \right). \tag{4.34}$$

We shall describe the Hecke operators in more detail later in Chaps. 7 and 9. For the moment we simply record the facts that each operator T_p is self-adjoint and commutes with the Laplacian, and that the Eisenstein series $E_{1/2+ir}(z)$ are eigenfunctions of the T_p. In fact, we have

$$T_p E_{1/2+ir} = (p^{ir} + p^{-ir})E_{1/2+ir}. \tag{4.35}$$

We thus obtain the following result.

Proposition 4.19 *For all $f \in \mathcal{E}(\Gamma \backslash \mathcal{H})$,*

$$T_p f = 2U_{\log p} f. \tag{4.36}$$

The two operators T_p and $U_{\log p}$ being self-adjoint, Proposition 4.19 implies the following corollary.

Corollary 4.20 *For all $f \in L^2(\Gamma \backslash \mathcal{H})$, $[T_p - 2U_{\log p}]f \in \overline{\mathcal{C}(\Gamma \backslash \mathcal{H})}$.*

To show that $\mathcal{C}(\Gamma \backslash \mathcal{H})$ is not reduced to $\{0\}$, it then remains to find a function $f \in L^2(\Gamma \backslash \mathcal{H})$ which does not satisfy (4.36). We shall now in fact show that there are "many" such functions, by constructing them "high in the cusp".

Given a positive real number R we write

$$\Omega_R = \Gamma_\infty \backslash \{z \mid \text{Im}(z) > R\}.$$

When R is strictly greater than 1 the projection of Ω_R to $\Gamma \backslash \mathcal{H}$ is injective. We can then think of the space $L^2(\Omega_R)$ as a Hilbert subspace of $L^2(\Gamma \backslash \mathcal{H})$. Given a non-zero integer n and a real number R strictly greater than 1, we write $V_{n,R}$ for the subspace of $C_c^\infty(\Omega_R)$ consisting of functions of the form $f(z) = h(y)e(nx)$.

Assume that $R > e^t$. Since the solutions to Eq. (4.29) propagate at a speed less than 1, for all $f \in V_{n,R}$, the function $U_t f$ ($t \geq 0$) has compact support contained in the set of points whose distance to Ω_R is smaller than t. The support of $U_t f$ is therefore contained in $\Omega_{e^{-t}R}$.

Each domain Ω_R is preserved by the action by translations of the group $\mathbf{R}/\mathbf{Z}$ and this induced an action

$$(s \cdot f)(x + iy) = f(x + s + iy), \quad s \in \mathbf{R}/\mathbf{Z},$$

on the space $C_0^\infty(\Omega_R)$. This action commutes with the Laplacian and thus with the operator U_t as well. Finally, the subspaces $V_{n,R}$ of $C_0^\infty(\Omega_R)$ correspond to the proper subspaces of this action relative to the characters $s \mapsto e(ns)$. Each operator U_t thus preserves the spaces $V_{n,R}$, for

$$U_t V_{n,R} \subset V_{n,e^{-t}R}, \quad R > e^t. \tag{4.37}$$

On the other hand, it follows directly from (4.34) that, for $R > p$, we have

$$T_p V_{n,R} \subset \begin{cases} V_{pn,R/p} & \text{if } p \nmid n \\ V_{pn,R/p} \oplus V_{n/p,pR} & \text{if } p \mid n. \end{cases} \tag{4.38}$$

Since the spaces $V_{n,R}$ and $V_{n',R'}$ are orthogonal for $n \neq n'$, it follows from (4.37) and (4.38) that $T_p V_{n,R}$ and $U_{\log p} V_{n,R}$ (where $n \neq 0, R > p$) are orthogonal.

The following theorem is now almost immediate.

Theorem 4.21 *For all $n \neq 0$, $R > p$, the map $(U_{\log p} - T_p)$ sends $V_{n,R}$ injectively into $\overline{C(\Gamma \backslash \mathcal{H})}$. In particular, there exists an infinity of linearly independent Maaß cusp forms for $\mathrm{SL}(2, \mathbf{Z})$ (and thus for every Fuchsian subgroup $\subset \mathrm{SL}(2, \mathbf{Z})$).*

Proof The operator T_p is injective on $V_{n,R}$: this can be seen by inspecting the action of T_p on the Fourier coefficients, see Lemma 7.6. But if $f \in V_{n,R}$ and $T_p f$ are non-zero, so is $(T_p - U_{\log p}) f$ as soon as $n \neq 0$ and R is taken strictly greater than p. □

4.4 Hyperbolic Periods of Eisenstein Series

We have already related the parabolic periodicity $E(z + 1, s) = E(z, s)$ of the Eisenstein series to the study of the Riemann zeta function. But $E(z, s)$, being Γ-invariant, a other periods as well; in this section we study its periodicity along closed primitive geodesics in $\Gamma \backslash \mathcal{H}$. This will allow us to show that certain L-functions defined by Hecke satisfy a functional equation.

4.4.1 Primitive Geodesics on the Modular Surface

A hyperbolic element $\delta \in \Gamma$ is conjugated in $\mathrm{SL}(2, \mathbf{R})$ to a diagonal matrix

$$\begin{pmatrix} t & 0 \\ 0 & t^{-1} \end{pmatrix}$$

where t is a real number of absolute value $|t| > 1$. The trace of δ is thus equal to $t + t^{-1}$ and one calls the number $N(\delta) = t^2$ the *norm* of δ. One says that δ is *primitive* if it is not a non-trivial power of another element in Γ. The axis (taken with its natural orientation) of a primitive hyperbolic element δ projects on a closed (oriented) geodesic of length $\log N(\delta)$ in the modular surface $\Gamma \backslash \mathcal{H}$. The different conjugates of δ in Γ give the same closed (oriented) geodesic. Every closed geodesic on $\Gamma \backslash \mathcal{H}$ is the projection of the axis of some hyperbolic element in Γ. All of the closed geodesics on $\Gamma \backslash \mathcal{H}$ are thus obtained in the way just described.

One can associate with a hyperbolic matrix

$$\delta = \begin{pmatrix} a & b \\ c & d \end{pmatrix} \in \mathrm{SL}(2,\mathbf{Z}) \quad (|a+d| > 2)$$

an integral quadratic form

$$Q_\delta(x,y) = cx^2 + (d-a)xy - by^2$$

of discriminant $D = (a+d)^2 - 4 > 0$.

Two matrices sharing the same trace are conjugate in $\mathrm{SL}(2,\mathbf{Z})$ if and only if the corresponding quadratic forms are $\mathrm{SL}(2,\mathbf{Z})$-equivalent, where $\mathrm{SL}(2,\mathbf{Z})$ acts on binary quadratic forms via the formula

$$(g, Q) \longmapsto Q \circ g^{-1}. \tag{4.39}$$

Conversely one may associate with every integral quadratic form

$$Q(x,y) = Ax^2 + Bxy + Cy^2$$

whose discriminant $D = B^2 - 4AC > 0$ is not a perfect square the oriented geodesic $\gamma(\theta_1, \theta_2)$ of $\mathcal{H}$ which joins the two (real) roots

$$\theta_1 = \frac{-B + \sqrt{D}}{2A} \quad \text{and} \quad \theta_2 = \frac{-B - \sqrt{D}}{2A} \tag{4.40}$$

of the degree two equation $Q(z, 1) = 0$. The subgroup of $\mathrm{SL}(2,\mathbf{R})$ fixing Q under the action (4.39) is, by definition, the special orthogonal group $\mathrm{SO}(Q)$; it is a Lie group of dimension 1, with two connected components (the stabilizers of the two sheets of the hyperbola $Q(x,y) = -1$).

One can show that

$$\mathrm{SO}(Q) = \left\{ \pm \begin{pmatrix} \frac{t-Bu}{2} & -Cu \\ Au & \frac{t+Bu}{2} \end{pmatrix} \; \middle| \; u, t \in \mathbf{R}, \; t^2 - Du^2 = 4 \right\}.$$

The intersection $\mathrm{SO}(Q, \mathbf{Z}) = \mathrm{SO}(Q) \cap \mathrm{SL}(2,\mathbf{Z})$ then contains all of the matrices such that t and u are integral; this is in fact an equality when A, B and C are relatively prime. It follows from the proof of Theorem 2.3 (see Exercise 2.17) that the group $\mathrm{SO}(Q, \mathbf{Z})$ is (discrete and) cocompact in $\mathrm{SO}(Q)$. Modulo $\pm I$ it is therefore an infinite cyclic group generated by

$$\delta = \delta_Q = \begin{pmatrix} \frac{t_0 - Bu_0}{2} & -Cu_0 \\ Au_0 & \frac{t_0 + Bu_0}{2} \end{pmatrix}, \tag{4.41}$$

where t_0 and u_0 are positive and together form a fundamental solution to the Pell equation $t^2 - Du^2 = 4$. (Note that we can always assume that this generator is of positive trace.)

The element $\delta_Q \in \Gamma$ is primitive and its axis is the geodesic $\gamma(\theta_1, \theta_2)$; the projection of this to $\Gamma \backslash \mathcal{H}$ is a closed oriented geodesic of length

$$2 \log\left(\frac{t_0 + \sqrt{D}u_0}{2}\right). \tag{4.42}$$

We put $\varepsilon_D = (t_0 + \sqrt{D}u_0)/2$.

We have therefore constructed a map

$$Q \longmapsto \delta_Q$$

which, to a quadratic form of positive discriminant D which is not a perfect square, associates a primitive hyperbolic element in Γ. Since the image of this map does not change if we replace Q by another quadratic form which is proportional to it by a positive scalar, we obtain more naturally a map from *primitive* quadratic forms, i.e., those such that $(A, B, C) = 1$, towards the primitive elements of Γ. The map which to δ associates the unique integral primitive quadratic form positively proportional to Q_δ is clearly an inverse map. We finally remark that these maps commute with the action of $\mathrm{PSL}(2, \mathbf{Z})$ by (4.39) on the space of quadratic forms and by conjugation on primitive elements. We deduce the following proposition.

Proposition 4.22 *The lengths of closed oriented geodesics in the modular surface $\Gamma \backslash \mathcal{H}$ are the numbers*

$$\{2 \log \varepsilon_D \mid D \in \mathbf{N}, \ D \equiv 0, 1 \ (\mathrm{mod} \ 4), \ \sqrt{D} \notin \mathbf{N}^2\}.$$

Each length appears with finite non-zero multiplicity equal to $h(D)$, the number of $\mathrm{SL}(2, \mathbf{Z})$-classes of primitive integral quadratic forms of discriminant D.

Remark 4.23 (A little bit of algebraic number theory) For every square free integer $d > 0$, we can form the real quadratic extension $\mathbf{Q}(\sqrt{d})/\mathbf{Q}$; let $\mathcal{O}_d$ be the ring of integers and $\alpha \mapsto \alpha'$ the Galois involution. The minimal polynomial of $\alpha = a + b\sqrt{d}, b \neq 0, a, b \in \mathbf{Q}$, is

$$(X - (a + b\sqrt{d}))(X - (a - b\sqrt{d})) = X^2 - 2aX + (a^2 - b^2d).$$

Thus α is an algebraic integer if and only if

$$2a \in \mathbf{Z}, \quad a^2 - b^2d \in \mathbf{Z}.$$

It follows easily from this that

$$\mathcal{O}_d = \begin{cases} \mathbf{Z}[\sqrt{d}] = \{m + n\sqrt{d} \mid m, n \in \mathbf{Z}\} & \text{if } d \equiv 2, 3 \pmod 4, \\ \{m + n(1 + \sqrt{d})/2 \mid m, n \in \mathbf{Z}\} & \text{if } d \equiv 1 \pmod 4. \end{cases}$$

The norm $N_d(x, y)$ given by

$$(x + y\sqrt{d})(x + y\sqrt{d})' = x^2 - dy^2, \qquad\qquad \text{if } d \equiv 2, 3 \pmod 4,$$
$$\left(x + y\tfrac{1+\sqrt{d}}{2}\right)\left(x + y\tfrac{1+\sqrt{d}}{2}\right)' = x^2 + xy + \tfrac{1-d}{4}y^2, \text{ if } d \equiv 1 \pmod 4,$$

then defines an integral quadratic form of discriminant

$$D = \begin{cases} 4d & \text{if } d \equiv 2, 3 \pmod 4, \\ d & \text{if } d \equiv 1 \pmod 4; \end{cases}$$

by definition this is in fact the *discriminant* of the number field $\mathbf{Q}(\sqrt{d})$.

In general, if Q is an integral quadratic form of discriminant Δ the discriminant D of the number field $\mathbf{Q}(\sqrt{\Delta})$ divides Δ and the quotient Δ/D is the square of an integer; this integer is equal to 1 if and only if the form Q is primitive.

A fractional ideal $\mathfrak{a}$ of $\mathbf{Q}(\sqrt{d})$ is an $\mathcal{O}_d$-module contained in $\mathbf{Q}(\sqrt{d})$. Two non-zero fractional ideals are *narrowly* equivalent if their quotient is a principal fractional ideal generated by a totally positive element of $\mathbf{Q}(\sqrt{d})$. In our case, an element $\mu \in \mathbf{Q}(\sqrt{d})$ is totally positive if it is of positive norm $N(\mu) = \mu\mu'$. Let (a_1, a_2) be a basis of $\mathfrak{a}$ viewed as a $\mathbf{Z}$-module. Permuting a_1 and a_2 if necessary we then have

$$\begin{vmatrix} a_1 & a_2 \\ a_1' & a_2' \end{vmatrix} = \sqrt{D}\,\mathrm{N}\mathfrak{a}, \qquad\qquad (4.43)$$

where $\mathrm{N}\mathfrak{a}$ denotes the norm of $\mathfrak{a}$, i.e., the multiplicative function which coincides with the index $[\mathcal{O}_d : \mathfrak{a}]$ on integral ideals. To show (4.43) we note that, upon multiplying $\mathfrak{a}$ by an integer if necessary, we can assume that $\mathfrak{a}$ is integral; writing $\mathcal{O}_d = \mathbf{Z} + \omega\mathbf{Z}$, we then have

$$\begin{vmatrix} a_1 & a_2 \\ a_1' & a_2' \end{vmatrix}^2 = [\mathcal{O}_d : \mathfrak{a}]^2 \begin{vmatrix} 1 & \omega \\ 1 & \omega' \end{vmatrix}^2 = \mathrm{N}\mathfrak{a}^2 D.$$

We put

$$Q_{a_1, a_2}(x, y) = \mathrm{N}\mathfrak{a}^{-1}(a_1 x + a_2 y)(a_1 x + a_2 y)'.$$

Since a_1, a_2 and $a_1 + a_2$ all lie in $\mathfrak{a}$ their norms are divisible by $\mathbb{N}\mathfrak{a}$. Thus

$$\mathbb{N}\mathfrak{a} | a_1 a_1', \; a_1 a_2' + a_1' a_2, \; a_2 a_2'$$

and the form Q_{a_1,a_2} is integral of discriminant D. A simple computation shows that the SL$(2, \mathbb{Z})$-equivalence class of Q_{a_1,a_2} depends only on the narrow equivalence class of $\mathfrak{a}$ and that the map $[\mathfrak{a}] \mapsto [Q_{a_1,a_2}]$ defines a bijection of this quotient onto the set of equivalence classes of integral binary quadratic forms of discriminant D. The number $h(D)$ is therefore equal to the *narrow class number* of the field $\mathbf{Q}(\sqrt{d})$. The term "narrow" reflects the fact that we consider only the oriented geodesics (or that we consider only the upper, and not also the lower, half-plane).

Starting from the primitive form $Q(x, y) = Ax^2 + Bxy + Cy^2$, and recalling the notation in (4.40), we can take as a fractional ideal the $\mathbf{Z}$-module generated by 1 and θ_2. This fractional ideal is *primitive* (not divisible by a rational integer $n > 1$) and of norm A^{-1}. Geometrically, the matrix

$$\kappa := \begin{pmatrix} 1 & -\theta_2 \\ 1 & -\theta_1 \end{pmatrix} \tag{4.44}$$

sends the geodesic $\gamma(\theta_1, \theta_2)$ onto $\gamma(0, \infty)$, the matrix $\kappa \delta \kappa^{-1}$ is therefore diagonal and equal to

$$\begin{pmatrix} \varepsilon_D & 0 \\ 0 & \varepsilon_D^{-1} \end{pmatrix}.$$

The various primitive elements δ_D hence correspond to the fundamental solution of the Pell equation $t^2 - Du^2 = 4$ (in other words, a fundamental unit in $\mathbf{Q}(\sqrt{d})$) viewed according to different $\mathbf{Z}$-bases of $\mathbf{Q}(\sqrt{d})$ which themselves correspond to the different narrow ideal classes.

An important case – the only case when the narrow class number of $\mathbf{Q}(\sqrt{d})$ is equal to 1 – is the case where the fractional ideal is $\mathcal{O}_d$. The corresponding quadratic form is N_d. For simplicity suppose that $d \equiv 2, 3 \pmod 4$; the associated geodesic is then $\gamma(-\sqrt{d}, \sqrt{d})$ and

$$\mathrm{SO}(Q_d) = \left\{ \pm \begin{pmatrix} x & dy \\ y & x \end{pmatrix} \; \middle| \; x, y \in \mathbf{R}, \; x^2 - dy^2 = 1 \right\}.$$

The intersection of this group with SL$(2, \mathbf{Z})$ consists of those matrices with x and y integral.

To conclude this subsection, let us show how to code the geodesics in $\Gamma \backslash \mathcal{H}$ in such a way as to recover the classical solution via continued fractions of the Pell-Fermat equation: $x^2 - dy^2 = 1$.

Coding Geodesics

Consider the geodesics

$$\gamma(p/q, p'/q') \quad (p, q, p', q' \in \mathbf{Z}, \ pq' - p'q = \pm 1, \ \infty = 1/0). \tag{4.45}$$

They do not intersect. To see this, first note that they are the images of $\gamma(0, \infty)$ by elements in $SL(2, \mathbf{Z})$. It then suffices to verify that if

$$g = \begin{pmatrix} a & b \\ c & d \end{pmatrix} \in SL(2, \mathbf{Z})$$

is different from $\pm I$ then $g(\gamma(0, \infty))$, when distinct from $\gamma(0, \infty)$, does not intersect it. This is obvious if a or b is zero, in which case g or $g \left(\begin{smallmatrix} 0 & -1 \\ 1 & 0 \end{smallmatrix} \right)$ is a parabolic element fixing 0 or ∞ and g maps $\gamma(0, \infty)$ onto itself or onto a disjoint geodesic. Now if $ab \neq 0$, the ratio

$$\frac{g(0)}{g(\infty)} = 1 - \frac{1}{ad}$$

is positive. The endpoints $p/q, p'/q'$ thus have the same sign so that $\gamma(p/q, p'/q')$ lies either strictly to the right or left of $\gamma(0, \infty)$.

We conclude from the above discussion that the images under $SL(2, \mathbf{Z})$ of the ideal geodesic triangle Δ_0 in $\mathcal{H}$ with vertices 0, 1 and ∞ define a triangulation of the hyperbolic plane $\mathcal{H}$. We call it the *Farey triangulation*. It is represented in Fig. 4.1, which was realized by Arnaud Chéritat.

Remark 4.24 We have $GL(2, \mathbf{Z}) = SL(2, \mathbf{Z}) \cup \sigma SL(2, \mathbf{Z})$ where

$$\sigma = \begin{pmatrix} -1 & 0 \\ 0 & 1 \end{pmatrix}.$$

Fig. 4.1 Farey's triangulation

One can therefore extend the action of $SL(2, \mathbf{Z})$ on $\mathcal{H}$ by having σ act on $\mathcal{H}$ by

$$\sigma : z \longmapsto -\bar{z}.$$

In this way the group $GL(2, \mathbf{Z})$ acts on $\mathcal{H}$ while preserving the Farey triangulation. The element σ reverses the orientation of $\mathcal{H}$, and the group $SL(2, \mathbf{Z})$ coincides with the subgroup of elements of $GL(2, \mathbf{Z})$ which preserve orientation.

An oriented geodesic γ in $\mathcal{H}$ defines a sequence of segments cut out by its intersection points with the Farey triangulation. Each such segment joins two sides of a triangle, and one can thus associate with it the vertex $p/q \in \mathbf{Q} \cup \{\infty\}$ at which these two sides intersect. An *excursion* of the geodesic γ is then defined to be the largest sequence of segments having some fixed vertex p/q. Since the group $SL(2, \mathbf{Z})$ preserves the Farey triangulation, it sends an excursion to an excursion. The *size* of an excursion is the number of segments composing it. Then to every oriented geodesic γ there corresponds a bi-infinite sequence $(\ldots, n_{-1}, n_0, n_1, n_2, \ldots)$, uniquely defined modulo shifts, which one can think of as the "coding" of γ. It is defined as the sequence of sizes of successive excursions of γ. This is a sequence of positive integers, which a priori can be finite on the right or left. This happens precisely when at least one endpoint lies in $\mathbf{Q} \cup \infty$, in which case we simply add zeros to make the sequence bi-infinite.[8]

With this background on geodesic excursions complete, we now return to the geodesic $\gamma(-\sqrt{d}, \sqrt{d})$ and its relation to Pell's equation. For simplicity, rather than consider the geodesic $\gamma(-\sqrt{d}, \sqrt{d})$, we prefer to look at its translate $\gamma_0 := \gamma(-\sqrt{d} + [\sqrt{d}], \sqrt{d} + [\sqrt{d}])$. This changes nothing when we pass to the quotient. Let us consider the sequence of excursions of the geodesic γ_0. Since $-\sqrt{d} + [\sqrt{d}] \in (-1, 0)$ and $\sqrt{d} + [\sqrt{d}] > 1$, we form the sequence of excursions beginning with that associated with the vertex $p/q = \infty$. Denote by b_0 its size. The first excursion is associated with the vertex b_0. The projective transformation corresponding to the matrix

$$\begin{pmatrix} b_0 & 1 \\ 1 & 0 \end{pmatrix} \in GL(2, \mathbf{Z})$$

preserves the Farey triangulation, reverses the orientation of $\mathcal{H}$ and sends the triple $(\infty, 0, 1)$ to $(b_0, \infty, b_0 + 1)$. The size of the excursion at b_0 is b_1. The next excursion of γ_0 is thus associated with the vertex

$$\begin{pmatrix} b_0 & 1 \\ 1 & 0 \end{pmatrix} \cdot b_1 = b_0 + \frac{1}{b_1}.$$

---

[8]If the geodesic is one of the $\gamma(p/q, p'/q')$ defining the triangulation, its coding is simply $(\ldots, 0, 1, 0, \ldots)$.

Iterating on n one sees that in general the n-th excursion is associated with the vertex $p_{n-1}/q_{n-1} = [b_0, b_1, \ldots, b_{n-1}, 0, \ldots]$, the rational number obtained by truncating the continued fraction expansion[9] of $\sqrt{d} + [\sqrt{d}]$ to order $n - 1$.

We now note that the primitive element $\delta \in \Gamma$ which generates the stabilizer of γ_0 in Γ preserves γ_0, the orientation of $\mathcal{H}$, as well as the Farey triangulation, but it acts non-trivially on γ_0. The sequence $(b_n)_{n \geq 0}$ is therefore periodic. We have just shown that the continued fraction expansion of $\sqrt{d}$ is periodic.

Denote by n_0 the period of the sequence $(b_n)_{n \geq 0}$. The generator of the stabilizer of γ_0 in $GL(2, \mathbf{Z})$ sends the point ∞ to p_{n_0-1}/q_{n_0-1}. The subgroup of $GL(2, \mathbf{Z})$ consisting of elements fixing the oriented geodesic $\gamma(-\sqrt{d}, \sqrt{d})$ is therefore generated by the elements

$$\pm \begin{pmatrix} x_0 & dy_0 \\ y_0 & x_0 \end{pmatrix}$$

where

$$\frac{x_0}{y_0} = \frac{p_{n_0-1}}{q_{n_0-1}} - [\sqrt{d}].$$

We deduce that

$$\delta = \begin{pmatrix} x_0 & dy_0 \\ y_0 & x_0 \end{pmatrix} \quad \text{or} \quad \begin{pmatrix} x_0 & dy_0 \\ y_0 & x_0 \end{pmatrix}^2$$

according to whether $x_0^2 - dy_0^2$ is equal to 1 or -1.

Example 4.25 In the case $d = 2$, we have

$$\sqrt{2} = 1 + \cfrac{1}{2 + \cfrac{1}{2 + \cdots}}.$$

The period n_0 is therefore equal to 1: $\sqrt{2} + [\sqrt{2}] = [2, 2, \ldots]$ and $p_0/q_0 = b_0 = 2$ so that $x_0/y_0 = 1$, i.e., $x_0 = y_0 = 1$, and

$$\delta = \begin{pmatrix} 1 & 2 \\ 1 & 1 \end{pmatrix}^2 = \begin{pmatrix} 3 & 4 \\ 2 & 3 \end{pmatrix}.$$

---

[9]Recall that the continued fraction expansion of β is

$$\beta = b_0 + \cfrac{1}{b_1 + \frac{1}{b_2 + \cdots}} = [b_0, b_1, \ldots],$$

where the (b_i) form a sequence of integers (a finite sequence, if $\beta \in \mathbf{Q}$) with $b_0 \in \mathbf{Z}$ and $b_i > 0$ for $i > 0$.

4.4.2 Hyperbolic Fourier Series of $E(z, s)$

We conserve the notation of the previous subsection. Let $\delta = \delta_Q \in \Gamma$ be a primitive hyperbolic element and write D for the discriminant of the corresponding quadratic form Q. We again write ε_D for the maximal eigenvalue of δ and κ for the matrix (4.44) which diagonalizes δ.

Since the Eisenstein series $E(z, s)$ are Γ-invariant, they are in particular δ-invariant and the function $z \mapsto E(\kappa^{-1}z, s)$ is invariant by the homothety $z \mapsto \varepsilon_D^2 z$:

$$E(\kappa^{-1}z, s) = E(\delta\kappa^{-1}z, s) = E(\kappa^{-1}(\varepsilon_D^2 z), s).$$

The restriction of the completed Eisenstein series $z \mapsto E^*(\kappa^{-1}z, s)$ $(\mathrm{Re}(s) > 1)$ to the geodesic $\gamma(0, \infty) = \{iv \mid v > 0\}$ can therefore be expanded into a Fourier series,

$$E^*(\kappa^{-1}(iv), s) = \sum_{k \in \mathbf{Z}} b_k(s) v^{i\pi k/\log \varepsilon_D} \tag{4.46}$$

where

$$b_k(s) = \frac{1}{2\log \varepsilon_D} \int_1^{\varepsilon_D^2} E^*(\kappa^{-1}(iv), s) v^{-i\pi k/\log \varepsilon_D} \frac{dv}{v}.$$

Since

$$\kappa^{-1}(iv) = \frac{-i\theta_1 v + \theta_2}{-iv + 1}$$

has imaginary part

$$\frac{v(\theta_2 - \theta_1)}{1 + v^2},$$

we have

$$E^*(\kappa^{-1}(iv), s) = \pi^{-s}\Gamma(s) \sum_{\substack{m,n \in \mathbf{Z} \\ (m,n) \neq (0,0)}} \frac{v^s(\theta_2 - \theta_1)^s}{((m\theta_1 + n)^2 v^2 + (m\theta_2 + n)^2)^s}.$$

We deduce that

$$b_k(s) = \frac{\pi^{-s}\Gamma(s)(\theta_2 - \theta_1)^s}{2\log \varepsilon_D}$$

$$\times \sum_{\beta \neq 0} N(\beta)^{-s} \left|\frac{\beta}{\beta'}\right|^{-i\pi k/\log \varepsilon_D} \int_{|\beta/\beta'|}^{\varepsilon_D^2|\beta/\beta'|} \left(\frac{v}{v^2 + 1}\right)^s v^{-i\pi k/\log \varepsilon_D} \frac{dv}{v},$$

where the sum runs over all of the non-zero elements of the $\mathbf{Z}$-module generated by 1 and θ_2: $\beta = m\theta_2 + n$ (and $\beta' = m\theta_1 + n$). Note then that these elements lie in an ideal $\mathfrak{b}$ of $\mathbf{Q}(\sqrt{D})$ whose class corresponds to the quadratic form Q and which moreover satisfies $\theta_2 - \theta_1 = \sqrt{D}N\mathfrak{b}$.

Let ε_0 be the positive fundamental unit in the quadratic field $\mathbf{Q}(\sqrt{D})$. Every unit in $\mathbf{Q}(\sqrt{D})$ is, up to multiplication by ± 1, an integer power of ε_0. Note that

$$\varepsilon_D = \varepsilon_0 \text{ or } \varepsilon_0^2$$

according[10] to whether ε_0 is of norm 1 or -1.

Given a $\beta \in \mathbf{Q}(\sqrt{D})$ we again write (β) for the principal (fractional) ideal generated by β. Two elements $\beta_1, \beta_2 \in \mathbf{Q}(\sqrt{D})$ generate the same ideal if and only if their quotient is a unit in $\mathbf{Q}(\sqrt{D})$. We have

$$\left| \frac{\varepsilon_0 \beta}{\varepsilon_0' \beta'} \right| = |\varepsilon_0^2| \left| \frac{\beta}{\beta'} \right|$$

and since $|\varepsilon_0^2|^{-i\pi/\log \varepsilon_D}$ is equal to 1 or -1 according to whether $\varepsilon_D = \varepsilon_0$ or ε_0^2, we find

$$\left| \frac{\varepsilon_0 \beta}{\varepsilon_0' \beta'} \right|^{-i\pi/\log \varepsilon_D} = \begin{cases} |\beta/\beta'|^{-i\pi/\log \varepsilon_D} & \text{if } \varepsilon_D = \varepsilon_0, \\ -|\beta/\beta'|^{-i\pi/\log \varepsilon_D} & \text{if } \varepsilon_D = \varepsilon_0^2. \end{cases}$$

The two cases can be treated in the same way, although the latter is slightly more delicate. Since we will need it in the following section, we content ourselves with dealing with this second case only. Assume then for the rest of this subsection that $\varepsilon_D = \varepsilon_0^2$.

In the expression for b_k we replace the sum over non-zero elements $\beta \in \mathfrak{b}$ by a sum over ideals (β) and integers $m \in \mathbf{Z}$. Furthermore, we separate the powers of ε_0 according to their parity. In this way we find

$$b_k(s) = \frac{\pi^{-s} \Gamma(s)(\sqrt{D}N\mathfrak{b})^s}{2 \log \varepsilon_D} \sum_{\substack{\mathfrak{b}|(\beta) \neq 0}} N(\beta)^{-s} \left| \frac{\beta}{\beta'} \right|^{-i\pi k/\log \varepsilon_D}$$

$$\times \left\{ \sum_{m \in \mathbf{Z}} \int_{\left| \frac{\beta' \varepsilon_0^{2m}}{\beta \varepsilon_0^{2m}} \right|}^{\varepsilon_D^2 \left| \frac{\beta' \varepsilon_0^{2m}}{\beta \varepsilon_0^{2m}} \right|} \left(\frac{v}{v^2 + 1} \right)^s v^{-i\pi k/\log \varepsilon_D} \frac{dv}{v} \right.$$

---

[10]We recover here the same alternative as at the end of the preceding subsection on the subject of solving the Pell equation by means of continued fractions.

$$+(-1)^k \sum_{m \in \mathbf{Z}} \int_{\beta' \varepsilon_0^{2m+1}}^{\varepsilon_D^2 \left| \begin{smallmatrix} \beta' \varepsilon_0^{2m+1} \\ \beta \varepsilon_0^{2m+1} \end{smallmatrix} \right|} \left(\frac{v}{v^2 + 1} \right)^s v^{-i\pi k/\log \varepsilon_D} \frac{dv}{v} \Bigg\} . \qquad (4.47)$$

Since $\varepsilon_D^2 = \varepsilon_0^4$ and using

$$\left| \frac{\beta' \varepsilon_0^{2m}}{\beta \varepsilon_0^{2m}} \right| = \varepsilon_0^{-4m} \left| \frac{\beta'}{\beta} \right| \quad \text{and} \quad \left| \frac{\beta' \varepsilon_0^{2m+1}}{\beta \varepsilon_0^{2m+1}} \right| = \varepsilon_0^{-4m-2} \left| \frac{\beta'}{\beta} \right|,$$

the two sums over $m \in \mathbf{Z}$ in (4.47) are both equal to

$$\int_0^\infty \left(\frac{v}{v^2 + 1} \right)^s v^{-i\pi k/\log \varepsilon_D} \frac{dv}{v}.$$

We deduce that $b_k(s) = 0$ for odd k and

$$b_k(s) = \frac{\pi^{-s} \Gamma(s) (\sqrt{D} \mathbb{N} \mathfrak{b})^s}{\log \varepsilon_D}$$

$$\times \sum_{\mathfrak{b} | (\beta) \neq 0} N(\beta)^{-s} \left| \frac{\beta}{\beta'} \right|^{-i\pi k/\log \varepsilon_D} \int_0^\infty \left(\frac{v}{v^2 + 1} \right)^s v^{-i\pi k/\log \varepsilon_D} \frac{dv}{v}$$

for even k. It remains to compute

$$\Gamma(s) \int_0^\infty \left(\frac{v}{v^2 + 1} \right)^s v^{-i\pi k/\log \varepsilon_D} \frac{dv}{v}$$

$$= \int_0^\infty \int_0^\infty t^s \left(\frac{v}{v^2 + 1} \right)^s e^{-t} v^{-i\pi k/\log \varepsilon_D} \frac{dv}{v} \frac{dt}{t}$$

$$= \int_0^\infty \int_0^\infty \left(\frac{t}{v + v^{-1}} \right)^s v^{-i\pi k/\log \varepsilon_D} e^{-t} \frac{dv}{v} \frac{dt}{t}$$

$$= \int_0^\infty \int_0^\infty \left(\frac{tv}{v + v^{-1}} \right)^{(s - i\pi k/\log \varepsilon_D)/2} \left(\frac{tv^{-1}}{v + v^{-1}} \right)^{(s + i\pi k/\log \varepsilon_D)/2}$$

$$\times e^{-tv/(v + v^{-1})} e^{-tv^{-1}/(v + v^{-1})} \frac{dv}{v} \frac{dt}{t}$$

$$= \int_0^\infty \tau_1^{(s - i\pi k/\log \varepsilon_D)/2} e^{-\tau_1} \frac{d\tau_1}{\tau_1} \int_0^\infty \tau_2^{(s + i\pi k/\log \varepsilon_D)/2} e^{-\tau_2} \frac{d\tau_2}{\tau_2}$$

$$= \Gamma\left(\frac{(s - i\pi k/\log \varepsilon_D)}{2} \right) \Gamma\left(\frac{(s + i\pi k/\log \varepsilon_D)}{2} \right).$$

Thus we finally obtain

$$E^*(\kappa^{-1}(iv), s) = \frac{(\sqrt{D}\mathbb{N}\mathfrak{b})^s}{2\pi^s \log \varepsilon_0}$$

$$\times \sum_{k \in \mathbb{Z}} \Gamma\left(\frac{s - i\pi k/\log \varepsilon_0}{2}\right) \Gamma\left(\frac{s + i\pi k/\log \varepsilon_0}{2}\right) L_\mathfrak{b}(s, \lambda^k) v^{i\pi k/\log \varepsilon_0},$$

where for all $\alpha \in \mathbf{Q}(\sqrt{D})$ we have set

$$\lambda^k(\alpha) = \left|\frac{\alpha}{\alpha'}\right|^{-i\pi k/\log \varepsilon_0}. \tag{4.48}$$

Note that, since $\lambda^k(\varepsilon\alpha) = \lambda^k(\alpha)$, for any unit ε of $\mathbf{Q}(\sqrt{D})$, this last formula defines a *Hecke character*.

Given a primitive fractional ideal $\mathfrak{b}$ we define the series

$$L_\mathfrak{b}(s, \lambda^k) = \sum_{\mathfrak{b}|(\beta) \neq 0} \lambda^k(\beta) N(\beta)^{-s}. \tag{4.49}$$

It follows for instance from the convergence properties of the Eisenstein series that the series (4.49) is absolutely convergent for $\mathrm{Re}(s) > 1$. The functional equation $E^*(z, s) = E^*(z, 1 - s)$ of the Eisenstein series implies the following theorem.

Theorem 4.26 *Let $\mathbf{Q}(\sqrt{d})$ be a real quadratic field of discriminant D and ε_0 its positive fundamental unit. Given a primitive fractional ideal $\mathfrak{b}$, let $L_\mathfrak{b}(s, \lambda^k)$ be the series (4.49) associated with the Hecke character (4.48) for a certain integer k. Then $L_\mathfrak{b}(s, \lambda^k)$ admits a meromorphic continuation to the complex s-plane with at most a simple pole at $s = 1$; it moreover satisfies the functional equation*

$$\Lambda_\mathfrak{b}(s, \lambda^k) := \left(\frac{\sqrt{D}\mathbb{N}\mathfrak{b}}{\pi}\right)^s \Gamma\left(\frac{s - i\pi k/\log \varepsilon_0}{2}\right) \Gamma\left(\frac{s + i\pi k/\log \varepsilon_0}{2}\right) L_\mathfrak{b}(s, \lambda^k)$$

$$= \Lambda_\mathfrak{b}(1 - s, \lambda^k).$$

4.5   Explicit Construction of Maaß Forms

We know that the completed Riemann zeta function is the Mellin transform of a theta series (see Exercise 1.16). This naturally leads one to wonder which object plays the role of this theta series in the case of the Hecke L-functions. This was the point of departure for the work of Maaß that we now explicate in the case of the number field $\mathbf{Q}(\sqrt{2})$ and for the L-function of Theorem 4.26.

Let us thus consider the totally real number field $\mathbf{Q}(\sqrt{2})$. Its discriminant D is equal to 8, its ring of integers $\mathbf{Z}[\sqrt{2}]$ is principal so that its class number is 1; the unique primitive ideal of $\mathbf{Q}(\sqrt{2})$ is therefore $\mathbf{Z}[\sqrt{2}]$. The unit group of $\mathbf{Q}(\sqrt{2})$ is generated by $\varepsilon_0 = 1 + \sqrt{2}$ and $\varepsilon_D = \varepsilon_0^2 = 3 + 2\sqrt{2}$.

Let k be an integer and λ^k the *Hecke character* given by (4.48). This is a homomorphism from the group of fractional ideals $\mathbf{Z}[\sqrt{2}]$ into $\mathbf{S}^1 \subset \mathbf{C}^*$. To simplify the notation we shall write

$$t_k = \frac{\pi k}{\log \varepsilon_0}.$$

Recall that the associated completed L-function is

$$\Lambda(s, \lambda^k) = (2\sqrt{2})^s \pi^{-s} \Gamma\left(\frac{s - it_k}{2}\right) \Gamma\left(\frac{s + it_k}{2}\right) L(s, \lambda^k),$$

where $L(s, \lambda^k)$ is given by (4.49). Let us compute its inverse Mellin transform

$$\frac{1}{2i\pi} \int_{(\sigma)} \Lambda(s, \lambda^k) y^{-s} ds.$$

We begin by calculating the Mellin transform of the K-Bessel function. This computation will be of use to us later.

Lemma 4.27 *The Mellin transform of K_ν is absolutely convergent as soon as* $\mathrm{Re}(s) > |\mathrm{Re}(\nu)|$, *and in this case we have*

$$\int_0^{+\infty} K_\nu(y) y^s \frac{dy}{y} = 2^{s-2} \Gamma\left(\frac{s + \nu}{2}\right) \Gamma\left(\frac{s - \nu}{2}\right). \tag{4.50}$$

Proof We use the integral expression

$$K_\nu(y) = \frac{1}{2} \int_0^{+\infty} \exp\left(-\frac{y}{2}(t + t^{-1})\right) t^\nu \frac{dt}{t}$$

which can be found in Appendix B. The right-hand side of (4.50) is then equal to

$$\frac{1}{2} \int_0^{+\infty} \int_0^{+\infty} \exp\left(-\frac{y}{2}(t + t^{-1})\right) t^\nu y^s \frac{dy\, dt}{y\ t}.$$

Using the change of variables $u = \frac{1}{2}ty$, $v = \frac{1}{2}t^{-1}y$, the above integral becomes

$$2^{s-2} \int_0^{+\infty} \int_0^{+\infty} e^{-u-v} u^{(s+\nu)/2} v^{(s-\nu)/2} \frac{du\, dv}{u\ v}.$$

This integral is absolutely convergent for $\mathrm{Re}(s) > |\mathrm{Re}(\nu)|$ and decomposes as a product of two Γ-functions. $\qquad\qquad\qquad\qquad\qquad\qquad\qquad\qquad\square$

By Mellin inversion we find that for $\sigma > |\mathrm{Re}(\nu)|$ and $y \in \mathbf{R}_+^*$,

$$K_\nu(y) = \frac{1}{2i\pi} \int_{(\sigma)} 2^{s-2} \Gamma\left(\frac{s+\nu}{2}\right)\Gamma\left(\frac{s-\nu}{2}\right) y^{-s} ds.$$

Moreover, for σ sufficiently large the sum (4.49) is absolutely convergent. We deduce that

$$\frac{1}{2i\pi} \int_{(\sigma)} (2\sqrt{2})^{-s} \Lambda(s, \lambda^k) y^{-s} ds$$

$$= \frac{1}{2i\pi} \int_{(\sigma)} \pi^{-s} \Gamma\left(\frac{s-it_k}{2}\right)\Gamma\left(\frac{s+it_k}{2}\right) L(s, \lambda^k) y^{-s} ds$$

$$= 4 \sum_{(\beta)\neq 0} \lambda^k(\beta) \frac{1}{2i\pi} \int_{(\sigma)} 2^{s-2} \Gamma\left(\frac{s-it_k}{2}\right)\Gamma\left(\frac{s+it_k}{2}\right) (2\pi N(\beta)y)^{-s} ds$$

$$= 4 \sum_{(\beta)\neq 0} \lambda^k(\beta) K_{it_k}(2\pi N(\beta)y).$$

This leads us to introduce the absolutely convergent series[11]

$$\phi_k(iy) = \sum_{(\beta)\neq 0} \lambda^k(\beta) \sqrt{y}\, K_{it_k}(2\pi N(\beta)y).$$

The above calculation implies that for $\mathrm{Re}(s)$ sufficiently large,

$$\int_0^{+\infty} \phi_k(iy) y^{s-1/2} \frac{dy}{y} = \frac{1}{4}(2\sqrt{2})^{-s} \Lambda(s, \lambda^k)$$

and, by Mellin inversion,

$$\phi_k(iy) = \frac{\sqrt{y}}{8i\pi} \int_{(\sigma)} (2\sqrt{2})^{-s} \Lambda(s, \lambda^k) y^{-s} ds, \qquad (4.51)$$

again for σ sufficiently large.

The functional equation of the Hecke L-function (Theorem 4.26) will then allow us to prove the following theorem, a particular case of an important theorem of Maaß.

---

[11]It is natural to include the factor of $\sqrt{y}$ in order to obtain a sum of Whittaker functions.

Theorem 4.28 (Maaß) *Let k be a non-zero integer. The series*

$$\phi_k(z) = \sum_{(\alpha) \neq 0} \lambda^k(\alpha) \sqrt{y} \, K_{it_k}(2\pi N(\alpha)y) \cos(2\pi N(\alpha)x),$$

where $t_k = \pi k / \log \varepsilon_0$ and $z = x + iy \in \mathcal{H}$, is absolutely convergent. The sum here runs over all non-zero ideals (α) of $\mathbf{Z}[\sqrt{2}]$. The series defines a non-zero Maaß cusp form in $\mathcal{C}(\Gamma(8) \backslash \mathcal{H})$ with eigenvalue $\lambda_k = 1/4 + t_k^2$.

Proof We begin by remarking that since the Whittaker functions $W_{it_k}(z) = 2\sqrt{y} K_{it_k}(2\pi y) e(x)$ are themselves Laplacian eigenfunctions with eigenvalue λ_k (see (3.11)) the same is true for ϕ_k.

Next we study the group $\Gamma_0(8)$. Recall (see Exercise 2.19) that the surface $\Gamma_0(8) \backslash \mathcal{H}$ is non-singular, has 4 cusps and is a degree 12 cover of the modular surface. Topologically, this surface is homeomorphic to a sphere minus 4 points. Its fundamental group is the free group on three generators, that we can take to correspond to the three loops encircling three distinct cusps. The parabolic elements

$$\begin{pmatrix} 1 & 1 \\ 0 & 1 \end{pmatrix}, \quad \begin{pmatrix} 1 & 0 \\ 8 & 1 \end{pmatrix} \text{ and } \begin{pmatrix} 5 & -2 \\ 8 & -3 \end{pmatrix}$$

generate the stabilizers of three distinct equivalence classes of cusps of $\Gamma_0(8)$. The group $\Gamma_0(8)$ is therefore the extension by $\{\pm 1\}$ of the free product of these three elements,

$$\Gamma_0(8) = \left\langle \begin{pmatrix} -1 & 0 \\ 0 & -1 \end{pmatrix}, \begin{pmatrix} 1 & 1 \\ 0 & 1 \end{pmatrix}, \begin{pmatrix} 1 & 0 \\ 8 & 1 \end{pmatrix}, \begin{pmatrix} 5 & -2 \\ 8 & -3 \end{pmatrix} \right\rangle.$$

We shall show that ϕ_k is almost $\Gamma_0(8)$-invariant, i.e., invariant up to a character. It suffices to do so for the last three generators. By construction, it is clear that ϕ_k is invariant by translation $z \mapsto z + 1$. The two lemmas that follow treat the two other generators.

Lemma 4.29 *We have*

$$\phi_k(z) = \phi_k(-1/8z).$$

In particular, ϕ_k is invariant under the action of $\begin{pmatrix} 1 & 0 \\ 8 & 1 \end{pmatrix}$.

Proof Shifting the integration contour in (4.51) with the help of the Phragmén-Lindelöf principle, and invoking the functional equation of $\Lambda(s, \lambda^k)$, we find

$$\phi_k(iy) = \frac{\sqrt{1/8y}}{8\pi i} \int_{\sigma - i\infty}^{\sigma + i\infty} (2\sqrt{2})^{-s} \Lambda(s, \lambda^k)(8y)^{-s} ds = \phi_k(i/8y).$$

Here we implicitly used the fact that k is non-zero and thus $\Lambda(s, \lambda^k)$ has no poles. When $k = 0$ one obtains an additional term coming from a residue. Note that ϕ_k is in particular non-zero.[12]

We have shown that $\phi_k(iy) = \phi_k(i/8y)$. The lemma can be deduced from this by observing that $f(z) = \phi_k(z) - \phi_k(-1/8z)$ is a Laplacian eigenfunction on $\mathcal{H}$ which, along with its partial derivative $\partial f/\partial x$, vanishes on the imaginary axis. An induction argument then shows that all partial derivatives of f vanish on the imaginary axis and thus, since f is real analytic, $f = 0$.

Finally, from the equality

$$\begin{pmatrix} 1 & 0 \\ -8 & 1 \end{pmatrix} = \begin{pmatrix} 0 & -1/\sqrt{8} \\ \sqrt{8} & 0 \end{pmatrix} \begin{pmatrix} 1 & 1 \\ 0 & 1 \end{pmatrix} \begin{pmatrix} 0 & 1/\sqrt{8} \\ -\sqrt{8} & 0 \end{pmatrix}$$

one deduces the invariance of ϕ_k under the action of $\begin{pmatrix} 1 & 0 \\ 8 & 1 \end{pmatrix}$. $\qquad\square$

The rest of the proof consists in proving the almost invariance of ϕ_k under the action of the remaining parabolic generator of $\Gamma_0(8)$. As in the above proof, the key point is a conjugacy relation between these parabolic generators within $SL(2, \mathbf{R})$.

Lemma 4.30 *We have*

$$\phi_k\left(\left(\begin{smallmatrix} 5 & -2 \\ 8 & -3 \end{smallmatrix}\right) z\right) = -\phi_k(z).$$

Proof An ideal (β) of even norm is divisible by the ideal $(\sqrt{2})$: $(\beta) = (\sqrt{2})(\alpha)$. Moreover, $\lambda^k(\sqrt{2}) = 1$, so that

$$\phi_k(z + 1/2) + \phi_k(z) = 2 \sum_{\substack{(\beta) \neq 0 \\ N(\beta) \text{ even}}} \lambda^k(\beta) \sqrt{y} K_{it_k}(2\pi N(\beta)y) \cos(2\pi N(\beta)x)$$

$$= 2 \sum_{(\alpha) \neq 0} \lambda^k(\sqrt{2}\,\alpha) \sqrt{y} K_{it_k}(2\pi N(\alpha)2y) \cos(2\pi N(\alpha)2x)$$

$$= \sqrt{2} \sum_{(\alpha) \neq 0} \lambda^k(\alpha) \sqrt{2y} K_{it_k}(2\pi N(\alpha)2y) \cos(2\pi N(\alpha)2x)$$

$$= \sqrt{2}\,\phi_k(2z),$$

and thus

$$\phi_k(z + 1/2) = \sqrt{2}\,\phi_k(2z) - \phi_k(z)$$
$$= \phi_k(z - 1/2). \tag{4.52}$$

---

[12]Indeed, one checks easily that the function $L(s, \lambda^k)$ is non-zero by writing it as an Euler product in the region of absolute convergence.

Now the matrix identity

$$\begin{pmatrix} 5 & -2 \\ 8 & -3 \end{pmatrix} = \begin{pmatrix} 1 & 1/2 \\ 0 & 1 \end{pmatrix} \begin{pmatrix} 1 & 0 \\ 8 & 1 \end{pmatrix} \begin{pmatrix} 1 & -1/2 \\ 0 & 1 \end{pmatrix},$$

implies that

$$\phi_k \left(\begin{pmatrix} 5 & -2 \\ 8 & -3 \end{pmatrix} z \right) = \phi_k \left(\begin{pmatrix} 1 & 0 \\ 8 & 1 \end{pmatrix} \begin{pmatrix} 1 & -1/2 \\ 0 & 1 \end{pmatrix} z + \frac{1}{2} \right).$$

We then deduce from (4.52) that

$$\phi_k \left(\begin{pmatrix} 5 & -2 \\ 8 & -3 \end{pmatrix} z \right) = \sqrt{2}\, \phi_k \left(2 \begin{pmatrix} 1 & 0 \\ 8 & 1 \end{pmatrix} \begin{pmatrix} 1 & -1/2 \\ 0 & 1 \end{pmatrix} z \right)$$

$$- \phi_k \left(\begin{pmatrix} 1 & 0 \\ 8 & 1 \end{pmatrix} \begin{pmatrix} 1 & -1/2 \\ 0 & 1 \end{pmatrix} z \right). \qquad (4.53)$$

But it follows from the invariance of ϕ_k under the action of $\begin{pmatrix} 1 & 0 \\ 8 & 1 \end{pmatrix}$ that

$$\phi_k \left(\begin{pmatrix} 1 & 0 \\ 8 & 1 \end{pmatrix} \begin{pmatrix} 1 & -1/2 \\ 0 & 1 \end{pmatrix} z \right) = \phi_k(z - 1/2). \qquad (4.54)$$

Next, Lemma 4.29 implies that

$$\sqrt{2}\, \phi_k \left(2 \begin{pmatrix} 1 & 0 \\ 8 & 1 \end{pmatrix} \begin{pmatrix} 1 & -1/2 \\ 0 & 1 \end{pmatrix} z \right)$$

$$= \sqrt{2}\, \phi_k \left(\begin{pmatrix} 1 & 0 \\ 4 & 1 \end{pmatrix} \begin{pmatrix} 2 & 0 \\ 0 & 1 \end{pmatrix} \begin{pmatrix} 1 & -1/2 \\ 0 & 1 \end{pmatrix} z \right)$$

$$= \sqrt{2}\, \phi_k \left(\begin{pmatrix} 1 & -1/2 \\ 0 & 1 \end{pmatrix} \begin{pmatrix} 0 & -1/\sqrt{8} \\ \sqrt{8} & 0 \end{pmatrix} \begin{pmatrix} 2 & 0 \\ 0 & 1 \end{pmatrix} \begin{pmatrix} 1 & -1/2 \\ 0 & 1 \end{pmatrix} z \right),$$

since

$$\begin{pmatrix} 1 & 0 \\ 4 & 1 \end{pmatrix} = \begin{pmatrix} 0 & -1/\sqrt{8} \\ \sqrt{8} & 0 \end{pmatrix} \begin{pmatrix} 1 & -1/2 \\ 0 & 1 \end{pmatrix} \begin{pmatrix} 0 & -1/\sqrt{8} \\ \sqrt{8} & 0 \end{pmatrix}.$$

Thus the identity (4.52) yields

$$\sqrt{2}\,\phi_k\left(\begin{pmatrix}1 & -1/2\\0 & 1\end{pmatrix}\begin{pmatrix}0 & -1/\sqrt{8}\\\sqrt{8} & 0\end{pmatrix}\begin{pmatrix}2 & 0\\0 & 1\end{pmatrix}\begin{pmatrix}1 & -1/2\\0 & 1\end{pmatrix}z\right)$$

$$= 2\phi_k\left(2\begin{pmatrix}0 & -1/\sqrt{8}\\\sqrt{8} & 0\end{pmatrix}\begin{pmatrix}2 & 0\\0 & 1\end{pmatrix}\begin{pmatrix}1 & -1/2\\0 & 1\end{pmatrix}z\right)$$

$$- \sqrt{2}\,\phi_k\left(\begin{pmatrix}0 & -1/\sqrt{8}\\\sqrt{8} & 0\end{pmatrix}\begin{pmatrix}2 & 0\\0 & 1\end{pmatrix}\begin{pmatrix}1 & -1/2\\0 & 1\end{pmatrix}z\right).$$

According to Lemma 4.29,

$$\phi_k\left(2\begin{pmatrix}0 & -1/\sqrt{8}\\\sqrt{8} & 0\end{pmatrix}\begin{pmatrix}2 & 0\\0 & 1\end{pmatrix}\begin{pmatrix}1 & -1/2\\0 & 1\end{pmatrix}z\right)$$

$$= \phi_k\left(\begin{pmatrix}0 & -1/\sqrt{8}\\\sqrt{8} & 0\end{pmatrix}\begin{pmatrix}1 & -1/2\\0 & 1\end{pmatrix}z\right)$$

$$= \phi_k(z - 1/2)$$

and

$$\phi_k\left(\begin{pmatrix}0 & -1/\sqrt{8}\\\sqrt{8} & 0\end{pmatrix}\begin{pmatrix}2 & 0\\0 & 1\end{pmatrix}\begin{pmatrix}1 & -1/2\\0 & 1\end{pmatrix}z\right) = \phi_k(2z - 1) = \phi_k(2z).$$

We deduce that

$$\sqrt{2}\,\phi_k\left(2\begin{pmatrix}1 & 0\\8 & 1\end{pmatrix}\begin{pmatrix}1 & -1/2\\0 & 1\end{pmatrix}z\right) = 2\phi_k(z - 1/2) - \sqrt{2}\,\phi_k(2z). \tag{4.55}$$

Finally, using (4.53), (4.54) and (4.55) we obtain

$$\phi_k\left(\begin{pmatrix}5 & -2\\8 & -3\end{pmatrix}z\right) = 2\phi_k(z - 1/2) - \sqrt{2}\,\phi_k(2z) - \phi_k(z - 1/2)$$

$$= -\phi_k(z), \quad \text{where we once again invoked (4.52).} \qquad \square$$

As the matrices

$$\begin{pmatrix}-1 & 0\\0 & -1\end{pmatrix}, \begin{pmatrix}1 & 1\\0 & 1\end{pmatrix}, \begin{pmatrix}1 & 0\\8 & 1\end{pmatrix} \text{ and } \begin{pmatrix}5 & -2\\8 & -3\end{pmatrix}$$

generate the group $\Gamma_0(8)$, we conclude that for all $\gamma \in \Gamma_0(8)$,

$$\phi_k(\gamma z) = \chi(\gamma)\phi_k(z), \tag{4.56}$$

where

$$\chi\left(\begin{pmatrix} a & b \\ c & d \end{pmatrix}\right) = \chi(d) = (-1)^{(d^2-1)/8} \pmod 2$$

$$= \begin{cases} 1 & \text{if } d \equiv \pm 1 \pmod 8 \\ -1 & \text{if } d \equiv \pm 3 \pmod 8. \end{cases}$$

Since the character χ is trivial on $\Gamma(8)$, we deduce finally from (4.56) that ϕ_k is $\Gamma(8)$-invariant.

It remains to verify that ϕ_k is cuspidal. Since the Bessel functions decay rapidly as y goes to infinity, the series defining ϕ_k vanishes at infinity. The same is true at 0 according to Lemma 4.29, at $1/2$ according to (4.52), and at $-1/4$ again from Lemma 4.29. But the points 0, $1/2$, $-1/4$, and ∞ represent the four distinct classes of cusps modulo $\Gamma_0(8)$. The form ϕ_k is therefore L^2 and vanishes at each cusp; this implies that it is cuspidal. □

The surface $\Gamma(8)\backslash\mathcal{H}$ admits then an explicit sequence of eigenvalues. No explicit non-zero eigenvalue is known for the modular surface, nor does one expect any explicit description.

4.6 Commentary and References

In this chapter we showed by two completely different methods that there exist "many" Maaß cusp forms. The first method – non constructive – is due to Lindenstrauss and Venkatesh [83]. Suitably extended this allows one to quantify "how many" and thereby obtain a "Weyl law". We shall provide a proof of this quantitative result in the next chapter by following the original approach of Selberg. The second method is constructive and is due to Maaß himself. Indeed, in [87] Maaß explicitly constructs Maaß cusp forms starting from quadratic extensions of $\mathbf{Q}$.

§ 4.1

The Eisenstein series $E(z, s)$ play a fundamental role in the two approaches. For the computation of their Fourier coefficients we followed the book of Bump [20, §1.6], the first chapter of which nicely completes this chapter.

§ 4.2

The principal source of this section is the book of Iwaniec [63].

The link between the Eisenstein series and the Prime Number Theorem is instructive. Note that Ikehara's theorem does not contain any arithmetic, see for example [75, p. 305] for a proof. The arithmetic is therefore hidden in the theorem of Hadamard-De la Vallée Poussin (Corollary 4.12). Nevertheless, it is possible to give a purely spectral proof of the analytic continuation and the functional equation of the Eisenstein series; the proof of Corollary 4.12 does not require any use of arithmetic. Recall that the Riemann Hypothesis states that the ζ function does not vanish in the half plane $\mathrm{Re}(s) > 1/2$. By working with non-arithmetic surfaces one can show that the spectral method above does not give more than Corollary 4.12. In other words, one can find zeros arbitrarily close to the line $\mathrm{Re}(s) = 1$. We refer the reader to [111] for more details on this question. Any improvement – at least via this method – of the theorem of Hadamard-De la Vallée Poussin would require taking into account the arithmetic of the surface (in this case the modular surface).

Iwaniec [63, Chap. 7] proves the analog of Theorem 4.16 for every finite area hyperbolic surface. The idea of the proof is the same. Here one has to introduce an Eisenstein series for every cusp. Moreover, the contour shift to $\sigma = 1/2$ in general passes by several poles of these Eisenstein series. These poles correspond to eigenvalues between 0 and $1/4$, the latter furnishing the *residual spectrum*. The number of residual eigenvalues appearing in the spectrum of surfaces obtained from finite index subgroups of $\mathrm{SL}(2, \mathbf{Z})$ is unbounded.[13] But these groups are not of congruence type. One can in fact show (see [63, Th. 11.3]) that for congruence subgroups of $\mathrm{SL}(2, \mathbf{Z})$ the residual spectrum is always $\{0\}$.

§ 4.3

The method that we use to show the existence of an infinity of Maaß cusp forms is not the original one of Selberg that we shall describe in the next chapter. We have instead followed the recent proof of Lindenstrauss and Venkatesh [83] which has the advantage of being directly generalizable to a very wide setting.

In the wave equation the term $u/4$ usually does not appear. It is nevertheless natural for the hyperbolic Laplacian, see the appendix to the first section of the book by Lax and Phillips [79]. We refer the reader to this text for the detailed theory of the automorphic wave equation.

It follows from Theorem 4.18 that the solutions to (4.29) (the waves) propagate at unit speed. This phenomenon is general to the so-called "hyperbolic"[14] equations, see [127, Th. 6.1].

---

[13]Starting from a genus 1 subgroup of $\mathrm{SL}(2, \mathbf{Z})$ one can indeed form cyclic covers of genus 1 and m cusps. As m gets large, it is possible to show (see [22]) that these covers have $m - 1$ small eigenvalues. On the other hand Huxley [60] (see also Otal [94] as well as Otal and Rosas [95]) proves that for a genus 1 surface the cuspidal spectrum is contained in $]1/4, +\infty[$.

[14]In this context the term "hyperbolic" has a different meaning from the hyperbolic geometry that we consider in this book.

§ 4.4

The idea to study hyperbolic periods of Eisenstein series is due to Siegel [123] and Zagier [143]. Wielonsky [140] extended their calculation to a much wider context, using the adelic language, thereby recovering a classical formula of Hecke concerning the zeta functions of number fields. For the classical theory of Hecke we refer the reader to the book of Lang [75, Chap. XIII]. We treat here the most elementary case – in the non-adelic set-up – of Wielonsky's formula by following, essentially, the presentation of Goldfeld in [46, §3.2], see also the beautiful book of Siegel [123].

A nice reference for coding geodesics on hyperbolic surfaces is the article of Series [7]. One can also consult the book by Dal'bo [30], which has more details.

Theorem 4.26 is a particular case of a theorem of Hecke [51, 52] (see as well [20, 75]).

§ 4.5

In [87], Maaß showed how to explicitly construct Maaß cusp forms by forming certain theta series associated with real quadratic extensions of $\mathbf{Q}$. Theorem 4.28 is a particular case of Maaß's theorem. The latter invokes the "converse theorem" of Hecke and Weil (see [20]) which gives a more conceptual framework for the calculations we present in this section. We have nevertheless preferred to restrict ourselves to the more "elementary" case of the extension $\mathbf{Q}(\sqrt{2})/\mathbf{Q}$. In the following paragraph we state a more general result that we shall not prove.

Maaß Forms and Algebraic Number Theory

Theorem 4.28 explicates a first link between Maaß forms and algebraic number theory. Maaß's theorem is more general: notably, it implies the following theorem that we shall admit, see [19] for an introduction to this theorem, to the Galois representations that we shall discuss below and more generally to the Langlands program.

Theorem 4.31 *Let K be a real quadratic field of discriminant D. Let η be a character of the ideal class group[15] of K. The series*

$$\phi_\eta(z) = \sum_{\mathfrak{a} \neq 0} \eta(\mathfrak{a}) \sqrt{y} \, K_0(2\pi N(\mathfrak{a})y) \cos(2\pi N(\mathfrak{a})x)$$

is absolutely convergent for $z = x + iy \in \mathcal{H}$. It defines an eigenfunction of the Laplacian with eigenvalue $1/4$ such that for all $\gamma = \left(\begin{smallmatrix} a & b \\ c & d \end{smallmatrix}\right) \in \Gamma_0(D)$,

$$\phi_\eta(\gamma z) = \chi_D(d)\phi_\eta(z),$$

---

[15] In the case of the field $\mathbf{Q}(\sqrt{2})$ such a character is necessarily trivial. We write it as λ^0.

where χ_D is the Dirichlet character associated with the quadratic extension $\mathbf{Q}(\sqrt{D})/\mathbf{Q}$. Moreover, ϕ_η is cuspidal if and only if the character η is not real.

In the latter case, the form ϕ_η defines a Maaß cusp form in $\mathcal{C}(\Gamma(D)\backslash\mathcal{H})$ of eigenvalue $1/4$! The Selberg conjecture is thus optimal when one allows for a finite level. The construction of Maaß is essentially the only known method for explicitly writing down Maaß cusp forms; one should note that the Fourier coefficients of such Maaß forms are algebraic. One conjectures that every Maaß form of eigenvalue $1/4$ should be "algebraic". More precisely, such a form should arise from an even Artin representation $\rho : \mathrm{Gal}(\overline{\mathbf{Q}}/\mathbf{Q}) \to \mathrm{GL}(2, \mathbf{C})$: we now briefly explain what we mean by this.

Galois Representations and Maaß Forms

Given a Galois extension K of $\mathbf{Q}$ and an irreducible representation[16] $\rho : \mathrm{Gal}(K/\mathbf{Q}) \to \mathrm{GL}(2, \mathbf{C})$, we can associate to each prime number p unramified in K the Frobenius conjugacy class Frob_p in $\mathrm{Gal}(K/\mathbf{Q})$. Artin then defines the function

$$
\begin{aligned}
L(s, \rho) &= \prod_p \det(I - \rho(\mathrm{Frob}_p)p^{-s})^{-1} \\
&= \prod_p (1 - \mathrm{tr}(\rho(\mathrm{Frob}_p))p^{-s} + \det(\rho(\mathrm{Frob}_p))p^{-2s})^{-1} \\
&=: \sum_{n=1}^{+\infty} \lambda_\rho(n)n^{-s}.
\end{aligned}
\tag{4.57}
$$

(One must pay attention to the definition of the local Euler factors at ramified primes.) According to a conjecture of Artin, verified by Langlands [77] and Tunnell [130] in a large number of cases, the function $L(s, \rho)$ extends meromorphically to the complex s-plane, and is in fact entire when ρ is non-trivial. When ρ is *even* – that is to say, that $\det(\rho)$ is 1 when we evaluate it on complex conjugation – the series

$$
\phi(z) = \sum_{n=1}^{+\infty} \lambda_\rho(n)y^{1/2}K_0(2\pi n y)\cos(2\pi n x)
\tag{4.58}
$$

should then define a Maaß form for a certain surface $Y(N) = \Gamma(N)\backslash\mathcal{H}$ of eigenvalue $1/4$. Then even representations with dihedral image in $\mathrm{PGL}(2, \mathbf{C})$ should correspond to real quadratic quadratic fields. We refer the reader to Taylor's article [128] or the book [19] for more details on these Galois theoretic aspects.

---

[16]See § 6.3.1 for a rapid introduction to the theory of representations of finite groups.

The use of a Hecke operator in the proof of the existence of Maaß cusp forms already suffices to show the link with arithmetic. We shall again see the essential role of arithmetic in the original proof of Selberg that we describe in the following section. It is therefore of no doubt – in the eyes of a specialist – that Maaß forms *are* of an arithmetic nature (despite the suspected transcendental nature of the Fourier coefficients of those Maaß forms not coming from even Galois representations). We shall furthermore see, in Chap. 7, the full interest of treating Maaß forms on the same footing as the more classical arithmetic objects such as holomorphic modular forms.

4.7 Exercises

Exercise 4.32 Show that the expression

$$u(z, t) = \frac{1}{4\pi} \int_{\mathcal{H}} \int_0^{+\infty} (e^{irt} + e^{-irt}) \omega_{r^2+1/4}(z, w) r \tanh(\pi r) f(w) \, dr d\mu(w)$$

defines a solution to Eq. (4.29).

Exercise 4.33 Let $\delta \in SL(2, \mathbf{Z})$ be a hyperbolic matrix. Show that the discriminant of the integral quadratic form Q_δ is not a perfect square.

Exercise 4.34 Let

$$Q(x, y) = Ax^2 + Bxy + Cy^2$$

be an integral quadratic form of discriminant D. Show that

$$SO(Q) = \left\{ \pm \begin{pmatrix} \frac{t-Bu}{2} & -Cu \\ Au & \frac{t+Bu}{2} \end{pmatrix} \mid u, t \in \mathbf{R}, \ t^2 - Du^2 = 4 \right\}.$$

Exercise 4.35 (Positive definite quadratic forms)

1. Show that the map $z \in \mathcal{H} \mapsto Q_z(u, v) = y^{-1} |uz + v|^2$ allows one to identify – in an $SL(2, \mathbf{R})$-equivariant way – the space of positive definite quadratic forms considered up to homothety with the half-plane $\mathcal{H}$.
2. Show that the series

$$\sum_{(m,n)\in\mathbf{Z}^2-\{(0,0)\}} \frac{1}{Q_z(m, n)^s}, \tag{4.59}$$

where s is a *complex* number with real part > 1, is absolutely convergent and equal to $2E(z, s)$.

3. Let d be a negative integer such that $-d$ is not a square. Consider the set $\mathcal{Q}_d$ of positive definite binary quadratic forms $Q(x, y) = ax^2 + bxy + cy^2 = [a, b, c]$ with $a, b, c \in \mathbf{Z}$, $b^2 - 4ac = d$, and $a > 0$. Show that to every equivalence class in $SL(2, \mathbf{Z}) \backslash \mathcal{Q}_d$ one can associate a point on the modular surface $SL(2, \mathbf{Z}) \backslash \mathcal{H}$. Deduce from this that the series

$$\zeta_Q(s) = \sum_{(m,n) \in \mathbf{Z}^2 - \{(0,0)\}} \frac{1}{Q(m, n)^s}$$

$$= \sum_{(m,n) \in \mathbf{Z}^2 - \{(0,0)\}} \frac{1}{(am^2 + bmn + cn^2)^s},$$

does not depend on the $SL(2, \mathbf{Z})$-equivalence class of Q.
4. Using the properness of the action of $SL_2(\mathbf{Z})$ on $\mathcal{H}$, show that the quotient $SL_2(\mathbf{Z}) \backslash \mathcal{Q}_d$ is finite. We call its cardinality $h(d)$ the class number.

Exercise 4.36 (Class number formula) We keep the notation of the preceding exercise but from now on we assume that $d = -p$ where p is a prime number > 3 and $p \equiv 3 \bmod 4$. We write $h = h(p)$ for the number of $SL(2, \mathbf{Z})$-equivalence classes of primitive integral positive definite quadratic forms $Q(x, y) = ax^2 + bxy + cy^2$ of discriminant $b^2 - 4ac = -p$. We fix a set S of h representatives of these primitive positive definite quadratic forms.

1. For every integer $n > 0$ and for all $Q \in S$, we put

$$R(n, Q) = \#\{(x, y) \in \mathbf{Z}^2 \mid Q(x, y) = n\}$$

and

$$R(n) = \sum_{Q \in S} R(n, Q).$$

By noting that $R(n)$ is equal to twice the number of norm n ideals of $\mathbf{Z}\left[\frac{1 + \sqrt{-p}}{2}\right]$ and by decomposing n in its prime factorization, show that

$$R(n) = 2 \sum_{m|n} \left(\frac{m}{p}\right).$$

2. Show that

$$(2\pi)^{-s} \Gamma(s) \zeta_Q(s) = \int_0^\infty y^s \left(\Theta_Q(iy) - 1\right) \frac{dy}{y},$$

where

$$\Theta_Q(z) = \sum_{(m,n)\in\mathbf{Z}^2} q^{am^2+bmn+cn^2}, \quad \text{with } q = e^{2\pi i z} \quad (z \in \mathcal{H}).$$

3. Using the Poisson summation formula, show that the function Θ_Q satisfies

$$\Theta_Q(z) = \frac{i}{2\sqrt{p}z}\Theta_Q\left(-\frac{1}{4pz}\right) \quad (z \in \mathcal{H}). \qquad (4.60)$$

4. Deduce that

$$\Theta_Q(iy) \sim \frac{1}{2\sqrt{p}y}, \quad \text{as } y \longrightarrow 0. \qquad (4.61)$$

5. Show that

$$\sum_{Q\in S} \Theta_Q(z) = h + \sum_{n=1}^{\infty} R(n)q^n.$$

6. Using the first question, deduce that

$$\sum_{Q\in S} \Theta_Q(z) = h + 2\sum_{m=1}^{\infty} \left(\frac{m}{p}\right) \frac{q^m}{1-q^m}.$$

7. Finally, letting $z = iy$ tend toward 0, deduce from (4.61) that one has

$$\frac{h}{2\sqrt{p}} = \frac{1}{\pi}\sum_{m=1}^{\infty} \left(\frac{m}{p}\right) \frac{1}{m}.$$

Chapter 5
The Trace Formula

In this chapter we derive the Selberg trace formula. We begin by describing it in a general framework which renders transparent the analogy with the Poisson summation formula recalled in the introduction. Having done so, the remaining work consists in explicating the general formula in the case of compact hyperbolic surfaces and then for the modular surface.

5.1 The Selberg Trace Formula I: General Framework

Let G be a locally compact topological group and Γ a cocompact discrete subgroup in G. Suppose that G is *unimodular*, meaning that every left-invariant measure on G is also right-invariant.

Henceforth we make the further assumption that G is a closed unimodular subgroup of $\mathrm{GL}(n, \mathbf{R})$. Let $\mathbf{H}$ be the Hilbert space

$$\mathbf{H} = L^2(\Gamma \backslash G).$$

We put a G-module structure on $\mathbf{H}$ by having G act by the right-regular representation R on $\mathbf{H}$ given by

$$(R(g)\varphi)(x) = \varphi(xg) \quad (x, g \in G, \ \varphi \in \mathbf{H}).$$

Let $f \in C_c^\infty(G)$. We put

$$R(f) = \int_G f(g)R(g)\, dg.$$

© Springer International Publishing Switzerland 2016
N. Bergeron, *The Spectrum of Hyperbolic Surfaces*, Universitext,
DOI 10.1007/978-3-319-27666-3_5

Then $R(f)$ defines an operator on $\mathbf{H}$ as follows:

$$
\begin{aligned}
(R(f)\varphi)(x) &= \int_G f(g)(R(g)\varphi)(x)\,dg \quad (\varphi \in \mathbf{H}) \\
&= \int_G f(g)\varphi(xg)\,dg \\
&= \int_G f(x^{-1}g)\varphi(g)\,dg,
\end{aligned}
$$

where the last equality results from the left-invariance of dg. Fubini's theorem then implies that

$$
\begin{aligned}
(R(f)\varphi)(x) &= \int_{\Gamma\backslash G} \left(\sum_{\gamma\in\Gamma} f(x^{-1}\gamma g)\varphi(\gamma g) \right) dg \\
&= \int_{\Gamma\backslash G} \left(\sum_{\gamma\in\Gamma} f(x^{-1}\gamma g) \right) \varphi(g)\,dg \\
&= \int_{\Gamma\backslash G} K(x,g)\varphi(g)\,dg,
\end{aligned}
$$

where

$$
K(x,y) = \sum_{\gamma\in\Gamma} f(x^{-1}\gamma y) \quad (x,y \in G).
$$

This sum – finite for all fixed x and y – is called the *Selberg kernel*.

The following lemma collects a few immediate properties of the Selberg kernel.

Lemma 5.1 *The Selberg kernel satisfies the following properties:*

1. $K(\gamma_1 x, \gamma_2 y) = K(x,y)$, $(\gamma_1, \gamma_2 \in \Gamma)$.
2. *The function K is C^∞ on the compact manifold $\Gamma\backslash G \times \Gamma\backslash G$.*
3. *The operator $R(f)$ is an integral operator with kernel $K(x,y)$.*

The operator $R(f)$ is therefore a Hilbert-Schmidt operator, see Theorem 3.27. We admit that it is also a trace class operator with

$$
\operatorname{tr} R(f) = \int_{\Gamma\backslash G} K(x,x)\,dx. \tag{5.1}
$$

One can deduce from all of the above a first formulation of the Selberg trace formula:

$$
\operatorname{tr} R(f) = \int_{\Gamma\backslash G} K(x,x)\,dx = \int_{\Gamma\backslash G} \sum_{\gamma\in\Gamma} f(x^{-1}\gamma x)\,dx, \tag{5.2}
$$

where $f \in C_c^\infty(G)$. This formula can take other more convenient forms, which we shall derive in the following sections. Denoting by $\{\gamma\}$ the conjugacy class of an element γ in Γ, and by $\Gamma_\gamma = \{\gamma_1 \in \Gamma \mid \gamma_1 \gamma \gamma_1^{-1} = \gamma\}$ the centralizer of γ in Γ, we have

$$\operatorname{tr} R(f) = \int_{\Gamma \backslash G} \sum_{\{\gamma\}} \sum_{\delta \in \Gamma_\gamma \backslash \Gamma} f(x^{-1} \delta^{-1} \gamma \delta x)\, dx.$$

The first sum runs over the set of conjugacy classes in Γ.

The function f being of compact support, it vanishes on all but finitely many conjugacy classes $\{\gamma\}$. Thus

$$\operatorname{tr} R(f) = \sum_{\{\gamma\}} \int_{\Gamma \backslash G} \sum_{\delta \in \Gamma_\gamma \backslash \Gamma} f(x^{-1} \delta^{-1} \gamma \delta x)\, dx.$$

Replacing the inner sum by an integral against a counting measure $d\delta$, we obtain

$$\operatorname{tr} R(f) = \sum_{\{\gamma\}} \int_{\Gamma \backslash G} \int_{\Gamma_\gamma \backslash \Gamma} f(x^{-1} \delta^{-1} \gamma \delta x)\, d\delta\, dx.$$

The change of variables $t = \delta x$ thus implies

$$\operatorname{tr} R(f) = \sum_{\{\gamma\}} \int_{\Gamma_\gamma \backslash G} f(t^{-1} \gamma t)\, dt.$$

Denote now by $G_\gamma = \{u \in G \mid \gamma u = u \gamma\}$ the centralizer of γ in G. Then the above equality can be written as

$$\operatorname{tr} R(f) = \sum_{\{\gamma\}} \int_{G_\gamma \backslash G} \int_{\Gamma_\gamma \backslash G_\gamma} f(t^{-1} u^{-1} \gamma u t)\, du\, dt,$$

where $u \in G_\gamma$. We finally obtain a new form of the Selberg trace formula, by putting $x = ut$. Namely,

$$\operatorname{tr} R(f) = \sum_{\{\gamma\}} \operatorname{vol}(\Gamma_\gamma \backslash G_\gamma) \int_{G_\gamma \backslash G} f(x^{-1} \gamma x)\, dx. \tag{5.3}$$

It is not clear that we may always make the above interchanges of sums and integrals. One must essentially verify that the integral

$$f_G(\gamma) = \int_{G_\gamma \backslash G} f(x^{-1} \gamma x)\, dx$$

is convergent for $f \in C_c^\infty(G)$. This is in fact the case, as can be deduce from Fubini's theorem and the fact that, if $G_1 \subset G_2 \subset G$ are unimodular groups, a right-invariant measure on $G_1 \backslash G$ can be decomposed as a product of right-invariant measures on $G_2 \backslash G$ and $G_1 \backslash G_2$. The integral $f_G(\gamma)$ is an *orbital integral*. This name is justified by the fact that the quotient $G_\gamma \backslash G$ is homeomorphic to the orbit $G \cdot \gamma = \{x^{-1}\gamma x \mid x \in G\}$ under the action of G on itself by conjugation. We know that the isotropy subgroup G_γ is unimodular, so that the measure dx on the quotient space is indeed G-invariant. The right-hand side of (5.3) is the *geometric side of the trace formula*[1]; to understand it one must understand these orbital integrals.

The subgroup Γ being cocompact, the representation R decomposes into a direct sum of irreducible unitary representations. Hence

$$R \cong \pi_1 \oplus \pi_1 \oplus \cdots \oplus \pi_2 \oplus \pi_2 \oplus \ldots$$

$$\cong m_1\pi_1 \oplus m_2\pi_2 \oplus \cdots \cong \bigoplus_{\pi \in \widehat{G}} m_\pi \pi,$$

where $m_\pi \in \{0, 1, 2, \ldots\}$ and $\widehat{G}$ is the *unitary dual* of G, i.e., the set of isomorphism classes of irreducible unitary representations of G. We can write this isomorphism as

$$R(g) \cong \bigoplus_{\pi \in \widehat{G}} m_\pi \pi(g) \quad (g \in G).$$

Integrating both sides against $f \in C_c(G)$, we obtain

$$R(f) \cong \bigoplus_{\pi \in \widehat{G}} m_\pi \pi(f) \qquad \text{and} \qquad \text{tr } R(f) = \sum_{\pi \in \widehat{G}} m_\pi \, \text{tr}(\pi(f)).$$

One can thus finally write a third version of the Selberg trace formula, whose left-hand side is called the *spectral side of the trace formula*,

$$\sum_{\pi \in \widehat{G}} m_\pi \, \text{tr}(\pi(f)) = \sum_{\{\gamma\}} \text{vol}(\Gamma_\gamma \backslash G_\gamma) \int_{G_\gamma \backslash G} f(x^{-1}\gamma x)dx, \qquad (5.4)$$

where $f \in C_c^\infty(G)$.

---

[1] Why one uses this terminology will become clear in the next section.

5.2 The Selberg Trace Formula II: The Case of Compact Surfaces

Let $S = \Gamma \backslash \mathcal{H}$ be a compact hyperbolic surface; recall that all of the non-trivial elements of $\overline{\Gamma} = \Gamma / \{\pm I\}$ are hyperbolic. To simplify, we assume in this section that $-I \notin \Gamma$.

5.2.1 The Pretrace Formula

Let $k : (-\infty, +\infty) \to \mathbf{C}$ be an even C^∞ function of compact support (or satisfying the decay condition at infinity (3.23)). Let f be an L^2 function on S. We can consider f as a function on $\mathcal{H}$, L^2 on compacta and automorphic with respect to Γ, i.e., $f(\gamma z) = f(z)$. Conversely, such a function obviously induces an L^2 function on S. Consider the operator

$$f \longmapsto \int_{\mathcal{H}} k(\cdot, w) f(w) \, d\mu(w) \tag{5.5}$$

on the Hilbert space $L^2(S)$. Here we make the same abuse of notation as in § 3.3: the kernel $k(z, w)$ is equal to $k(\rho(z, w))$.

Let D be a fundamental domain for Γ whose boundary is of zero measure. The images of the translations of D by Γ tesselate the plane $\mathcal{H}$, and we can write

$$\int_{\mathcal{H}} k(z, w) f(w) \, d\mu(w) = \int_D K(z, w) f(w) \, d\mu(w),$$

where as usual we write $K(z, w) = \sum_{\gamma \in \Gamma} k(z, \gamma w)$. We have already verified that the function K is well-defined, C^∞, and bi-automorphic, i.e., $K(\gamma_1 z, \gamma_2 w) = K(z, w)$ for all $\gamma_1, \gamma_2 \in \Gamma$. The Hilbert-Schmidt theorem (Theorem 3.27) thus implies that the operator

$$f \longmapsto \int_S K(z, w) f(w) \, d\mu(w) \tag{5.6}$$

is a compact operator on $L^2(S)$. The action of K on a function $f \in L^2(S)$ can be calculated with the help of (5.5), if one lifts f to the hyperbolic plane $\mathcal{H}$.

Let $\varphi_0, \varphi_1, \varphi_2, \ldots$ be a complete orthonormal system of (real) eigenfunctions of the Laplacian on S, corresponding to the sequence of eigenvalues $0 = \lambda_0 < \lambda_1 \leqslant \lambda_2 \leqslant \cdots$

By viewing the functions φ_j as automorphic Laplacian eigenfunctions on $\mathcal{H}$, Theorem 3.7 implies

$$\int_{\mathcal{H}} k(z,w)\varphi_j(w)\,d\mu(w) = h(r_j)\varphi_j(z),$$

where r_j is always one of the two roots of $1/4 + r_j^2 = \lambda_j$.

This can be rewritten as

$$\int_S K(z,w)\varphi_j(w)\,d\mu(w) = h(r_j)\varphi_j(z).$$

In other words, the functions φ_j are also eigenfunctions of the operator (5.6) and, since the system $\{\varphi_j\}$ is complete,

$$K(z,w) = \sum_j h(r_j)\varphi_j(z)\varphi_j(w), \tag{5.7}$$

where the convergence is in $L^2(S \times S)$. We have therefore found that

$$\sum_{\gamma \in \Gamma} k(z,\gamma w) = \sum_{j \geq 0} h(r_j)\varphi_j(z)\varphi_j(w). \tag{5.8}$$

Lemma 5.2 *The operator (5.5) is trace class.*

Proof We can apply the Laplacian to the function $z \mapsto K(z,w)$ and thereby obtain another automorphic kernel ΔK. It follows from (5.7) that

$$(\Delta K)(z,w) = \sum_j h(r_j)\lambda_j\varphi_j(z)\varphi_j(w),$$

where the convergence is in $L^2(S \times S)$. Since the kernel ΔK is continuous, the associated operator is a Hilbert-Schmidt operator and thus

$$\sum_j |h(r_j)\lambda_j|^2 < +\infty. \tag{5.9}$$

Since the sum $\sum_j \lambda_j^{-2}$ converges, one finally deduces from the Cauchy-Schwarz inequality and from (5.9) that the series

$$\sum_j h(r_j)$$

is absolutely convergent. $\square$

Remark 5.3 Lemma 5.2 can also be obtained as a consequence of the rapid decay of the function h. According to the proof of the spectral theorem (Theorem 3.32) the series

$$\sum_{j=1}^{+\infty} \lambda_j^{-\sigma} \varphi_j(z) \varphi_j(w)$$

is absolutely uniformly convergent on $S \times S$ as soon as $\sigma > 1$. We find that Eq. (5.7) is moreover valid in the sense of pointwise convergence.

As in § 3.2 it will be more convenient to replace the function k by the function U given by (3.25). The equality

$$\sum_{\gamma \in \Gamma} U(\cosh \rho(z, \gamma w)) = \sum_{j \geq 0} h(r_j) \varphi_j(z) \varphi_j(w), \tag{5.10}$$

where h is given by Lemma 3.9, is called the *pretrace formula*.

5.2.2   The Geometric Side of the Trace Formula

Put $z = w$ in either of the two forms of the pretrace formula and integrate it along S, so as to eliminate the eigenfunctions. We prefer to work with the formula (5.10) and this over the universal cover $\mathcal{H}$. We then integrate over a fundamental domain D of Γ. We obtain

$$\sum_{j} h(r_j) = AU(1) + {\sum_{\gamma}}' \int_D U(\cosh \rho(z, \gamma z)) \, d\mu(z), \tag{5.11}$$

where A is the (hyperbolic) area of D and the symbol $\sum'$ means that the sum runs over non-identity elements $\gamma \in \Gamma$.

We now group the terms of the right-hand side of (5.11) by conjugacy classes in Γ. Thus we let $\{\gamma\}$ be the conjugacy class of γ in Γ. We are led to consider sums of the form

$$\sum_{\gamma \in \{\gamma_0\}} \int_D U(\cosh \rho(z, \gamma z)) \, d\mu(z).$$

A typical term of this last sum is

$$\int_D U(\cosh \rho(z, \gamma_1^{-1} \gamma_0 \gamma_1 z)) \, d\mu(z) = \int_D U(\cosh \rho(\gamma_1 z, \gamma_0 \gamma_1 z)) \, d\mu(z)$$

$$= \int_{\gamma_1 D} U(\cosh \rho(z, \gamma_0 z)) \, d\mu(z).$$

On the other hand, if $\gamma_1^{-1}\gamma_0\gamma_1 = \gamma_2^{-1}\gamma_0\gamma_2$ then $\gamma_1\gamma_2^{-1} \in \Gamma_{\gamma_0}$, the centralizer of γ_0 in Γ; in other words, $\gamma_2 = \gamma'\gamma_1$ with $\gamma' \in \Gamma_{\gamma_0}$.

We can therefore replace the sum on the right-hand side of (5.11) by

$$\sum{}^* \int_{D_\gamma} U(\cosh \rho(z, \gamma z)) \, d\mu(z), \tag{5.12}$$

where the sum now runs over a set of representatives of conjugacy classes other than the identity class in Γ and $D_\gamma = \bigcup_{\gamma_1} \gamma_1 D$, where γ_1 describes the set of representatives of the orbits of the left-action of Γ_γ on Γ. It is immediate that D_γ is a fundamental domain for the (non-compact) surface $\Gamma_\gamma \backslash \mathcal{H}$, and that the integral in (5.12) is independent of the choice of such a fundamental domain (at least if the boundary is not too pathological, for example piecewise C^1). We must now make a convenient choice of fundamental domain for Γ_γ. In order to do so we should better understand the group Γ_γ.

5.2.3 Contribution of Hyperbolic Elements

Each element $\gamma \in \Gamma - \{I\}$ is hyperbolic since Γ is torsion-free (and so does not contain any elliptic elements) and cocompact (and so does not contain any parabolic elements); there exists then a unique γ-invariant geodesic a_γ in $\mathcal{H}$, the *axis* of γ. Write $\ell(\gamma)$ for the *displacement distance* of γ, i.e., the real number ℓ satisfying

$$\rho(z, \gamma z) = \ell \text{ for all } z \in a_\gamma, \quad \rho(z, \gamma z) > \ell \text{ for all } z \in \mathcal{H} - a_\gamma.$$

If a_γ is parametrized by arclength and if the orientation of a_γ goes from z to γz ($z \in a_\gamma$), we have

$$\gamma(a_\gamma(t)) = a_\gamma(t + \ell(\gamma)), \quad t \in \mathbf{R}.$$

The displacement distance is given by

$$\ell(\gamma) = 2 \operatorname{arccosh}(\operatorname{tr}(\gamma)/2). \tag{5.13}$$

Since conjugate elements have the same displacement distance, the latter notion is a well-defined invariant of a conjugacy class in Γ. In fact it corresponds to the length of the closed geodesic associated to γ in S.

We shall call an element $\gamma \in \Gamma - \{I\}$ *primitive* if it cannot be written as a non-trivial power of an element in Γ. The corresponding geodesic is called a *prime geodesic*. Such a geodesic cannot be obtained by winding m times, with $m \geqslant 2$, around a shorter geodesic.

Lemma 5.4 *For all $\gamma \in \Gamma - \{I\}$ there exists a unique primitive element $\delta \in \Gamma$ such that $\gamma = \delta^m$ for a certain $m \geqslant 1$. The elements δ^n, $n \in \mathbf{Z}$, are pairwise non-conjugate in Γ and the centralizer of γ in Γ is*

$$\Gamma_\gamma = \{\delta^n \mid n \in \mathbf{Z}\}.$$

Proof Let a_γ be the axis of Γ. Denote by Z the subgroup of Γ which fixes the axis a_γ. Then Z acts freely and properly discontinuously on a_γ. The restriction $Z_{|a_\gamma}$ is therefore a discrete subgroup of $\mathbf{R}$ and there exists $\delta \in Z$ such that $\ell(\delta) > 0$ and $\ell(\delta) \leqslant \ell(\alpha)$ for all $\alpha \in Z - \{I\}$. It follows from all of this that for all $\alpha \in Z$, there exists an integer $n \in \mathbf{Z}$ such that $\alpha_{|a_\gamma} = (\delta_{|a_\gamma})^n$. Since $(\delta_{|a_\gamma})^n = (\delta^n)_{|a_\gamma}$ and since Γ acts freely, we deduce that $\alpha = \delta^n$. In particular, there exists a unique $m \in \mathbf{Z}$ such that $\gamma = \delta^m$ and, replacing δ by δ^{-1} if necessary, we can assume that $m > 0$. The uniqueness of δ comes from the fact that such an element naturally preserves the axis a_γ and thus belongs to the group Z.

It remains to show that the group Z is nothing other than the centralizer of γ in Γ. But if $\alpha \in \Gamma_\gamma$, by the uniqueness of the axis of γ it is clear that α preserves a_γ and thus lies in Z. $\qquad\square$

Fix then a generator δ of Γ_γ such that $\gamma = \delta^m$ for an integer $m > 0$. Conjugating δ in $SL(2, \mathbf{R})$ if necessary, we can assume that δ acts on $\mathcal{H}$ by a homothety $z \mapsto pz$ for some positive $p \neq 1$. Then, replacing δ by its inverse if necessary, we can assume that $p > 1$. Then $\log p$ is the hyperbolic distance from i to pi, and is therefore the displacement distance $\ell(\delta)$ of δ. The group Γ_γ is the cyclic group generated by δ, and a fundamental domain for its action on $\mathcal{H}$ is given by the horizontal strip $1 < y < p$. We thus obtain

$$\int_{D_\gamma} U(\cosh \rho(z, \gamma z)) \, d\mu(z) = \int_1^p \int_{-\infty}^{+\infty} U(\cosh \rho(z, p^m z)) \frac{dx \, dy}{y^2}$$

$$= \int_1^p \int_{-\infty}^{+\infty} U\left(1 + 2\left(\ell|z|/y\right)^2\right) y^{-2} dx \, dy \quad (\text{where } 2\ell = |p^{m/2} - p^{-m/2}|)$$

$$= \left(\int_1^p y^{-1} dy\right) \int_{-\infty}^{+\infty} U(1 + 2\ell^2(x^2 + 1)) \, dx$$

$$= \frac{\log p}{\ell} \int_{\ell^2}^{+\infty} \frac{U(1 + 2u)}{\sqrt{u - \ell^2}} du$$

$$= \frac{\log p}{2\ell} g(m \log p)$$

$$= |p^{m/2} - p^{-m/2}|^{-1} g(m \log p) \log p,$$

where as usual (see (3.29))

$$g(u) = \sqrt{2} \int_{|u|}^{+\infty} \frac{k(\rho) \sinh \rho}{\sqrt{\cosh \rho - \cosh u}} \, d\rho.$$

Remark 5.5 The function g is the Fourier transform of the function h given by Lemma 3.9. Using the Abel transform, we can moreover recover the function U (or k) from the function g via the formula

$$U(1 + 2u) = -\frac{1}{\pi} \int_{u}^{+\infty} (v - u)^{-1/2} dq(v),$$

where

$$q(v) = \frac{1}{2} g \left(2 \log(\sqrt{v + 1} + \sqrt{v}) \right).$$

If U (or k) is of compact support, the function $g : (-\infty, +\infty) \to \mathbf{C}$ is an even C^∞ function of compact support. The converse holds as well.

5.2.4   The Trace Formula

We can now finally prove the *Selberg trace formula*. In this formula, if γ is a closed geodesic on S, we denote by $N_\gamma = e^{\ell(\gamma)}$ the *norm* of γ and by $\Lambda(\gamma)$ the length $\Lambda(\gamma) = \ell(\gamma_0)$, where γ_0 is the unique oriented prime geodesic satisfying $\gamma = \gamma_0^m$ for an integer $m \geq 1$.

Theorem 5.6 *Let S be a compact hyperbolic surface and $G(S)$ the set of closed oriented geodesics on S. For each $j \in \mathbf{N}$ we fix a complex number r_j so that the $\lambda_j = 1/4 + r_j^2$ describe the Laplacian spectrum on S. Let $g : (-\infty, +\infty) \to \mathbf{C}$ be an even C^∞ function of compact support and $h = \widehat{g}$ its Fourier transform. Then,*

$$\sum_{i=0}^{+\infty} h(r_i) = \frac{\text{area}(S)}{4\pi} \int_{-\infty}^{+\infty} rh(r) \tanh(\pi r) \, dr + \sum_{\gamma \in G(S)} \frac{\Lambda(\gamma)}{N_\gamma^{1/2} - N_\gamma^{-1/2}} g(\log N_\gamma).$$

The series on both sides are absolutely convergent.

Proof Since g is of compact support, there exists a kernel k of compact support such that g comes from k via (3.30). The function h is the Fourier transform of g and so lies in $\mathcal{S}(\mathbf{R})$; the function h is also the Selberg transform of k, see Lemma 3.9. We can apply formula (5.11).

Since g is of compact support and h is of rapid decay, the sum and integral on the right-hand side present no convergence issues. Moreover, Lemma 5.2 implies that the sum on the left-hand side is itself absolutely convergent.

It follows from § 5.2.3 that in (5.11) the sum

$$\sideset{}{'}\sum_{\gamma} \int_D U(\cosh \rho(z, \gamma z)) \, d\mu(z)$$

is equal to

$$\sum_{\gamma \in G(S)} \frac{\Lambda(\gamma)}{N_\gamma^{1/2} - N_\gamma^{-1/2}} g(\log N_\gamma).$$

The area A is the area of the surface S. Then

$$U(1) = -\frac{1}{\pi} \int_0^{+\infty} \frac{q'(v)}{\sqrt{v}} \, dv$$

$$= -\frac{1}{2\pi} \int_0^{+\infty} \frac{g'(u)}{\sinh u/2} \, du.$$

Now, since g is even so is h and

$$g(u) = \frac{1}{2\pi} \int_{-\infty}^{+\infty} e^{-iru} h(r) \, dr = \frac{1}{\pi} \int_0^{+\infty} (\cos ru) h(r) \, dr, \quad u \in \mathbf{R}.$$

Since h lies in the Schwartz space, we can differentiate g under the integration. Doing so, we find

$$g'(u) = -\frac{1}{\pi} \int_0^{+\infty} rh(r)(\sin ru) \, dr.$$

The function $(r, u) \mapsto (\sinh(u/2))^{-1} rh(r) \sin ru$ is integrable on the product space $(0, +\infty) \times (0, +\infty)$ and one obtains

$$U(1) = \frac{1}{2\pi^2} \int_0^{+\infty} rh(r) \int_0^{+\infty} \frac{\sin ru}{\sinh u/2} \, du \, dr.$$

Now,

$$\frac{1}{\sinh(u/2)} = \frac{2}{e^{u/2} - e^{-u/2}} = 2e^{-u/2} \sum_{n \geq 0} e^{-nu}$$

so that

$$\int_0^{+\infty} \frac{\sin ru}{\sinh u/2}\, du = 2 \sum_{n\geq 0} \int_0^{+\infty} e^{-(2n+1)u/2} \sin(ru)\, du$$

$$= 2 \sum_{n\geq 0} \frac{4r}{4r^2 + (2n+1)^2}$$

$$= \sum_{n\in\mathbf{Z}} \frac{4r}{4r^2 + (2n+1)^2}.$$

The Fourier transform of $f_y(x) = e^{-y|x|}$, where y is a positive parameter, is $\widehat{f_y}(\xi) = 2y/(\xi^2 + y^2)$. The Poisson summation formula then implies

$$\sum_{n\in\mathbf{Z}} \frac{r}{r^2 + (n+1/2)^2} = \sum_{n\in\mathbf{Z}} \pi e^{-r2\pi|n|} e^{i\pi n}$$

$$= \pi \frac{1 - e^{-2\pi r}}{1 + e^{-2\pi r}} = \pi \tanh(\pi r).$$

Finally, we obtain

$$U(1) = \frac{1}{2\pi} \int_0^{+\infty} rh(r) \tanh(\pi r)\, dr. \qquad \square$$

Let $\varepsilon > 0$. Put

$$\mathcal{B}_\varepsilon = \{r \in \mathbf{C} \mid |\operatorname{Im} r| < \tfrac{1}{2} + \varepsilon\}$$

and assume that $h : \mathcal{B}_\varepsilon \to \mathbf{C}$ is an even holomorphic function with the decay property

$$h(r) = O((1 + |r|^2)^{-1-\varepsilon}) \text{ uniformly on } \mathcal{B}_\varepsilon.$$

Then h and its Fourier transform

$$g(u) = \frac{1}{2\pi} \int_{-\infty}^{+\infty} e^{-iru} h(r)\, dr, \quad u \in \mathbf{R}$$

form what is called an *admissible* pair of test function. We leave as an exercise the task of verifying that the Selberg trace formula is valid for every admissible pair (h, g).

Remark 5.7 Note that the spectral theorem remains valid when the group Γ contains elliptic elements; the proof is identical but for the fact that one must work with Γ-invariant functions rather than functions on the quotient. Similarly, the trace formula can be formulated when the group Γ has torsion, by taking into account elliptic terms. The idea is the same as for the hyperbolic terms but the calculations are a bit more delicate, see Iwaniec [63]. Allowing for these modifications, one obtains the following more general theorem.

Theorem 5.8 (Selberg trace formula) *Let Γ be a cocompact discrete subgroup of* SL(2, **R**). *For each $j \in$ **N** we fix a complex number r_j in such a way that $\lambda_j = 1/4 + r_j^2$ describes the spectrum of the hyperbolic Laplacian on $\Gamma \backslash \mathcal{H}$. Let (h, g) be an admissible pair of test functions. Then*

$$\sum_{j=0}^{+\infty} h(r_j) = \frac{A_\Gamma}{4\pi} \int_{-\infty}^{+\infty} r h(r) \tanh(\pi r)\, dr$$

$$+ \sum_P \operatorname{arccosh}(t/2) \sum_{\ell=1}^{+\infty} \frac{g\left(2\ell \operatorname{arccosh}(t/2)\right)}{\sinh\left(\ell \operatorname{arccosh}(t/2)\right)}$$

$$+ \sum_E \frac{1}{m} \sum_{\ell=1}^{m-1} \frac{1}{\sin(\ell\pi/m)} \int_{-\infty}^{+\infty} h(r) \frac{e^{-2\ell\pi r/m}}{1 + e^{-2\pi r}}\, dr.$$

Here P (resp., E) denotes the set of primitive hyperbolic (resp., elliptic) conjugacy classes in Γ, t denotes the displacement distance of an element in P, m the order of an element in E, and A_Γ the area of a fundamental domain for the action of Γ on $\mathcal{H}$.

5.3 The Selberg Trace Formula III: The Case of SL(2, **Z**)

We consider here the case of the modular group $\Gamma = $ SL(2, **Z**). Fix as usual an even C^∞ function $k : $ **R** $\to$ **C** of compact support and write K for the associated automorphic kernel. Let D be the usual fundamental domain for the action of Γ on $\mathcal{H}$; recall that D is non-compact.

Consider the kernel $K(z, w)$ as a function of z, constraining w to lie in a fixed compact subset. We calculate (see Theorem 3.7)

$$\langle K(\cdot, w), u_j \rangle = h(r_j)\overline{u_j(w)}$$

and

$$\langle K(\cdot, w), E(\cdot, 1/2 + ir) \rangle = h(r)\overline{E(w, 1/2 + ir)},$$

where h is the Selberg transform of k. Theorem 4.16 then allows one to decompose $K(z, w)$ according to the spectrum of Γ. We deduce that

$$K(z, w) = \sum_j h(r_j) u_j(z) \overline{u_j(w)}$$
$$+ \frac{1}{4\pi} \int_{-\infty}^{+\infty} h(r) E(z, 1/2 + ir) \overline{E(w, 1/2 + ir)} \, dr, \qquad (5.14)$$

where the right-hand side converges absolutely and uniformly on compacta. As long as k is non-zero, the integral

$$\int_D K(z, z) \, d\mu(z)$$

diverges. The operator T_k whose kernel is K is therefore not of trace class. The Plancherel formula (Theorem 4.16) nevertheless implies that

$$\int_D |K(z, w)|^2 d\mu(z) = \sum_j |h(r_j) u_j(w)|^2$$
$$+ \frac{1}{4\pi} \int_{-\infty}^{+\infty} |h(r) E(w, 1/2 + ir)|^2 dr. \qquad (5.15)$$

With the help of this equality and the theory of Eisenstein series we shall isolate the non-compact part of K. Then, by subtracting this contribution, we shall obtain a trace class operator.

Theorem 4.16 decomposes the space $L^2(\Gamma \backslash \mathcal{H})$ into a direct sum of three Δ-invariant subspaces,

$$\mathbf{C} \cdot 1 \oplus \mathcal{C}(\Gamma \backslash \mathcal{H}) \oplus \mathrm{Im}(E)$$

where E as usual denotes the Eisenstein transform, see § 4.2.5. For any function $f \in C_c^\infty((0, +\infty))$, it follows from Theorem 3.7 that

$$T_k(E(f)) = E(hf).$$

The subspace $\mathrm{Im}(E)$ is in particular left invariant by T_k, and $-T_k$ being a continuous self-adjoint operator – the same is true of its orthogonal complement

$$\mathrm{Im}(E)^{\perp} = \overline{\mathbf{C} \cdot 1 \oplus \mathcal{C}(\Gamma \backslash \mathcal{H})}.$$

We shall show that the restriction of T_k to $\mathrm{Im}(E)^{\perp}$ defines a compact operator. To do this, we introduce the kernel K_{cont} which corresponds to the action of T_k on $\mathrm{Im}(E)$.

5.3.1 The Kernel K_{cont}

Let

$$K_{\text{cont}}(z, w) = \frac{1}{4\pi} \int_{-\infty}^{+\infty} h(r) E(z, 1/2 + ir) \overline{E(w, 1/2 + ir)} \, dr.$$

The Plancherel formula relative to the spectral decomposition given by Theorem 4.16 implies that

$$\int_D |K_{\text{cont}}(z, w)|^2 d\mu(w) = \frac{1}{4\pi} \int_{-\infty}^{+\infty} |h(r) E(z, 1/2 + ir)|^2 dr$$

$$\leq \int_D |K(z, w)|^2 d\mu(w),$$

the last equality resulting from (5.15). There are therefore no convergence problems in the definition of K_{cont} and the associated integral operator is well-defined. Its action on functions is given by the following proposition.

Proposition 5.9 *Let* $f \in C_c^{\infty}((0, +\infty))$ *and* $u \in \text{Im}(E)^{\perp}$. *Then*

$$\int_D K_{\text{cont}}(z, w)(Ef)(w) \, d\mu(w) = \frac{1}{2\pi} \int_0^{+\infty} h(r) f(r) E(z, 1/2 + ir) \, dr$$

and

$$\int_D K_{\text{cont}}(z, w) u(w) \, d\mu(w) = 0.$$

Proof The statement follows from the Plancherel formula associated with the spectral decomposition given by Theorem 4.16. □

We deduce from Proposition 5.9 that the action of K_{cont} on $\text{Im}(E)$ is identical to that of K, while the action of K_{cont} on $\text{Im}(E)^{\perp}$ is trivial. The trace of K on $\text{Im}(E)^{\perp}$ is therefore the same as the trace of $K - K_{\text{cont}}$ on the entire space. This leads us naturally to the study of the integral operator with kernel $K - K_{\text{cont}}$.

5.3.2 The Kernel $K - K_{\text{cont}}$

In this subsection we prove the following theorem.

Theorem 5.10 *The integral operator with kernel* $K - K_{\text{cont}}$ *is a Hilbert-Schmidt operator (and so in particular is compact).*

Proof It is natural to truncate the Eisenstein series present in the definition of K_{cont}. Let $Y > 0$ and write

$$K_{\text{cont}}^{Y,1}(z, w) = \frac{1}{4\pi} \int_{-\infty}^{+\infty} h(r) E^Y(z, 1/2 + ir) \overline{E^Y(w, 1/2 + ir)} \, dr,$$

where E^Y denotes the truncated Eisenstein series, see § 4.2.3. The kernel K_{cont} then decomposes as a sum of three terms

$$K_{\text{cont}}(z, w) = K_{\text{cont}}^{Y,1}(z, w) + K_{\text{cont}}^{Y,2}(z, w) + K_{\text{cont}}^{Y,3}(z, w),$$

where

$$K_{\text{cont}}^{Y,2}(z, w)$$
$$= \frac{1}{4\pi} \int_{-\infty}^{+\infty} \delta_Y(z) \delta_Y(w) \left(\text{Im}(z)^{1/2+ir} + \varphi(1/2 + ir) \, \text{Im}(z)^{1/2-ir} \right)$$
$$\cdot \overline{\left(\text{Im}(w)^{1/2+ir} + \varphi(1/2 + ir) \, \text{Im}(w)^{1/2-ir} \right)} h(r) \, dr$$

with $\delta_Y(z) = 1$ if $z \in D$ has imaginary part $\geq Y$ and $\delta_Y(z) = 0$ if $z \in D$ has imaginary part $< Y$. In the following lemmas we study the contributions of each of these three terms.

Lemma 5.11 *The integral operator with kernel $K_{\text{cont}}^{Y,1}$ is a Hilbert-Schmidt operator.*

Proof It suffices to bound its norm in $L^2(D \times D)$. We begin by observing

$$\left(\int_{D \times D} |K_{\text{cont}}^{Y,1}(z, w)|^2 d\mu(z) \, d\mu(w) \right)^{1/2}$$
$$\leq \frac{1}{4\pi} \int_{-\infty}^{+\infty} |h(r)|^2 \int_D |E^Y(z, 1/2 + ir)|^2 d\mu(z) \, dr.$$

As σ tends toward $1/2$, we have

$$\varphi(\sigma + ir) = \varphi(1/2 + ir) + (\sigma - 1/2)\varphi'(1/2 + ir) + o(\sigma - 1/2)$$

and

$$\varphi(\sigma + ir)\varphi(\sigma - ir) = 1 + (2\sigma - 1)\varphi'(1/2 + ir)\varphi(1/2 + ir)^{-1} + o(\sigma - 1/2).$$

(In the last expression we used the fact that $\varphi(s)\varphi(1 - s) = 1$ in order to replace $\varphi(1/2 - ir)$ by $\varphi(1/2 + ir)^{-1}$.) By passing to the limit $\sigma \to 1/2$, Lemma 4.10

implies that

$$\int_D |E^Y(z, 1/2 + ir)|^2 d\mu(z) = \frac{1}{2ir} \left(\varphi(1/2 - ir)Y^{2ir} - \varphi(1/2 + ir)Y^{-2ir} \right)$$
$$+ 2 \log Y - \frac{\varphi'}{\varphi}(1/2 + ir)$$
$$= -\frac{\varphi'}{\varphi}(1/2 + ir) + O_Y(1),$$

for any real r. The function h being of rapid decay, it suffices to show that the integral

$$\frac{1}{4\pi} \int_{-\infty}^{+\infty} \frac{-\varphi'}{\varphi}(1/2 + ir) |h(r)|^2 dr \qquad (5.16)$$

converges. Now $\varphi(s) = \xi(2 - 2s)/\xi(2s)$ and Proposition 4.7 implies that $|\log \varphi(1/2 + ir)| = O(|r|^{1+\varepsilon})$ for all $\varepsilon > 0$. In particular

$$\int_{-T}^{T} \frac{-\varphi'}{\varphi}(1/2 + ir)\, dr = O(T^2)$$

for T large enough and a simple integration by parts finally implies that the integral (5.16) converges.   □

Lemma 5.12 *We have*

$$K_{\text{cont}}^{Y,2}(z, w) = \delta_Y(z)\delta_Y(w)K_0(z, w) + K_{\text{cont}}^{Y,4}(z, w),$$

where $K_{\text{cont}}^{Y,4}$ is the kernel of a Hilbert-Schmidt operator and

$$K_0(z, w) = \int_{-\infty}^{+\infty} k(z, w + t)dt.$$

Moreover, the integral $\int_D K_{\text{cont}}^{Y,4}(z, z)\, d\mu(z)$ is absolutely convergent and tends to 0 as $Y \to +\infty$.

Proof Suppose that $y = \text{Im}(z)$ and $\eta = \text{Im}(w)$ are both greater than Y. By expanding the kernel $K_{\text{cont}}^{Y,2}(z, w)$ we obtain four terms, two of which give

$$\frac{1}{4\pi}(y\eta)^{1/2} \int_{-\infty}^{+\infty} \left((y/\eta)^{ir} + (\eta/y)^{ir} \right) h(r)\, dr = \frac{1}{2\pi}(y\eta)^{1/2} \int_{-\infty}^{+\infty} (y/\eta)^{ir} h(r)\, dr$$

since h is even. But

$$K_0(z, w) = \int_{-\infty}^{+\infty} k(z, w + t)dt = \int_{-\infty}^{+\infty} k(y, \eta + t)dt$$

$$= \int_{-\infty}^{+\infty} U\left(1 + \frac{(y-\eta)^2 + t^2}{2y\eta}\right) dt$$

$$= \int_{-\infty}^{+\infty} U\left(\cosh(\log(y/\eta)) + \frac{t^2}{2y\eta}\right) dt$$

$$= (y\eta)^{1/2}\sqrt{2} \int_{-\infty}^{+\infty} U(\cosh(\log(y/\eta)) + \xi^2) d\xi$$

$$= (y\eta)^{1/2}\sqrt{2} \int_{\cosh(\log(y/\eta))} \frac{U(u)}{\sqrt{u - \cosh(\log(y/\eta))}} du$$

$$= (y/\eta)^{1/2}\sqrt{2} \int_{\log(y/\eta)} \frac{k(\rho)\sinh\rho}{\sqrt{\cosh\rho - \cosh(\log(y/\eta))}} d\rho$$

$$= (y/\eta)^{1/2} g(\log(y/\eta))$$

$$= (y\eta)^{1/2} \frac{1}{2\pi} \int_{-\infty}^{+\infty} h(r)(y/\eta)^{ir} dr.$$

The first two terms above thus contribute

$$\delta_Y(z)\delta_Y(w)K_0(z, w).$$

The two other terms can be treated in the same way. Consider, for example,

$$\frac{1}{4\pi} \int_{-\infty}^{+\infty} \varphi(1/2 + ir)(y\eta)^{1/2 - ir} h(r) dr. \tag{5.17}$$

Let ε be a positive real number strictly less than $1/2$ such that the function φ stays bounded in the strip $1/2 \leq \mathrm{Re}(s) \leq 1/2 + \varepsilon$. Since h is of rapid decay, we can shift the integration contour to $\mathrm{Im}(r) = -\varepsilon$ to find that the integral (5.17) is $O((y\eta)^{1/2 - \varepsilon})$. The last two terms thus produce a contribution of $O(\delta_Y(\eta)\delta_Y(y)(y\eta)^{1/2 - \varepsilon})$ and the lemma follows. □

Lemma 5.13 *The integral operator with kernel $K_{\mathrm{cont}}^{Y,3}$ is of Hilbert-Schmidt type. Moreover, the integral $\int_D K_{\mathrm{cont}}^{Y,3}(z, z)\, d\mu(z)$ converges absolutely and tends to 0 as $Y \to +\infty$.*

Proof We treat all the terms present in $K_{\mathrm{cont}}^{Y,3}$ in the same way. Consider, for example,

$$\delta_Y(z) \int_{-\infty}^{+\infty} \mathrm{Im}(z)^{1/2 - ir} \varphi(1/2 + ir)\overline{E^Y(w, 1/2 + ir)} h(r) dr. \tag{5.18}$$

The proof of Lemma 5.11 implies that this integral converges absolutely; we can moreover shift the contour integral in

$$\delta_Y(z) \int_{\text{Im}(r)=0} h(r) \, \text{Im}(z)^{1/2-ir} \varphi(1/2 + ir) E^Y(w, 1/2 - ir) \, dr$$

to, for example, $\text{Im}(r) = -3/2$. In doing so we encounter the pole at $s = 1$ of the Eisenstein series E. Note nevertheless that $\varphi(s)E(w, 1-s) = E(w, s)$ and thus that $\varphi(s)E^Y(w, 1-s) = E^Y(w, s)$. In particular, the function $s \mapsto \varphi(s)E^Y(w, 1-s)$ has no poles in the half-plane $\text{Re}(s) \geq 1/2$. In light of this, the integral (5.18) becomes

$$\delta_Y(z) \int_{\text{Im}(r)=-3/2} h(r) \, \text{Im}(z)^{1/2-ir} E^Y(w, 1/2 + ir) \, dr.$$

But $E^Y(w, s) = O(1)$ along the line $\text{Re}(s) = 2$ and the integral is then $O(1/\text{Im}(z))$. In particular, the kernel is Hilbert-Schmidt and the integral $\int_D K_{\text{cont}}^{Y,3}(z, z) \, d\mu(z)$ converges absolutely and tends to 0 as $Y \to +\infty$. □

To conclude, observe that it follows from the proof of Proposition 3.25 that

$$K_Y(z, w) := K(z, w) - \delta_Y(z)\delta_Y(w)K_0(z, w)$$

is a Hilbert-Schmidt kernel. According to Lemmas 5.11, 5.12, and 5.13, the decomposition

$$K - K_{\text{cont}} = K_Y - (K_{\text{cont}}^{Y,1} + K_{\text{cont}}^{Y,3} + K_{\text{cont}}^{Y,4})$$

is a sum of Hilbert-Schmidt kernels. This ends the proof of Theorem 5.10. □

5.3.3 The Spectral Side of the Trace Formula

Proposition 5.14 *The operator with kernel $K - K_{\text{cont}}$ is of trace class.*

Proof Since the function h is of rapid decay, it suffices to show the following lemma. □

Lemma 5.15 *We have*

$$|\{j \mid |r_j| \leq T\}| = O(T^2),$$

for T large enough. Here $\lambda_j = 1/4 + r_j^2$ runs through the (discrete) spectrum of Δ in $\text{Im}(E)^\perp$.

Proof The sum (5.15) converges absolutely and uniformly on compacta. We deduce that for all $Y > 0$,

$$\sum_j |h(r_j)|^2 \int_{D^Y} |u_j(w)|^2 d\mu(w) \leq \int_{D^Y} \int_D |K(z,w)|^2 d\mu(z)\, d\mu(w). \qquad (5.19)$$

It remains then to choose a kernel k in such a way as to bound the number of r_j of absolute value less than a given constant T.

Take for k the characteristic function of a small ball centered at 0. In fact, it shall be more convenient in what follows to use a function of the form $U(\cosh \rho) = k(\rho)$, and we take for U the characteristic function of a small ball centered at 1 and of radius δ. (Such a kernel is not smooth, if one prefers one can always approximate it by compactly supported C^∞ functions.) We begin by evaluating the Selberg transform h of k.

Recall that

$$h(r) = \int_{\mathcal{H}} k(i,z) y^s d\mu(z),$$

where $s = 1/2 + ir$. Taking $s = 0$, we obtain

$$h(i/2) = \int_{\mathcal{H}} k(i,z)\, d\mu(z) = 2\pi\delta$$

(here $2\pi\delta$ is the area of a hyperbolic disk whose radius r is given by

$$\sinh(r/2) = \sqrt{\frac{1}{2}(\cosh r - 1)} = \sqrt{\frac{\delta}{2}}\,).$$

Recall furthermore that

$$\frac{1}{2}(\cosh \rho(z,i) - 1) = u(z,i) = \frac{|z-i|^2}{4y},$$

where $y = \operatorname{Im}(z)$. If

$$\frac{1}{2}(\cosh \rho(z,i) - 1) \leq \frac{\delta}{2},$$

we then have $|y - 1| \leq \sqrt{2y\delta}$. Suppose that $\delta \leq 1/4$. Then $y \leq 2$ and so $|y - 1| \leq 2\sqrt{\delta}$. Moreover, for all $z \in \mathbf{C}$ of modulus less than 1, we have $|e^z - 1| \leq 2|z|$. We deduce that for δ sufficiently small and $|s| \leq (4\sqrt{2\delta})^{-1}$,

$$|y^s - 1| \leq 4|s|\sqrt{2\delta}.$$

From this it follows that

$$|h(r) - h(i/2)| \leq 4|1/2 + ir| \sqrt{2\delta} h(i/2),$$

and finally

$$2\pi\delta \leq |h(r)| \leq 6\pi\delta,$$

if $|1/2 + ir| \leq (8\sqrt{2\delta})^{-1}$.

Now, we have

$$\int_D |K(z, w)|^2 d\mu(z) = \frac{1}{4} \sum_{\gamma, \gamma' \in \Gamma} \int_D k(\gamma'z, w) k(\gamma'z, \gamma w) d\mu(z)$$

$$= \frac{1}{2} \sum_{\gamma \in \Gamma} \int_{\mathcal{H}} k(z, w) k(z, \gamma w) d\mu(z). \tag{5.20}$$

By definition of the kernel k, we must have $u(z, w) \leq \delta/2$ and $u(z, \gamma w) \leq \delta/2$, which implies, via the triangle inequality for the hyperbolic metric and the inequality $\cosh(a + b) \leq 2 \cosh a \cosh b - 1$, that $u(w, \gamma w) \leq \delta(\delta + 2)$. Put

$$N_\delta(w) = |\{\gamma \in \Gamma \mid u(w, \gamma w) \leq \delta(\delta + 2)\}|.$$

Applying the Cauchy-Schwarz inequality to (5.20) and recalling that k is a characteristic function, we obtain the inequality

$$\int_D |K(z, w)|^2 d\mu(z) \leq \frac{1}{2} N_\delta(w) \int_{\mathcal{H}} k(i, z)^2 d\mu(z) = \frac{1}{2} N_\delta(w) h(i/2).$$

We shall prove in Lemma 5.16 that

$$N_\delta(w) = O(\sqrt{\delta} Y + 1)$$

for $\delta \in (0, 1]$ and $w \in D^Y$. Keeping only the first sum in the left-hand side of inequality (5.19) and restricting the summation to r_j such that $|1/2 + ir_j| \leq (8\sqrt{2\delta})^{-1}$ we finally see that

$$(2\pi\delta)^2 \sum_{|1/2 + ir_j| < (8\sqrt{2\delta})^{-1}} \int_{D^Y} |u_j(w)|^2 d\mu(w) = O(\delta^{3/2} Y + \delta).$$

Putting $T = \delta^{-1/2}$ this gives

$$\sum_{|r_j| \leq T} \int_{D^Y} |u_j(w)|^2 d\mu(w) = O(T^2 + TY). \tag{5.21}$$

But the $u_j, j \geqslant 1$, are cuspidal and thus decay rapidly in the cusp. More precisely,

$$\int_{D-D^Y} |u_j(w)|^2 d\mu(w) = O(|r_j|Y^{-2}).$$

(5.22)

We shall admit this last estimate, sending the interested reader to the references at the end of this chapter. The lemma is then deduced from (5.21) and (5.22) by taking $T = Y$.   □

Lemma 5.16 *Let $z = x + iy \in D$ and $\delta \in (0, 1]$. We have*

$$N_\delta(z) = O(\sqrt{\delta}y + 1),$$

where the implied constant is universal.

Proof If $u(\gamma z, z) \leqslant \delta(\delta + 2)$ then $u(\gamma z, z) \leqslant 3\delta$. Since δ is smaller than 1, the orbit $\Gamma \cdot z$ contains only a finite number – independent of z – of points whose imaginary part is smaller than y. Thus

$$N_\delta(z) = |\{k \in \mathbf{Z} \mid u(z, z + k) \leqslant 3\delta\}| + O(1).$$

Finally, $u(z, z + k) = k^2/4y^2$ and we obtain the stated bound.   □

It follows from Proposition 5.14 and Lemmas 5.12 and 5.13 that

$$\sum_j h(r_j) = \int_D (K(z, z) - K_{\text{cont}}(z, z)) \, d\mu(z)$$

$$= \lim_{Y \to +\infty} \int_D (K_Y(z, z) - K_{\text{cont}}^{Y,1}(z, z)) \, d\mu(z),$$

(5.23)

where the eigenvalues $\lambda_j = 1/4 + r_j^2$ form the discrete spectrum of Δ in $\text{Im}(E)^\perp$. The left-hand side of (5.23) is thus analogous to the spectral side in the trace formula for a compact surface. We now endeavor to understand the various terms on the right-hand side.

We begin with the term involving $K_{\text{cont}}^{Y,1}$:

$$\int_D K_{\text{cont}}^{Y,1}(z, z) \, d\mu(z) = \frac{1}{4\pi} \int_{-\infty}^{+\infty} h(r) \int_D |E^Y(z, 1/2 + ir)|^2 d\mu(z) \, dr.$$

The inner integral is calculated in the proof of Lemma 5.11 and we obtain

$$\frac{1}{4\pi} \int_{-\infty}^{+\infty} h(r) \left(2 \log Y - \frac{\varphi'}{\varphi}(1/2 + ir) \right)$$

$$+ \frac{1}{2ir} \left(\varphi(1/2 - ir)Y^{2ir} - \varphi(1/2 + ir)Y^{-2ir} \right) \Bigg) dr$$

$$= \frac{1}{4\pi} \int_{-\infty}^{+\infty} Y^{2ir} \left(\frac{\varphi(1/2 - ir) - \varphi(1/2 + ir)}{2ir} \right) h(r)\, dr$$

$$+ \frac{1}{4\pi} \int_{-\infty}^{+\infty} \varphi(1/2 + ir) \left(\frac{\sin(r \log Y)}{2r} \right) h(r)\, dr$$

$$+ \frac{1}{2\pi} \int_{-\infty}^{+\infty} h(r)\, dr \log Y - \frac{1}{4\pi} \int_{-\infty}^{+\infty} h(r) \frac{\varphi'}{\varphi}(1/2 + ir)\, dr.$$

The first integral tends to 0 as Y goes to $+\infty$ according to the Riemann-Lebesgue lemma. Since $\int_{-\infty}^{+\infty} (\sin(x)/x)\, dx = \pi$, a simple change of variables shows that the second integral tends toward

$$\frac{1}{4} \varphi(1/2)h(0)$$

as $Y \to +\infty$. We therefore obtain the following proposition.

Proposition 5.17 *We have*

$$\int_D K_Y(z, z)\, d\mu(z) = \sum_j h(r_j) + \frac{1}{4\pi} \int_{-\infty}^{+\infty} \frac{-\varphi'(s)}{\varphi(s)} (1/2 + ir)h(r)\, dr$$

$$+ \frac{1}{4} h(0)\varphi(1/2) + g(0) \log Y + o(1),$$

as $Y \to +\infty$. (Here as in the rest of the text, $h = \widehat{g}$.)

5.3.4   The Geometric Side of the Trace Formula: Parabolic Term Contribution

It remains to calculate

$$\int_D K_Y(z, z)\, d\mu(z) = \int_{D^Y} K(z, z)\, d\mu(z) + \int_{\mathrm{Im}(z) \geqslant Y} (K(z, z) - K_0(z, z))\, d\mu(z).$$

The proof of Proposition 3.25 implies that the second integral is $o(1)$ as Y goes to $+\infty$. We have only to calculate

$$\int_{D^Y} K(z, z)\, d\mu(z).$$

But

$$K(z, z) = \sum_{\gamma \in \Gamma} k(z, \gamma z)$$

and for $\mathrm{Im}(z)$ sufficiently large, we have $k(z, \gamma z) \neq 0$ if and only if $\gamma \in \Gamma_\infty$. We thus find

$$\sum_{\gamma \in \Gamma} \int_{D^Y} k(z, \gamma z)\, d\mu(z)$$

$$= \sum_{\gamma | \{\gamma\} \cap \Gamma_\infty = \varnothing} \int_D k(z, \gamma z)\, d\mu(z) + \sum_{\gamma | \{\gamma\} \cap \Gamma_\infty \neq \varnothing} \int_{D^Y} k(z, \gamma z)\, d\mu(z).$$

The first side can be treated as in the compact case. Any (non-trivial) parabolic conjugacy class is a power of the conjugacy class of

$$\gamma_0 = \begin{pmatrix} 1 & 1 \\ 0 & 1 \end{pmatrix}.$$

We can therefore write

$$\sum_{\gamma | \{\gamma\} \cap \Gamma_\infty \neq \varnothing} \int_{D^Y} k(z, \gamma z)\, d\mu(z)$$

$$= \int_{D^Y} k(z, z)\, d\mu(z) + \sum_{\ell \in \mathbf{Z}^*} \sum_{\tau \in \Gamma_\infty \backslash \Gamma} \int_{D^Y} k(z, \tau^{-1} \gamma_0^\ell \tau z).$$

The first term can be treated as in the compact case. Since the area of D is equal to $\pi/3$, it tends to

$$\frac{1}{12} \int_{-\infty}^{+\infty} r h(r) \tanh(\pi r)\, dr$$

as Y goes to $+\infty$. Then for every integer $\ell \geqslant 1$,

$$\sum_{\tau \in \Gamma_\infty \backslash \Gamma} \int_{D^Y} k(z, \tau^{-1} \gamma_0^\ell \tau z) = \int_{\Gamma_\infty \backslash \mathcal{H}^Y} k(z, \gamma_0^\ell z)\, d\mu(z)$$

$$= \int_{\Gamma_\infty \backslash \mathcal{H}^Y} k(z, z + \ell)\, d\mu(z),$$

where

$$\mathcal{H}^Y = \{z \in \mathcal{H} \mid \mathrm{Im}(z) \leqslant Y\}.$$

This is the contribution of the parabolic terms to the trace formula; we calculate it in the following proposition.

Proposition 5.18 *We have*

$$\sum_{\ell=1}^{+\infty} \int_{\Gamma_\infty \backslash \mathcal{H}^Y} k(z, z + \ell) \, d\mu(z)$$

$$= g(0) \log Y - g(0) \log 2 + \frac{1}{4} h(0) - \frac{1}{2\pi} \int_{-\infty}^{+\infty} h(r) \frac{\Gamma'}{\Gamma} (1 + ir) \, dr + o(1)$$

as $Y \to +\infty$.

Proof We begin by remarking that given an integer $\ell \neq 0$, a change of variables analogous to the one made in the proof of Lemma 5.12 implies that

$$\int_{\Gamma_\infty \backslash \mathcal{H}^Y} k(z, z + \ell) \, d\mu(z) = \int_0^1 \int_0^Y k(\rho(z, z + \ell)) \, d\mu(z)$$

$$= \int_0^Y k \left(\arg \cosh(1 + \ell^2/2y^2) \right) y^{-2} dy$$

$$= \frac{1}{|\ell| \sqrt{2}} \int_{\arg \cosh(1+\ell^2/2Y^2)}^{+\infty} \frac{k(\rho) \sinh \rho}{\sqrt{\cosh \rho - 1}} d\rho.$$

We deduce that

$$\sum_{\ell=1}^{+\infty} \int_{\Gamma_\infty \backslash \mathcal{H}^Y} k(z, z + \ell) \, d\mu(z)$$

$$= \sqrt{2} \int_{\arg \cosh(1+1/2Y^2)}^{+\infty} \frac{k(\rho) \sinh \rho}{\sqrt{\cosh \rho - 1}} \left(\sum_{1 \leq \ell < \sqrt{2} Y \sqrt{\cosh \rho - 1}} \frac{1}{\ell} \right) d\rho$$

$$= \sqrt{2} \int_{\arg \cosh(1+1/2Y^2)}^{+\infty} \frac{k(\rho) \sinh \rho}{\sqrt{\cosh \rho - 1}} \left(\log(\sqrt{2} Y \sqrt{\cosh \rho - 1}) + \gamma \right.$$

$$+ O(Y^{-1}(\cosh \rho - 1)^{-1/2}) \right) d\rho$$

$$= L(Y) + O(Y^{-1} \log Y),$$

where

$$L(Y) = \sqrt{2} \int_0^{+\infty} \frac{k(\rho) \sinh \rho}{\sqrt{\cosh \rho - 1}} \left(\log(Y \sqrt{2(\cosh \rho - 1)}) + \gamma \right) d\rho$$

and γ is the *Euler constant*. It remains then to explicate $L(Y)$. Recall first of all that

$$\sqrt{2}\int_0^{+\infty} \frac{k(\rho)\sinh\rho}{\sqrt{\cosh\rho-1}}d\rho = g(0).$$

Thus,

$$L(Y) = g(0)(\log(2Y)+\gamma) + 2\int_0^{+\infty} k(\rho)\cosh(\rho/2)\log(\sinh(\rho/2))d\rho.$$

But the Abel inversion formula implies that

$$k(\rho) = -\frac{1}{\sqrt{2}\,\pi}\int_\rho^{+\infty} \frac{dg(t)}{\sqrt{\cosh t-\cosh\rho}}$$

$$= -\frac{1}{2\pi}\int_\rho^{+\infty} \frac{dg(t)}{\sqrt{\sinh^2(t/2)-\sinh^2(\rho/2)}}$$

from which we deduce that

$$2\int_0^{+\infty} k(\rho)\cosh(\rho/2)\log(\sinh(\rho/2))d\rho$$

$$= -\frac{1}{\pi}\int_0^{+\infty}\int_\rho^{+\infty} \frac{\cosh(\rho/2)\log(\sinh(\rho/2))}{\sqrt{\sinh^2(t/2)-\sinh^2(\rho/2)}}dg(t)d\rho$$

$$= -\frac{1}{2\pi}\int_0^{+\infty}\left(\int_0^t \frac{\log(\sinh^2(\rho/2))}{\sqrt{\frac{\sinh^2(t/2)}{\sinh^2(\rho/2)}-1}}\frac{d(\sinh^2(\rho/2))}{\sinh^2(\rho/2)}\right)dg(t)$$

$$= -\frac{1}{2\pi}\int_0^{+\infty}\left(\int_0^v \frac{\log u}{\sqrt{u(v-u)}}du\right)dq(v),$$

where $u = \sinh^2(\rho/2)$, $v = \sinh^2(t/2)$, and $g(t) = q(v)$. But

$$-\frac{1}{2\pi}\int_0^{+\infty}\left(\int_0^v \frac{\log u}{\sqrt{u(v-u)}}du\right)dq(v)$$

$$= -\frac{1}{2\pi}\int_0^{+\infty}\left(\int_0^1 \frac{\log(wv)}{\sqrt{w(1-w)}}dw\right)dq(v)$$

$$= \frac{1}{2\pi}q(0)\int_0^1 \frac{\log w}{\sqrt{w(1-w)}}dw - \frac{1}{2\pi}\left(\int_0^1 \frac{dw}{\sqrt{w(1-w)}}\right)\left(\int_0^{+\infty}\log(v)dq(v)\right)$$

$$= -g(0)\log 2 - \int_0^{+\infty}\log(\sinh(t/2))\,dg(t).$$

In this way,

$$L(Y) = g(0)(\log Y + \gamma) - \int_0^{+\infty} \log(\sinh(t/2)) \, dg(t). \tag{5.24}$$

We can either content ourselves with the expression (5.24) or rewrite it in the stated form

$$L(Y) = g(0) \log Y - g(0) \log 2 + \frac{1}{4}h(0) - \frac{1}{2\pi} \int_{-\infty}^{+\infty} h(t)\frac{\Gamma'}{\Gamma}(1 + it)dt,$$

see [63, §10.3]. □

5.3.5   The Trace Formula

The hyperbolic and elliptic terms contribute to the "trace" $\int_{D^Y} K(z, z) \, d\mu(z)$ in a similar way as for the compact case. By passing to the limit $Y \to +\infty$ Propositions 5.17 and 5.18 imply the trace formula for the group SL(2, **Z**).

Theorem 5.19 *Let* $\Gamma = $ SL(2, **Z**)*. Let* (h, g) *be an admissible pair of functions. Let* $\lambda_j = 1/4 + r_j^2$, $j \in \mathbb{N}$, *be the set of (discrete) eigenvalues of the hyperbolic Laplacian in* $L^2(\Gamma \backslash \mathcal{H})$, *counted with multiplicity. Then*

$$\sum_{j=0}^{+\infty} h(r_j) = \frac{1}{12} \int_{-\infty}^{+\infty} rh(r) \tanh(\pi r) \, dr$$

$$+ \sum_P \operatorname{arccosh} (t/2) \sum_{\ell=1}^{+\infty} \frac{g\,(2\ell \operatorname{arccosh} (t/2))}{\sinh\,(\ell \operatorname{arccosh} (t/2))}$$

$$+ \frac{1}{2} \int_{-\infty}^{+\infty} h(r)\frac{e^{-\pi r}}{1 + e^{-2\pi r}}dr + \frac{1}{3}\sum_{\ell=1}^{2} \frac{1}{\sin\,(\ell\pi/3)} \int_{-\infty}^{+\infty} h(r)\frac{e^{-2\ell\pi r/3}}{1 + e^{-2\pi r}}dr$$

$$+ g(0) \log(\pi/2) - \frac{1}{2\pi} \int_{-\infty}^{+\infty} h(r)\left[\frac{\Gamma'}{\Gamma}(1/2 + ir) + \frac{\Gamma'}{\Gamma}(1 + ir)\right] dr$$

$$+ 2\sum_{n=1}^{+\infty} \frac{\Lambda(n)}{n}g(2 \log n).$$

Here P denotes the set of primitive hyperbolic conjugacy classes in PSL(2, **Z**), *t the displacement distance of an element in P, and*

$$\Lambda(n) = \begin{cases} \log(p) & \text{if } n = p^k \text{ with } p \text{ prime and } k \in \mathbf{N}^*, \\ 0 & \text{otherwise.} \end{cases}$$

Proof There are exactly two elliptic conjugacy classes in PSL(2, **Z**) one of order 2 and the other of order 3. The terms involving $\log Y$ in Propositions 5.17 and 5.18 cancel each other out. Passing to the limit $Y \to +\infty$, these two propositions thus imply the equality

$$\sum_{j=0}^{+\infty} h(r_j) = \frac{1}{12} \int_{-\infty}^{+\infty} r h(r) \tanh(\pi r) \, dr$$

$$+ \sum_{P} \operatorname{arccosh}(t/2) \sum_{\ell=1}^{+\infty} \frac{g\left(2\ell \operatorname{arccosh}(t/2)\right)}{\sinh\left(\ell \operatorname{arccosh}(t/2)\right)} + \frac{1}{2} \int_{-\infty}^{+\infty} h(r) \frac{e^{-\pi r}}{1 + e^{-2\pi r}} dr$$

$$+ \frac{1}{3} \sum_{\ell=1}^{2} \frac{1}{\sin\left(\ell\pi/3\right)} \int_{-\infty}^{+\infty} h(r) \frac{e^{-2\ell\pi r/3}}{1 + e^{-2\pi r}} dr - \frac{1}{2\pi} \int_{-\infty}^{+\infty} h(r) \frac{\Gamma'}{\Gamma}(1 + ir) \, dr$$

$$+ \frac{1}{4\pi} \int_{-\infty}^{+\infty} \frac{\varphi'}{\varphi}(1/2 + ir) h(r) \, dr - \log(2) g(0) + \frac{1}{4}(1 - \varphi(1/2)) h(0). \qquad (5.25)$$

Recall now that we have

$$\varphi(s) = \pi^{2s-1} \frac{\Gamma(1-s)}{\Gamma(s)} \frac{\zeta(2-2s)}{\zeta(2s)} \quad \text{and} \quad \frac{\zeta'(s)}{\zeta(s)} = -\sum_{n=1}^{+\infty} \frac{\Lambda(n)}{n^s}$$

for $\operatorname{Re}(s) > 1$. A simple calculation with the logarithmic derivative of φ then implies that

$$\frac{1}{4\pi} \int_{-\infty}^{+\infty} \frac{\varphi'}{\varphi}(1/2 + ir) h(r) \, dr$$

$$= g(0) \log(\pi) - \frac{1}{2\pi} \int_{-\infty}^{+\infty} h(r) \frac{\Gamma'}{\Gamma}(1/2 + ir) \, dr + 2 \sum_{n=1}^{+\infty} \frac{\Lambda(n)}{n} g(2 \log n).$$

This concludes the proof of the theorem.   □

Remark 5.20 We can in the same way develop a trace formula for an arbitrary Fuchsian group of the first kind, see [63]. The case of the group $\Gamma_0(N)$ is particularly interesting: one has the following theorem.

Theorem 5.21 *Let $N \geq 1$ be a square-free integer and let (h, g) be an admissible function pair. Denote by $\lambda_j = 1/4 + r_j^2$, $j \in \mathbb{N}$, the set of (discrete) eigenvalues of the hyperbolic Laplacian in $L^2(\Gamma_0(N)\backslash\mathcal{H})$, counted with multiplicity. Then*

$$\sum_{j=0}^{+\infty} h(r_j) = \frac{A_N}{4\pi} \int_{-\infty}^{+\infty} rh(r) \tanh(\pi r)\, dr$$

$$+ \sum_P \operatorname{arccosh}(t/2) \sum_{\ell=1}^{+\infty} \frac{g\,(2\ell \operatorname{arccosh}(t/2))}{\sinh\,(\ell \operatorname{arccosh}(t/2))}$$

$$+ \sum_E \sum_{\ell=1}^{m-1} \frac{1}{m \sin\,(\ell\pi/m)} \int_{-\infty}^{+\infty} h(r) \frac{e^{-2\ell\pi r/m}}{1 + e^{-2\pi r}} dr$$

$$+ 2^{\omega(N)} \left\{ g(0) \log(\pi/2) - \frac{1}{2\pi} \int_{-\infty}^{+\infty} h(r) \left[\frac{\Gamma'}{\Gamma}(1/2 + ir) + \frac{\Gamma'}{\Gamma}(1 + ir) \right] dr \right.$$

$$\left. + 2\sum_{n=1}^{+\infty} \frac{\Lambda(n)}{n} g(2\log n) - \sum_{p|N,\, p \text{ premier}} \sum_{k=0}^{+\infty} \frac{\log p}{p^k} g(2k\log p) \right\}.$$

Here A_N is the area of a fundamental domain for the action of $\Gamma_0(N)$ on $\mathcal{H}$, P (resp. E) is the set of primitive hyperbolic (resp. elliptic) conjugacy classes in $\Gamma_0(N)/\{\pm I\}$, t is the displacement distance of an element in P, m is the order of an element in E, and $\omega(N)$ is the number of prime divisors of N.

5.4 Applications

We now give two types of applications of the trace formulas explicated above.

5.4.1 The Weyl Law

Let $S = \Gamma\backslash\mathcal{H}$ be a compact hyperbolic surface. Let us apply the Selberg trace formula to the admissible pair $h(r) = e^{-\delta r^2}$ and $g(x) = (4\pi\delta)^{-1/2} e^{-x^2/4\delta}$, where δ

is a positive parameter. According to Theorem 5.6 we have

$$\sum_{j=0}^{+\infty} e^{-\delta r_j^2} = \frac{\text{area}(S)}{4\pi} \int_{-\infty}^{+\infty} re^{-\delta r^2} \tanh(\pi r)\, dr$$

$$+ (4\pi\delta)^{-1/2} \sum_{\gamma \in G(S)} \frac{\Lambda(\gamma)}{N_\gamma^{1/2} - N_\gamma^{-1/2}} e^{-\ell(\gamma)^2/4\delta}. \qquad (5.26)$$

Lemma 5.22 *The number of closed geodesics on S of length $\leqslant L$ is $O(e^L)$ for L large enough.*

Proof Fix a base point $p \in \mathcal{H}$ and consider the Γ-orbit of p in $\mathcal{H}$. For each point in this orbit one may construct the corresponding Dirichlet fundamental domain, thereby obtaining a tiling of $\mathcal{H}$. For a real number $L > 0$, let $N(L)$ be the number of fundamental domains meeting the hyperbolic disk of radius $L + 2\,\text{diam}(S)$ centered at p. Note that the number of closed geodesics on S of length less than L is bounded above by $N(L)$. Since the area of the disk is $O(e^L)$, we have $N(L) = O(e^L)$ and Lemma 5.22 follows. $\qquad\square$

According to Lemma 5.22, the second term in the right-hand side of (5.26) is bounded by $O(\delta^{-1/2}e^{-\ell_0^2/8\delta})$, where ℓ_0 is the length of the smallest geodesic, and converges then to 0 as $\delta \to 0$. An integration by parts gives

$$\int_{-\infty}^{+\infty} re^{-\delta r^2} \tanh(\pi r)\, dr = \frac{1}{\delta} + O(1).$$

Hence one obtains

$$\sum_{j=0}^{+\infty} e^{-\delta r_j^2} = \frac{\text{area}(S)}{4\pi\delta}(1 + O(1)) \qquad (5.27)$$

for δ small enough.

Tauberian theorems then allow one to relate the asymptotic expansion of the partial sums of a series to the asymptotic expansion of its Abel transform. In particular, a Tauberian theorem of Karamata (see the references at the end of this chapter) allows one to deduce from (5.27) the following theorem – known as a *Weyl law*.

Theorem 5.23 *As $T \to +\infty$,*

$$|\{j \mid 0 \leqslant \lambda_j \leqslant T\}| \sim \frac{\text{area}(S)}{4\pi} T.$$

When $\Gamma = \mathrm{SL}(2, \mathbf{Z})$ the trace formula (5.25) replaces Theorem 5.6. Although we skip the details here, formula (5.25) allows one to deduce

$$\sum_{j=0}^{+\infty} e^{-\delta r_j^2} - \frac{1}{4\pi} \int_{-\infty}^{+\infty} \frac{\varphi'}{\varphi}(1/2 + ir) e^{-\delta r^2} dr$$

$$\sim \frac{\mathrm{area}(S)}{4\pi} \int_{-\infty}^{+\infty} r e^{-\delta r^2} \tanh(\pi r) dr,$$

as δ tends toward 0. (Recall that $\varphi(s) = \xi(2 - 2s)/\xi(2s)$, where ξ is the completed Riemann zeta function.) Just as in the proof of Theorem 5.23, it follows from the theorem of Karamata that

$$|\{j \geq 0 \mid \lambda_j \leq T\}| - \frac{1}{4\pi} \int_{-T}^{T} \frac{\varphi'}{\varphi}(1/2 + ir) dr \sim \frac{T^2}{12}, \qquad (5.28)$$

as $T \to +\infty$. Taking into account the explicit expression for $\varphi(s)$ in this case and in particular the fact that $\varphi(s)$ is a holomorphic function of order 1, we have, for any $\varepsilon > 0$,

$$\left| \frac{1}{4\pi} \int_{-T}^{T} \frac{\varphi'}{\varphi}(1/2 + ir) dr \right| = O(T^{1+\varepsilon}),$$

for T large enough. We thus obtain the following result.

Theorem 5.24 Let λ_j denote the (discrete) eigenvalues of the hyperbolic Laplacian in $L^2(\mathrm{SL}(2, \mathbf{Z})\backslash\mathcal{H})$, counted with multiplicity. Then

$$|\{j \mid 0 \leq \lambda_j \leq T\}| \sim \frac{T}{12},$$

as $T \to +\infty$.

5.4.2 The Prime Geodesic Theorem

We return now to the case of a compact surface $S = \Gamma\backslash\mathcal{H}$ and write $\pi(x)$ for the number of prime geodesics γ on S such that $N_\gamma = e^{\ell(\gamma)} \leq x$. Let

$$s_j = \frac{1}{2} + \sqrt{\frac{1}{4} - \lambda_j}, \quad \text{for } \lambda_j \leq \frac{1}{4},$$

where as usual the λ_j are the Laplacian eigenvalues on S. Selberg [115] highlighted the analogy between his trace formula and the "explicit formulas" in number theory relating prime numbers to zeros of the Riemann zeta function. In this analogy the prime numbers correspond to the lengths of prime geodesics. He developed this analogy by proving the following theorem, which is reminiscent of the prime number theorem recalled in the preceding chapter.

Theorem 5.25 *For x large enough,*

$$\pi(x) = \text{li}(x) + \sum_{1 > s_j > 3/4} \text{li}(x^{s_k}) + O(x^{3/4}/\log x).$$

Here li denotes the *logarithmic integral*:

$$\text{li}(x) = \int_2^x \frac{d\tau}{\log \tau} \sim x/\log x, \quad x \longrightarrow \infty.$$

Proof We shall give asymptotic expansions of various auxiliary functions closely related to $\pi(x)$. We begin by considering the function

$$H(T) = \sum_{\ell(\gamma) \leq T} \frac{\Lambda(\gamma)}{N_\gamma^{1/2} - N_\gamma^{-1/2}} 2 \cosh\left(\frac{1}{2} \log N_\gamma\right)$$

$$= \sum_{\ell(\gamma) \leq T} \Lambda(\gamma)(1 + N_\gamma^{-1})(1 - N_\gamma^{-1})^{-1},$$

where $N_\gamma = e^{\ell(\gamma)}$ and $\Lambda(\gamma)$ as usual denotes the length of the unique oriented prime geodesic γ_0 such that $\gamma = \gamma_0^m$ for an integer $m \geq 1$. Put $E_T(\alpha) = \alpha^{-1} e^{T\alpha}$.

Lemma 5.26 *For T large enough we have*

$$H(T) = E_T(1) + \sum_{1 > s_j > 3/4} E_T(s_j) + O(e^{3T/4}).$$

We admit momentarily the lemma and show how it can be used to deduce the theorem. It is clear that

$$H(T) = \sum_{\ell(\gamma) \leq T} \Lambda(\gamma)(1 + N_\gamma^{-1})(1 + O(N_\gamma^{-1}))$$

$$= \sum_{\ell(\gamma) \leq T} \Lambda(\gamma) + O\left(\sum_{\ell(\gamma) \leq T} \Lambda(\gamma) N_\gamma^{-1}\right). \tag{5.29}$$

Put

$$\psi(T) = \sum_{\ell(\gamma) \leq T} \Lambda(\gamma).$$

Since only a finite number of N_γ are less than a fixed constant and since – according to Lemma 5.26 – $H(T)$ tends to infinity with T, it follows from (5.29) that

$$H(T) = \psi(T) + o(\psi(T)) \tag{5.30}$$

as T tends toward $+\infty$. Lemma 5.26 implies that $\psi(T) \sim E_T(1)$ as T tends toward $+\infty$. We deduce that

$$\sum_{\ell(\gamma)\leqslant T} \Lambda(\gamma)N_\gamma^{-1} = \int_0^T e^{-x}d\psi(x) = O(T).$$

Lemma 5.26 and identity (5.29) then imply

$$\psi(T) = E_T(1) + \sum_{1>s_j>3/4} E_T(s_j) + O(e^{3T/4}). \tag{5.31}$$

Consider now the function

$$\vartheta(T)\left(= \sum_{\substack{\ell(\gamma)\leqslant T \\ \gamma \text{ prime}}} \Lambda(\gamma)\right) = \sum_{\substack{\ell(\gamma)\leqslant T \\ \gamma \text{ prime}}} \ell(\gamma).$$

It is immediate that

$$\psi(T) = \vartheta(T) + \vartheta(T/2) + \cdots + \vartheta(T/k), \tag{5.32}$$

where k is of order T since (5.32) is satisfied as soon as $k \geqslant T/\ell_1$, where ℓ_1 denotes the length of the shortest closed geodesic on S.

It follows from (5.31) and (5.32) that

$$\vartheta(T) = E_T(1) + \sum_{1>s_j>3/4} E_T(s_j) + O(e^{3T/4}). \tag{5.33}$$

Now observe that $\pi(x) = \int_\delta^{\log x} T^{-1}d\vartheta(T)$ if $\delta < \ell_1$; hence it follows from (5.33) that

$$\pi(x) = \text{li}(x) + \sum_{1>s_j>3/4} \text{li}(x^{s_k}) + O(x^{3/4}/\log x), \tag{5.34}$$

where $\text{li}(x) = \int_2^x dt/\log t$. This concludes the proof of the theorem. $\square$

Proof of Lemma 5.26 We now apply the Selberg trace formula to well-chosen test functions. We begin by defining a family of functions g_T^ε.

1. Let $\chi_{[-T,T]}$ be the characteristic function of the interval $[-T,T]$ in $\mathbf{R}$ and g_T the function defined by $g_T(x) = 2\cosh(x/2)\chi_{[-T,T]}(x)$.
2. Let φ be an even non-negative C^∞ function with support contained in $[-1,1]$ and such that $\int_{-1}^1 \varphi(x)\,dx = 1$. Given a real number $\varepsilon > 0$, write φ_ε for $\varphi_\varepsilon(x) = \varepsilon^{-1}\varphi(x/\varepsilon)$. The function φ_ε is supported in $[-\varepsilon,\varepsilon]$ and we have $\int_{-\varepsilon}^\varepsilon \varphi_\varepsilon(x)\,dx = 1$; the functions φ_ε thus approximate the Dirac mass at 0 as ε tends to 0.
3. Finally, we put

$$g_T^\varepsilon(x) = (g_T * \varphi_\varepsilon)(x) = 2\int_{-\infty}^{+\infty} \cosh((x-y)/2)\chi_{[-T,T]}(x-y)\varphi_\varepsilon(y)\,dy.$$

For any $\varepsilon, T > 0$, the function g_T^ε is even, C^∞, and of compact support. We can therefore apply the Selberg trace formula to it. The corresponding function h in the trace formula (the Fourier transform of g_T^ε) shall be denoted h_T^ε.

For the Fourier transform $\widehat{f}(r) = \int_{-\infty}^{+\infty} e^{-irx} f(x)\,dx$, we have $\widehat{f_1 * f_2} = \widehat{f_1}\widehat{f_2}$ and $2\pi\,\widehat{\widehat{f}} = f$ if f is even. A direct calculation then shows that

$$\begin{aligned}
\widehat{g}_T(r) &= \int_{-\infty}^{+\infty} e^{-irx} g_T(x)\,dx \\
&= 2\int_{-T}^{T} e^{-irx}\cosh(x/2)\,dx \\
&= \int_{-T}^{T} e^{(1/2-ir)x} + e^{-(1/2+ir)x}\,dx \\
&= S(1/2 + ir) + S(1/2 - ir),
\end{aligned}$$

where $S(w) = 2w^{-1}\sinh(Tw)$ with the convention $S(0) = 2T$. The function h_T^ε is then given by

$$h_T^\varepsilon(r) = (S(1/2 + ir) + S(1/2 - ir))\widehat{\varphi}_\varepsilon(r).$$

We now define functions H_ε, which will be approximations to $H(T)$, by letting

$$H_\varepsilon(T) = \sum_{\gamma \in G(S)} \frac{\Lambda(\gamma)}{N_\gamma^{1/2} - N_\gamma^{-1/2}} g_T^\varepsilon(\log N_\gamma).$$

Observe that for all $\varepsilon > 0$, $H_\varepsilon(T - \varepsilon) \leqslant H(T) \leqslant H_\varepsilon(T + \varepsilon)$. On the other hand, the trace formula implies

$$H_\varepsilon(T) = \sum_{i=0}^{+\infty} h_T^\varepsilon(r_i) - \frac{\text{area}(S)}{4\pi} \int_{-\infty}^{+\infty} r h_T^\varepsilon(r) \tanh(\pi r)\, dr$$

$$= \sum_i^* h_T^\varepsilon(r_i) + \int_0^{+\infty} h_T^\varepsilon(r)\, dm(r), \tag{5.35}$$

where $\sum^*$ denotes the finite sum over the purely imaginary r_j, and $dm(r)$ is the measure on $[0, +\infty)$ given by $dN(r) - \frac{\text{area}(S)}{2\pi} r \tanh(\pi r)\, dr$, with $N(r) = \sum_{0 < r_i \leqslant r} 1$.

Now put $\varepsilon = e^{-T/4}$. Note that $\widehat{\varphi}_\varepsilon(x) = \widehat{\varphi}(\varepsilon x) = 1 + O(\varepsilon x)$ as $\varepsilon \to 0$ with x fixed. Each term of the sum $\sum_i^* h_T^\varepsilon(r_i)$ corresponding to one of the $s_j \in (3/4, 1]$ is thus of the form

$$(2s_j^{-1} \sinh(s_j T)) \widehat{\varphi}_\varepsilon(r_j) = E_T(s_j) + O(\varepsilon e^{s_j T}).$$

We conclude that

$$\sum_i^* h_T^\varepsilon(r_i) = E_T(1) + \sum_{1 > s_j > 3/4} E_T(s_j) + O(\varepsilon e^{3T/4}). \tag{5.36}$$

It remains to control the term $\int_0^{+\infty} h_T^\varepsilon(r) dm(r)$ in (5.35). We begin by remarking that $\widehat{\varphi}(\rho) = O((1 + |\rho|)^{-2})$ and $|\widehat{g}_T(r)| \leqslant O((1 + r)^{-1} e^{T/2})$. It then follows from the expression for h_T^ε that for each real number r,

$$|h_T^\varepsilon(r)| = O\big(e^{T/2}(1 + r)^{-1}(1 + \varepsilon r)^{-2}\big).$$

This gives

$$\left| \int_0^{+\infty} h_T^\varepsilon dm(r) \right| = O\left(e^{T/2} \int_0^{+\infty} (1 + r)^{-1}(1 + \varepsilon r)^{-2} |dm(r)| \right).$$

To bound this last integral we break it up as $\int_0^{+\infty} = \int_0^{1/\varepsilon} + \int_{1/\varepsilon}^{+\infty}$ and integrate by parts (so as to replace $dm(r)$ by $m(r)$). According to the Weyl law, $N(r)$ is $O(r^2)$, which implies an upper bound on the integral of the form $O(\varepsilon^{-1} e^{T/2}) = O(e^{3T/4})$. In other words

$$\int_0^{+\infty} h_T^\varepsilon(r) dm(r) = O(e^{3T/4}). \tag{5.37}$$

Combining (5.36) and (5.37), we thus find

$$H_\varepsilon(T) = E_T(1) + \sum_{1 > s_j > 3/4} E_T(s_j) + O(e^{3T/4}). \tag{5.38}$$

But

$$H_\varepsilon(T - \varepsilon) \leqslant H(T) \leqslant H_\varepsilon(T + \varepsilon)$$

and, for fixed $\alpha \in (3/4, 1]$,

$$E_{T \pm \varepsilon}(\alpha) = \alpha^{-1} e^{(T \pm \varepsilon)\alpha} = \alpha^{-1} e^{T\alpha}(1 + O(\varepsilon))$$

$$= E_T(\alpha) + O(e^{3T/4}).$$

It then follows from (5.38) that

$$H(T) = E_T(1) + \sum_{1 > s_j > 3/4} E_T(s_j) + O(e^{3T/4}),$$

which completes the proof of Lemma 5.26.   □

5.5   Comments and References

§ 5.1

Since they are semisimple, it is not difficult to verify that the classical groups $SL(n, \mathbf{R})$, $U(n, \mathbf{R})$, $SO(p, q, \mathbf{R})$ are unimodular. A theorem of Borel [16] then states that each of these groups contains a discrete cocompact subgroup.

One can show that under the general hypotheses of the first section, the operator $R(f)$ is a trace class operator, see [69, Lem. 10.11]. Moreover, Eq. (5.1) is satisfied.

In the comments on § 3.9 we noted that the proof of the first part of the spectral theorem can be made to show that the representation R decomposes as a direct sum of irreducible unitary representations.

The trace formula as presented in this section is due to the work of Selberg [115]. The exercise 5.30 shows that for $G = \mathbf{R}$ and $\Gamma = \mathbf{Z}$ one recovers the Poisson summation formula. The exercise 5.32 makes the link between the Frobenius reciprocity theorem in the representation theory of finite groups. One can find a nice introduction to this subject in the book of Serre [119]; one can also refer to § 6.3.1 for a rapid introduction as well as to acquaint oneself with the relevant notation.

The principal application of the trace formula that Selberg gave was to the group $SL(2, \mathbf{R})$, as in § 5.2 and 5.3. We could not however conclude these comments without mentioning the Arthur trace formula. This immense generalization to any reductive group – after passing to the adeles – of the original trace formula is an essential tool in the realization of the "Langlands program"; for a masterful introduction – by Arthur himself – see [4].

The Selberg trace formula is intimately linked to Weil's "explicit formulae" in number theory. We shall now try to make this more precise:

Let $\mathbf{A}$ be the ring of adeles of $\mathbf{Q}$, see for example [20, 92] for a definition. We can take for G an adelic group $G(\mathbf{A})$ and Γ equal to $G(\mathbf{Q})$. By an appropriate choice of function on $G(\mathbf{A})$, one can make sense of the Selberg trace formula for the pair $(G(\mathbf{A}), G(\mathbf{Q}))$. This is the setting of the Arthur trace formula.

In [45] Goldfeld considers the group $G(\mathbf{A}) = \mathbf{A} \rtimes I_0$, where

$$I_0 = \{x \in \mathbf{A}^* \mid |x| = \textstyle\prod_p |x_p|_p = 1\}$$

is the group of norm 1 *ideles*, which acts on $\mathbf{A}$ by left-multiplication. The multiplication in the semi-direct product is given by $(x, x')(y, y') = (xy, x'y + y')$, with $(x, x'), (y, y') \in \mathbf{A} \rtimes I_0$. The following *Weil explicit formula* then follows from the Selberg trace formula:

$$\sum_{\substack{\rho = 1/2 + i\gamma,\, \zeta(\rho) = 0 \\ \mathrm{Re}(\rho) \in [0,1]}} h(\gamma) = h(i/2) + h(-i/2) - g(0) \log \pi$$

$$+ \frac{1}{2\pi} \int_{-\infty}^{+\infty} h(r) \frac{\Gamma'}{\Gamma}\left(\frac{1}{4} + \frac{1}{2}ir\right) dr - 2 \sum_{n=1}^{+\infty} \frac{\Lambda(n)}{\sqrt{n}} g(\log n),$$

where g is a function of compact support,

$$h(r) = \int_{-\infty}^{+\infty} g(u) e^{iru} du$$

and

$$\Lambda(n) = \begin{cases} \log(p) & \text{if } n = p^k \text{ with } p \text{ prime and } k \in \mathbf{N}^*, \\ 0 & \text{otherwise.} \end{cases}$$

§5.3

In this section we have mostly followed the notes of Cohen and Sarnak [28] on the trace formula. Another reference is [63] which treats the case of a general Fuchsian group of the first kind. In particular, the integral[2] (5.16) converges for an arbitrary Fuchsian group of the first kind, see Iwaniec [63, §10.2]. In the special case studied in this section the function $\varphi(s) = \xi(2 - 2s)/\xi(2s)$ is explicit, which simplifies the proof.

The asymptotic expansion (5.22) that we admitted in the proof of Lemma 5.15 should not be surprising since the u_j are rapidly decreasing. One can deduce it from [63, Th. 3.2].

The Selberg trace formula for $SL(2, \mathbf{Z})$ is very close to the Weil explicit formula, with the prime geodesic theorem extending this analogy.

§5.4

The Tauberian theorem of Karamata – which generalizes a celebrated Tauberian theorem of Hardy and Littlewood – is proved in [139].

The Weyl law holds in fact on any compact Riemannian manifold, see [8].

---

[2]The function φ is then replaced by a matrix-valued function called the scattering matrix.

We have not given the details of the proof of Theorem 5.24. For this one can consult [63, p. 158–159].

Theorem 5.24 is a quantitative version of the existence of an infinite number of linearly independent Maaß cusp forms on the modular surface. It is due to Selberg and is one of the first spectacular consequences of the trace formula. The version given here – without error terms – can also be obtained from the method of Lindenstrauss and Venkatesh described in the preceding chapter. This latter method has the advantage of rather easily extending to a much more general setting, whereas the approach by the Arthur trace formula is much more difficult to generalize in an efficient manner.

When Γ is an arbitrary Fuchsian group of the first kind, the analog of (5.28) remains true but one cannot in general control the continuous spectrum. Phillips and Sarnak [98] in fact conjecture that for "generic" Γ there exists only a finite number of Maaß cusp forms.

The proof of Theorem 5.25 that we reproduce in this section is due to Randol [25].

It is natural to conjecture that the $3/4$ in Theorem 5.25 can be replaced by $1/2$. One can think of this conjecture as an analog of the Riemann hypothesis. It would in fact follow from the Selberg conjecture that for any congruence surface and large enough x one has

$$\pi(x) = \mathrm{li}(x) + O(x^{1/2}/\log x).$$

The proof of Theorem 5.25 extends to the modular surface $\mathrm{SL}(2, \mathbf{Z})\backslash\mathcal{H}$ through the trace formula (5.25), see [110]. Applying Proposition 4.22 we may then deduce the following theorem (we adopt here the notation of § 4.4.1).

Theorem 5.27 *We have*

$$\sum_{\varepsilon_D \leq x} h(D) = \mathrm{li}(x^2) + O(x^{3/2}/\log x),$$

the sum running over all non-square positive integers D congruent to 0 or 1 mod 4 such that $\varepsilon_D \leq x$.

Gauß understood that the asymptotic behavior of the narrow class number $h(D)$ of real quadratic fields is a subtle thing to grasp. Nevertheless, Siegel [122] succeeded in showing that

$$\sum_{D \leq x} h(D) \log \varepsilon_D = \frac{\pi^2 x^{3/2}}{18\zeta(3)} + O(x \log x),$$

thereby confirming a conjecture of Gauß. Here the sum runs over non-square integers D congruent to 0 or 1 mod 4 such that $D \leq x$. By way of comparison note that an Abel summation allows one to deduce from Theorem 5.27 the following corollary.

Corollary 5.28 *We have*

$$\sum_{\varepsilon_D \leq x} h(D) \log \varepsilon_D = \frac{x^2}{2} + O(x^{3/2}),$$

the sum running over all non-square integers congruent to 0 or 1 mod 4 and such that $\varepsilon_D \leq x$.

Pushing through this method Sarnak succeeds, in [110], in giving an asymptotic to the mean value of the narrow class number $h(D)$ ordered by the size of $\log \varepsilon_D$.

5.6 Exercises

Exercise 5.29

1. Show that the transitive action of $SL(2, \mathbf{R})$ on $\mathcal{H}$ realizes $\mathcal{H}$ as a homogeneous space:

$$\mathcal{H} = SL(2, \mathbf{R})/SO(2).$$

2. Using the measure μ on $\mathcal{H} = SL(2, \mathbf{R})/SO(2)$ and the angular measure on the rotation group $SO(2)$ for the plane, show that the group $SL(2, \mathbf{R})$ is unimodular.

Exercise 5.30 Let $G = \mathbf{R}$ and $\Gamma = \mathbf{Z}$. Let $f \in C_c^\infty(\mathbf{R})$. Recall that $\widehat{\mathbf{R}} = \{\pi_\lambda = e(\lambda \cdot) \mid \lambda \in \mathbf{R}\}$. Deduce from

$$\operatorname{tr} \pi_\lambda(f) = \int_{\mathbf{R}} f(y) e(\lambda y) \, dy = \widehat{f}(-\lambda),$$

that the Selberg trace formula is nothing other than the Poisson summation formula (for f).

Exercise 5.31 Show that the Selberg trace formula is valid for every admissible pair (h, g).

Exercise 5.32 Let G be a finite group and Γ any subgroup of G. Let $\pi \in \widehat{G}$ be arbitrary and define $f(x) = \operatorname{tr} \pi(x)$. The group G equipped with the discrete topology is a locally compact group such that the quotient $\Gamma \backslash G$ is compact, and we can take as an invariant measure on G the counting measure.

1. Mimicking the calculations for the trace of the operator $R(f)$, show that

$$\operatorname{tr} \pi(f) = \operatorname{tr} \left(\frac{1}{|G|} \sum_{x \in G} f(x) \pi(x) \right).$$

Using the orthogonality of characters deduce from the trace formula that

$$m_\pi = \frac{1}{|\Gamma|} \sum_{\gamma \in \Gamma} \operatorname{tr} \pi(\gamma).$$

2. Deduce the following particular case of *Frobenius reciprocity*:

$$\dim_{\mathbf{C}} \operatorname{Hom}_\Gamma(1_\Gamma, \pi_{|\Gamma}) = \dim_{\mathbf{C}} \operatorname{Hom}(\operatorname{Ind}_\Gamma^G(1_\Gamma), \pi).$$

Here 1_Γ denotes the trivial representation (of dimension 1) of Γ and $\operatorname{Ind}_\Gamma^G(1_\Gamma)$ is the representation $(g \cdot f)(x) = f(g^{-1}x)$ in the space of function $G/\Gamma \to \mathbf{C}$.

Exercise 5.33 (Elliptic elements contribution) Let Γ be a discrete cocompact subgroup of $SL(2, \mathbf{R})$ and let γ be a non-trivial elliptic element in Γ.

1. Show that there exists a unique primitive elliptic element $\delta \in \Gamma$ such that $\gamma = \delta^{\ell}$ for a certain integer $1 \leqslant \ell \leqslant m$, where m is the order of δ.
2. Show that δ generates the centralizer of γ in Γ.
3. Show that, after conjugating δ in G if necessary, one can assume that $\delta = k(\theta)$ with $\theta = \frac{\pi}{m}$. Deduce that the (hyperbolic) angular sector S of angle 2θ and vertex $i \in \mathcal{H}$ is a fundamental domain for the action of the centralizer of γ on $\mathcal{H}$.
4. Show that the contribution of the conjugacy class of $\gamma = \delta^{\ell}$ to the trace formula is

$$\int_S k(z, k(\theta\ell)z)d\mu(z) = \frac{1}{m}\int_{\mathcal{H}} k(z, k(\theta\ell)z)d\mu(z).$$

5. Show that this last expression is equal to

$$\frac{\pi}{m}\int_0^{+\infty} k\left(u\sin^2\frac{\pi\ell}{m}\right)(u+1)^{-1/2}du.$$

6. Finally, show that this integral can be simply expressed in terms of the function g as

$$\frac{1}{m}\int_0^{+\infty} \frac{g(r)\cosh(r/2)}{\cosh r - \cos(2\pi\ell/m)}dr.$$

Chapter 6
Multiplicity of λ_1 and the Selberg Conjecture

We fix positive integers a and b and denote by $\Gamma = \Gamma_{a,b}$ the group constructed in § 2.2. Given a prime p let $\Gamma(p)$ be the subgroup consisting of images of elements $x \in \mathcal{O}^1 = D_{a,b}(\mathbf{Z})^1$ such that $x - 1 \in pD_{a,b}(\mathbf{Z})$. The Selberg conjecture (conjecture 3.37) implies in particular that the first non-zero eigenvalue $\lambda_1(\Gamma(p)\backslash\mathcal{H})$ is greater than or equal to $1/4$. The goal of this chapter is to prove the following approximation to the Selberg conjecture.

Theorem 6.1 *If p is a sufficiently large prime,*

$$\lambda_1(\Gamma(p)\backslash\mathcal{H}) \geqslant \min\left(\frac{5}{36}, \lambda_1(\Gamma\backslash\mathcal{H})\right).$$

The idea of the proof is provide upper and lower bounds on the multiplicity of $\lambda_1(\Gamma(p)\backslash\mathcal{H})$ by two different methods so as to deduce a contradiction whenever p is large and λ_1 is smaller than $5/36$ and at the bottom of the spectrum of Γ. We obtain the upper bound with the aid of the trace formula; this is the subject of the first two sections. The lower bound comes from exploiting the fact that the Galois groups of the congruence covers $\Gamma/\Gamma(p)$ are isomorphic to $\mathrm{PSL}(2, \mathbf{Z}/p\mathbf{Z})$. If λ_1 is smaller than the bottom of the spectrum of Γ, the group $\mathrm{PSL}(2, \mathbf{Z}/p\mathbf{Z})$ acts non-trivially on the corresponding eigenspace and the multiplicity is minorized by the dimension of a non-trivial representation of $\mathrm{PSL}(2, \mathbf{Z}/p\mathbf{Z})$. It remains to classify the irreducible representations of $\mathrm{PSL}(2, \mathbf{Z}/p\mathbf{Z})$.

6.1 Point Counting in Arithmetic Lattices

Given a prime p, we propose here to study the growth of the number of points in the orbit of $i \in \mathcal{H}$ under the action of the group $\Gamma(p)$ in hyperbolic balls as both the radius and the prime p tend toward infinity. We begin by remarking that a simple

© Springer International Publishing Switzerland 2016
N. Bergeron, *The Spectrum of Hyperbolic Surfaces*, Universitext,
DOI 10.1007/978-3-319-27666-3_6

calculation using the formula

$$\cosh \rho(z, w) = 1 + \frac{|z - w|^2}{2 \operatorname{Im} z \operatorname{Im} w}$$

implies that if

$$A = \begin{pmatrix} \alpha & \beta \\ \gamma & \delta \end{pmatrix} \in G = \mathrm{SL}(2, \mathbf{R}),$$

then

$$2 \cosh \rho(i, Ai) = \alpha^2 + \beta^2 + \gamma^2 + \delta^2 = \operatorname{tr}({}^t\!AA).$$

Write $\|.\|$ for the norm on G defined by $\|A\| = (\operatorname{tr}({}^t\!AA))^{1/2}$. We seek to bound

$$N(T, \Gamma(p)) := \sum_{\gamma \in \Gamma(p), \, \|\gamma\| \le T} 1$$

as $T \in \mathbf{R}$ and the prime p tend toward infinity.

Theorem 6.2 *For all real $\varepsilon > 0$ and for T and p sufficiently large, we have*

$$N(T, \Gamma(p)) = O_\varepsilon\Big(\frac{T^{2+\varepsilon}}{p^3} + \frac{T^{1+\varepsilon}}{p^2} + 1\Big).$$

Proof By definition of the group $\Gamma(p) = \Gamma_{a,b}(p)$, after replacing T by a multiple depending only on the constants a and b if necessary, we must bound the number of integers x_0, x_1, x_2, x_3 satisfying the following conditions:

$$|x_0|, |x_1|, |x_2|, |x_3| \le T, \tag{6.1}$$

$$x_0^2 - ax_1^2 - bx_2^2 + abx_3^2 = 1, \tag{6.2}$$

$$x_0 \equiv 1 \pmod{p}, \tag{6.3}$$

$$x_1 \equiv x_2 \equiv x_3 \equiv 0 \pmod{p}. \tag{6.4}$$

Conditions (6.2) and (6.4) imply that

$$x_0^2 \equiv 1 \pmod{p^2}.$$

Condition (6.3) implies that if $p \ne 2$ then

$$x_0 \equiv 1 \pmod{p^2}. \tag{6.5}$$

It is immediate that there exist $O\left(T/p^2\right)$ choices of $x_0 \neq 1$ satisfying (6.1) and (6.5). Then there exist $O\left(1 + T/p\right)$ choices of x_3 satisfying (6.1) and (6.4).

Finally, for every choice of $x_0 \neq 1$ and x_3 as above, we put $\xi = x_0^2 + abx_3^2 - 1$. Condition (6.1) implies that ξ is $O(T^2)$. To each pair of integers (x_1, x_2) satisfying $ax_1^2 + bx_2^2 = \xi$ we associate a factorization of the ideal (ξ) in the ring of integers of the number field $K = \mathbf{Q}[\sqrt{a}, \sqrt{b}, \sqrt{-1}]$:

$$(x_1\sqrt{a} + ix_2\sqrt{b})(x_1\sqrt{a} - ix_2\sqrt{b}) = (\xi) \tag{6.6}$$

– a product of two ideals. Lemma 6.3 which follows implies that there exists at most $O_\varepsilon(T^\varepsilon)$ possible such decompositions. Moreover, the map sending a pair (x_1, x_2) to the ideal $(x_1\sqrt{a} + ix_2\sqrt{b})$ is injective up to units. Consider therefore a unit u such that

$$x_1\sqrt{a} + ix_2\sqrt{b} = u(x_1'\sqrt{a} + ix_2'\sqrt{b})$$

with $|x_1|, |x_2|, |x_1'|, |x_2'| \leqslant T$. Then for every embedding $\sigma : K \to \mathbf{C}$ we have

$$\log|\sigma(u)| = O(\log T).$$

Now according to the Dirichlet unit theorem, the logarithms of the units form a lattice in a suitable Euclidean space. From this it follows that there are at most $O_\varepsilon(T^\varepsilon)$ units satisfying the above bounds. The number of pairs (x_1, x_2) is therefore $O_\varepsilon(T^\varepsilon)$ and we have

$$N(T, \Gamma(p)) = O_\varepsilon\left(\frac{T^{1+\varepsilon}}{p^2}\left(\frac{T^{1+\varepsilon}}{p} + 1\right) + 1\right)$$

$$= O_\varepsilon\left(\frac{T^{2+\varepsilon}}{p^3} + \frac{T^{1+\varepsilon}}{p^2} + 1\right). \qquad \square$$

Lemma 6.3 *The number of divisors of the ideal (ξ) is $O_\varepsilon(N(\xi)^\varepsilon)$ for every $\varepsilon > 0$. Here $N(\xi)$ denotes the norm of ξ.*

Proof Denote by d the function which to an ideal in the ring of integers of the number field K associates the number of its divisors. Let ε be a positive real number. We write f for the function which sends an ideal I to the number $d(I)/N(I)^\varepsilon$, where $N(I)$ is the norm of I. One easily sees that f is multiplicative on relatively prime ideals and, if $\mathfrak{p}$ is a prime ideal, then

$$f(\mathfrak{p}^m) = \frac{m+1}{N(\mathfrak{p})^{\varepsilon m}} \longrightarrow 0 \text{ as } N(\mathfrak{p})^m \longrightarrow +\infty.$$

In particular, there exists constants A and B such that

- $|f(\mathfrak{p}^m)| < A$ for every prime ideal $\mathfrak{p}$ and all integers m;
- $|f(\mathfrak{p}^m)| < 1$, for $N(\mathfrak{p}^m) > B$.

By decomposing an arbitrary ideal into its prime factors, we deduce that f is bounded.   □

6.2 Multiplicity of the First Eigenvalue

We let $m(\lambda_1, \Gamma(p))$ denote the multiplicity of the first non-zero eigenvalue of the Laplacian on $S_p := \Gamma(p)\backslash\mathcal{H}$. The following theorem allows one to bound $m(\lambda_1, \Gamma(p))$ as a function of p under the hypothesis that $\lambda_1(S_p) < 1/4$ for sufficiently large p.

Theorem 6.4 *Suppose that for sufficiently large p, one has $\lambda_1(S_p) \leqslant s(1-s)$ where $s \in (1/2, 1)$. Then for all real $\varepsilon > 0$ and for p sufficiently large we have*

$$m(\lambda_1, \Gamma(p)) = O_\varepsilon(p^{6(1-s)+\varepsilon}).$$

Proof for compact S_p Theorem 6.2 furnishes an upper bound for $N(T, \Gamma(p))$ for sufficiently large T and p. To prove Theorem 6.4 we link $N(T, \Gamma(p))$ to the spectrum of the hyperbolic Laplacian on S_p. For this we consider a smooth compactly supported point-pair invariant $k(z, w)$ with associated automorphic kernel $K(z, w)$. The surface S_p being compact, Eq. (5.7) and Remark 5.3 imply that the series

$$K(z, w) = \sum_j h(r_j)\varphi_j(z)\varphi_j(w)$$

converges in $L^2(S_p \times S_p)$, absolutely and uniformly on $S_p \times S_p$. Here the $\varphi_j \in L^2(\Gamma(p)\backslash\mathcal{H})$ form an orthonormal system of Laplacian eigenfunctions with corresponding eigenvalues $\lambda_j = 1/4 + r_j^2 = s_j(1 - s_j)$ and h denotes the Selberg transform of k. We deduce in particular that

$$\int_{S_p \times S_p} |K(z, w)|^2 d\mu(z)\, d\mu(w) = \sum_j |h(r_j)|^2 \tag{6.7}$$

$$\geqslant m(\lambda_1, \Gamma(p))|h(r_1)|^2.$$

We must now choose a good kernel k and apply formula (6.7). Consider then the kernel

$$k_R(z, w) = \chi_R(\rho(z, w)),$$

where R is a positive real number to be chosen a posteriori and χ_R is the characteristic function of the interval $[0, R]$. The kernel k_R is not C^∞, but this is not a problem here (we could instead approximate it by a smooth kernel). By approximating k_R by a compactly supported piecewise affine function equal to 1 on the interval $[0, R]$, we can show (see [63, (12.8)]) that[1]:

$$h_R(r) = \pi^{1/2} \frac{\Gamma(s - 1/2)}{\Gamma(s + 1)} e^{sR} + O(e^{R/2}), \tag{6.8}$$

as $R \to +\infty$ (where $\lambda = 1/4 + r^2 = s(1 - s)$ and $s \in (1/2, 1)$ as usual).

It is natural to introduce the kernel

$$\kappa_R(z, w) = \int_{\mathcal{H}} k_R(z, x) k_R(x, w)\, d\mu(x) \tag{6.9}$$

and the associated automorphic kernel

$$\mathcal{K}_R(z, w) = \sum_{\gamma \in \Gamma(p)} \kappa_R(z, \gamma w).$$

We thus have

$$\mathcal{K}_R(z, w) = \int_{S_p} K_R(z, x) K_R(x, w)\, d\mu(x)$$

and

$$\int_{S_p \times S_p} |K_R(z, w)|^2 d\mu(z)\, d\mu(w) = \int_{S_p} \mathcal{K}_R(z, z)\, d\mu(z)$$

$$= \sum_{\gamma \in \Gamma(p)} \sum_{\delta \in \Gamma(p) \backslash \Gamma} \int_{S_1} \kappa_R(\delta z, \gamma \delta z)\, d\mu(z)$$

$$= [\Gamma : \Gamma(p)] \sum_{\gamma \in \Gamma(p)} \int_{S_1} \kappa_R(z, \gamma z)\, d\mu(z),$$

---

[1] Note that the expression (3.19) immediately implies that there exists a constant $c \neq 0$ such that

$$h_R(r) = \left(\int_{B(i,R)} \omega_\lambda(i, z)\, d\mu(z) \right)^2 \sim c e^{sR},$$

as $R \to +\infty$. This asymptotic is sufficient for what follows.

since $\Gamma(p)$ is normal in Γ. Lemma 6.6 which follows and the compactness of S_1 allows one to estimate the last integral: for any real number $R > 0$ and for all $\gamma \in \Gamma$,

$$\int_{S_1} \kappa_R(z, \gamma z)\, d\mu(z) = \begin{cases} O(e^R e^{-\rho(i,\gamma i)/2}) & \text{if } \rho(i, \gamma i) \leqslant 2(R + \mathrm{diam}(S_1)) \\ 0 & \text{else.} \end{cases}$$

Let $\delta = 2\ \mathrm{diam}(S_1)$. We thus obtain

$$\int_{S_p \times S_p} |K_R(z, w)|^2 d\mu(z)\, d\mu(w)$$

$$= O\left([\Gamma : \Gamma(p)] e^R \sum_{\gamma \in \Gamma(p),\ \rho(i,\gamma i) \leqslant 2R+\delta} e^{-\rho(i,\gamma i)/2}\right). \qquad (6.10)$$

Since the constant δ depends only on S_1, an integration by parts allows one to deduce from Theorem 6.2 that for all $\varepsilon > 0$,

$$\sum_{\gamma \in \Gamma(p),\ \rho(i,\gamma i) \leqslant 2R+\delta} e^{-\rho(i,\gamma i)/2} = \int_0^{2R+\delta} e^{-t/2} d\left(\sum_{\gamma \in \Gamma(p),\ \rho(i,\gamma i) \leqslant t} 1\right)$$

$$= O\left(\int_0^{2R} e^{-t/2} N(e^{t/2}, \Gamma(p)) dt\right)$$

$$= O_\varepsilon\left(\int_0^{2R} \left(\frac{e^{(1+\varepsilon)t}}{p^3} + \frac{e^{(1+\varepsilon)t/2}}{p^2} + 1\right) e^{-t/2} dt\right)$$

$$= O_\varepsilon\left(\int_0^{2R} \left(\frac{e^{(1/2+\varepsilon)t}}{p^3} + \frac{e^{\varepsilon t/2}}{p^2} + e^{-t/2}\right) dt\right)$$

$$= O_\varepsilon\left(\frac{e^{(1+2\varepsilon)R}}{p^3} + \frac{e^{\varepsilon R}}{p^2} + 1\right),$$

for sufficiently large R. We deduce from (6.7), (6.8) and (6.10) that for all $\varepsilon > 0$,

$$\mathrm{m}(\lambda_1, \Gamma(p)) = O_\varepsilon\left([\Gamma(p) : \Gamma] e^{(1-2s)R}\left(\frac{e^{(1+2\varepsilon)R}}{p^3} + \frac{e^{\varepsilon R}}{p^2} + 1\right)\right), \qquad (6.11)$$

for sufficiently large R.

Lemma 6.5 *If p is sufficiently large, the quotient $\Gamma/\Gamma(p)$ is isomorphic to the finite group* $\mathrm{SL}(2, \mathbf{F}_p)$.

Proof We can suppose that a and b are prime to p. Thus

$$D_{a,b}(\mathbf{Z})/p D_{a,b}(\mathbf{Z}) = D_{a,b}(\mathbf{F}_p).$$

We admit the fact (see the references at the end of this chapter) that the equation

$$x_0^2 - ax_1^2 - bx_2^2 = 0$$

has a non-trivial solution modulo p. Exchanging the roles of a and b if necessary we can assume that there exist $x_0, x_1 \in \mathbf{F}_p$ such that

$$b = x_0^2 - ax_1^2$$

modulo p. Consider the map Φ from $D_{a,b}(\mathbf{F}_p)$ to $\mathcal{M}_2(\mathbf{F}_p)$ defined on basis elements of $D_{a,b}(\mathbf{F}_p)$ by

$$
1 \longmapsto \begin{pmatrix} 1 & 0 \\ 0 & 1 \end{pmatrix}, \quad i \longmapsto \begin{pmatrix} 0 & 1 \\ a & 0 \end{pmatrix},
$$
$$
j \longmapsto \begin{pmatrix} x_0 & -x_1 \\ ax_1 & -x_0 \end{pmatrix}, \quad k \longmapsto \begin{pmatrix} ax_1 & -x_0 \\ ax_0 & -ax_1 \end{pmatrix}.
$$
(6.12)

One can verify that $\Phi(i)^2 = \Phi(a)$, $\Phi(j)^2 = \Phi(b)$ and $\Phi(i)\Phi(j) = -\Phi(j)\Phi(i)$ and that the matrices in (6.12) are linearly independent mod p. The map Φ then induces an algebra isomorphism from $D_{a,b}(\mathbf{F}_p)$ onto $\mathcal{M}_2(\mathbf{F}_p)$ and therefore an isomorphism between $D_{a,b}(\mathbf{F}_p)^1$ and $\mathrm{SL}(2, \mathbf{F}_p)$.

Finally, as in the case of $\mathrm{SL}(2, \mathbf{Z})$, we can deduce from the approximation theorem [92, Th. 5.2.10] for quaternion algebras that the map $D_{a,b}(\mathbf{Z})^1 \to D_{a,b}(\mathbf{F}_p)^1$ induced by reduction mod p is surjective for sufficiently large p. The composition with Φ then identifies the quotient $\Gamma/\Gamma(p)$ with the group $\mathrm{SL}(2, \mathbf{F}_p)$. □

From now on we assume that p is sufficiently large for the conclusion of Lemma 6.5 to hold. In particular,

$$[\Gamma : \Gamma(p)] = p^3(1 - p^{-2})$$

and (6.11) can be rewritten as

$$\mathrm{m}(\lambda_1, \Gamma(p)) = O_\varepsilon \left(e^{(1-2s)R}(e^{(1+2\varepsilon)R} + pe^{\varepsilon 2R} + p^3) \right). \tag{6.13}$$

Finally, taking R such that $e^R = p^3$, we obtain the claimed upper bound. □

Lemma 6.6 *Fix $R > 0$. Then for $z, w \in \mathcal{H}$,*

$$|\kappa_R(z, w)| \leqslant 4\pi e^{R - \rho(z,w)/2},$$

as long as $\rho(z, w) \leqslant 2R$; otherwise this expression is equal to 0.

Proof It follows immediately from the definition of κ_R that

$$\kappa_R(z, w) = \mathrm{area}(B(z, R) \cap B(w, R)),$$

where we consider the hyperbolic area and $B(z, R)$ denotes the hyperbolic ball of center z and hyperbolic radius R. In particular, it is clear that $F_R(z, w) = 0$ if $\rho(z, w) > 2R$. Now suppose that $\rho(z, w) \leqslant 2R$ and write $E = B(z, R) \cap B(w, R)$. We can assume that $z = i$ and that w lies on the vertical half-geodesic going from i to infinity, so that it suffices to study the function $\kappa_R(\rho) = \kappa_R(\rho(i, w))$. In polar coordinates about i relative to the upward pointing vector based at i, denote by $c(r, \theta)$ the distance of the point with coordinates (r, θ) to the point with coordinates $(\rho, 0)$. Hyperbolic trigonometry (see (1.15)) states that

$$\cosh c(r, \theta) = \cosh \rho \cosh r - \cos \theta \sinh \rho \sinh r.$$

Consider the angle $\alpha \in [0, \pi/2]$ defined by the relation

$$\cos \alpha = \frac{\cosh \rho \cosh R - \cosh R}{\sinh \rho \sinh R}$$

$$= \frac{\cosh R(\cosh \rho - 1)}{\sinh \rho \sinh R} = \frac{\tanh(\rho/2)}{\tanh R}.$$

Denote by v the point with polar coordinates $(\rho/2, 0)$. For a point $p \in \partial B(i, R) \cap B(w, R)$ let (r, θ) be its polar coordinates. We have $c(r, \theta) \leqslant R$ and $\theta \in [-\alpha, \alpha]$. By considering the triangle with vertices i, v and (r, θ) – colored in grey in Fig. 6.1 – we find

$$\cosh \rho(v, p) = \cosh(\rho/2) \cosh R - \cos \theta \sinh(\rho/2) \sinh R$$

$$\leqslant \cosh(\rho/2) \cosh R - \cos \alpha \sinh(\rho/2) \sinh R$$

$$\leqslant \cosh(\rho/2) \cosh R - \left(\frac{\tanh(\rho/2)}{\tanh R} \right) \sinh(\rho/2) \sinh R$$

$$\leqslant \frac{\cosh R}{\cosh(\rho/2)}.$$

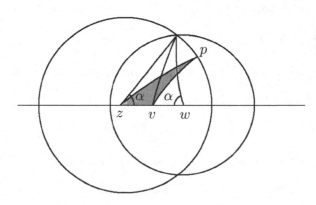

Fig. 6.1 The triangle (zpv)

By symmetry in z and w, we finally deduce that the set E is included in a hyperbolic ball of radius T (here $B(v, T)$) defined by the relation

$$\cosh T = \frac{\cosh R}{\cosh(\rho/2)}.$$

The area of such a ball is

$$4\pi \sinh^2(T/2) = 2\pi(\cosh T - 1) = 2\pi \left(\frac{\cosh R}{\cosh(\rho/2)} - 1 \right) \leqslant 4\pi e^{R-\rho/2}.$$

This concludes the proof of Lemma 6.6. $\qquad\qquad\qquad\qquad\qquad\qquad\qquad\qquad\square$

In the above proof we assumed that S_p was compact. It remains then to treat the case where Γ is the modular group.

Proof when S_p is non-compact In the above proof we used the compactness of S on two occasions. We now show how to bypass this hypothesis in each of the two cases.

Let k be a point-pair invariant of the form κ_R as before. Let K again denote the associated automorphic kernel and h the Selberg transform of k. As in the compact case, the integral operator with kernel K is a positive operator on $L^2(\Gamma(p)\backslash\mathcal{H})$. The orthogonal projection onto the subspace $L^2(\Gamma(p)\backslash\mathcal{H})$ and orthogonal to the subspace generated by the Laplacian eigenfunctions with eigenvalue λ_1 in $\mathcal{C}(\Gamma(p)\backslash\mathcal{H})$ is also a positive operator. The operator obtained by composing these two commuting operators is therefore itself positive; it is moreover an integral operator with kernel

$$B(z, w) = K(z, w) - \sum_{\lambda_j=\lambda_1} h(\lambda_1)\varphi_j(z)\overline{\varphi_j(w)}.$$

By considering (for fixed z) the sequence of functions

$$f_n(w) = \begin{cases} 1 & \text{if } \rho(z, w) \leqslant 1/n, \\ 0 & \text{if } \rho(z, w) > 1/n, \end{cases}$$

we have

$$B(z, z) = \lim_{n\to+\infty} \langle Bf_n, f_n \rangle \geqslant 0.$$

In other words,

$$K(z, z) - \sum_{\lambda_j=\lambda_1} h(\lambda_1)|\varphi_j(z)|^2 \geqslant 0. \qquad\qquad (6.14)$$

Compared to the compact case, the situation is complicated by the presence of cusps: the integral $\int_{S_p} K(z,z)\,d\mu(z)$ is divergent. We again alleviate this problem by truncating a fundamental domain for $\Gamma(p)$ in $\mathcal{H}$.

The fundamental domain

$$D = \{z = x + iy \in \mathcal{H} \mid |x| \leqslant 1/2 \text{ and } |z| > 1\}$$

for Γ decomposes, for all $Y > 1$, as a union

$$D = D^Y \cup P(Y)$$

where $P(Y)$ is the fundamental domain of a cusp

$$P(Y) = \{z = x + iy \mid |x| \leqslant 1/2 \text{ and } y > Y\}$$

based on the horocycle

$$h_Y = \{z = x + iy \mid y = Y\}.$$

A cuspidal region subdivides into a *collar* of width $\log 2$ and the closure of a contiguous cuspidal region. More precisely (and again at the level of the fundamental domain),

$$P(Y) = C(Y) \cup \overline{P(2Y)},$$

where

$$C(Y) = \{z = x + iy \mid |x| \leqslant 1/2 \text{ and } Y < y < 2Y\}$$

is the subset of points of $P(Y)$ lying between the horocycles h_Y and h_{2Y} (see Fig. 6.2).

Fig. 6.2 The collar $C(Y)$

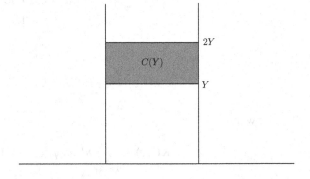

We obtain an analogous decomposition of $\Gamma(p)\backslash\mathcal{H}$ by considering the covering map $\Gamma(p)\backslash\mathcal{H} \to \Gamma\backslash\mathcal{H}$. Fix a real number Y strictly greater than 1. Let Ω_p be the compact subset of $\Gamma(p)\backslash\mathcal{H}$ obtained by taking the union of the heart – by which we mean the preimage of D^Y – in $\Gamma(p)\backslash\mathcal{H}$ and the collars of different cusps. Note that Ω_p depends on the choice of Y. We are going to compare the L^2-norm of a Laplacian eigenfunction on $\Gamma(p)\backslash\mathcal{H}$ with the L^2 norm of its restriction to Ω_p.

Let f then be a Laplacian eigenfunction in $\mathcal{C}(\Gamma(p)\backslash\mathcal{H})$ with eigenvalue λ. Let $P(Y)$ be the fundamental domain of a cusp that we'll take to be of the form

$$P(Y) = \{z = x + iy \mid 0 < x < q, \; y > Y\}$$

so that its collar is

$$C(Y) = \{x + iy \mid 0 < x < q, \; Y < y < 2Y\}.$$

Lemma 6.7 *Let $s_0 > 1/2$ be a real number. Then there exists a constant $C_{s_0} > 0$ such that for all $Y > 0$, $s \in (s_0, 1)$ and any function f as above with $\lambda = s(1 - s)$, we have*

$$\frac{\int_Y^{2Y} \int_0^q |f(z)|^2 d\mu(z)}{\int_{2Y}^{+\infty} \int_0^q |f(z)|^2 d\mu(z)} \geq C_{s_0}.$$

Proof Conjugating the group $\Gamma(p)$ in G if necessary, we can assume that $q = 1$. The function f then satisfies $f(z + 1) = f(z)$, and we can therefore expand it in its Fourier series

$$f(z) = \sum_n f_n(y)e(nx),$$

where

$$f_n(y) = \int_0^1 f(z)e(-nx).$$

Since f is C^∞, the Fourier series above converges absolutely and uniformly on compacta, and since $f \in \mathcal{C}(\Gamma(p)\backslash\mathcal{H})$, the constant term f_0 is constantly equal to 0.

Since f has eigenvalue λ, the function $F(y) = f_n(y/(2\pi n))$ is a solution to the differential equation (3.10). Since moreover f is cuspidal, it is bounded (by definition); in particular $f(z) = o(e^{2\pi y})$ as $y \to \infty$. It follows from the general solution of equation (3.10) that the function $z \mapsto f_n(y)e(nx)$ is a multiple of the Whittaker function $z \mapsto W_s(nz)$ (see (3.11)). We define $\hat{f}(n)$ by the identity

$$f_n(y)e(nx) = \hat{f}(n)W_s(nz), \quad \forall z = x + iy \in \mathcal{H}.$$

We then have the following expansion

$$f(z) = \sum_{n \neq 0} \hat{f}(n) W_s(nz).\tag{6.15}$$

Parseval's identity takes the form

$$\sum_{n \neq 0} |\hat{f}(n) W_s(iny)|^2 = \int_0^1 |f(x+iy)|^2 dx,$$

for all $y > 0$. Thus

$$\int_Y^\infty \int_0^1 |f(z)|^2 d\mu(z) = \sum_{n \neq 0} |\hat{f}(n)|^2 \int_Y^\infty |W_s(iny)|^2 y^{-2} dy.\tag{6.16}$$

We put $\nu = s - 1/2$. According to the expression (3.11), Lemma 6.7 is a consequence of the following result regarding Bessel functions, the proof of which we omit: for $Y > 0$ and $\nu \in (\nu_0, 1/2)$ we have

$$\frac{\int_Y^{2Y} (K_\nu^2(x)/x)\, dx}{\int_{2Y}^\infty (K_\nu^2(x)/x)\, dx} \geq C_{\nu_0} > 0.\tag{6.17}$$

(Note that the Bessel function K_ν is real for ν real.) $\square$

Let us come back to the proof of Theorem 6.4. As in the compact case one deduces from (6.8) that there exists a constant $c > 0$ such that

$$h_R(r_1) \geq c e^{2s_1 R}$$

as $R \to +\infty$. On integrating inequality (6.14) over Ω_p we get

$$\int_{\Omega_p} \sum_{\lambda_j = \lambda_1} |\varphi_j(z)|^2 d\mu(z) = O\left(e^{-2s_1 R} \int_{\Omega_p} K(z,z)\, d\mu(z)\right).$$

Swapping the sum and integral, we may apply Lemma 6.7 to the functions φ_j to arrive at (modifying the constant if necessary)

$$\int_{\Gamma(p)\backslash \mathcal{H}} \sum_{\lambda_j = \lambda_1} |\varphi_j(z)|^2 d\mu(z) = O\left(\int_{\Omega_p} \sum_{\lambda_j = \lambda_1} |\varphi_j(z)|^2 d\mu(z)\right).$$

Since

$$\mathrm{m}(\lambda_1, \Gamma(p)) = \int_{\Gamma(p)\backslash \mathcal{H}} \sum_{\lambda_j = \lambda_1} |\varphi_j(z)|^2 d\mu(z),$$

we have therefore shown that

$$\mathrm{m}(\lambda_1, \Gamma(p)) = O\left(e^{-2s_1 R} \int_{\Omega_p} K(z, z) \, d\mu(z)\right). \tag{6.18}$$

Finally, similarly as in the compact case but substituting the compactness of D^Y for that of S, one shows that

$$\int_{\Omega_p} K(z, z) \, d\mu(z) = O\left([\Gamma(p) : \Gamma] e^R \sum_{\gamma \in \Gamma(p), \, \rho(i, \gamma i) \leqslant 2R} e^{-\rho(i, \gamma i)/2}\right).$$

The conclusion of the proof now proceeds precisely as in the compact case.   □

6.3 Representation Theory of PSL(2, **Z**/p**Z**)

The goal of this section is to prove the following theorem.

Theorem 6.8 *Let $p \geqslant 5$ be a prime. The dimension of a non-trivial linear representation of* PSL(2, **Z**/p**Z**) *is* $\geqslant (p-1)/2$.

For the proof we shall use a few basic facts in the representation theory of finite groups. We shall thus begin by recalling the necessary results.

6.3.1 Review of the Representation Theory of Finite Groups

We start by recalling that a *representation* of a finite group G is a pair (π, V) where V is a complex vector space V and π is a homomorphism from G into GL(V). The dimension of the representation (π, V) is the dimension of V. We shall often simply write π for a representation of G. A subspace $W \subset V$ is called an *invariant subspace* if for all $g \in G$ one has $\pi(g)(W) = W$. A representation (π, V) with $V \neq \{0\}$ is said to be *irreducible* if it admits no non-trivial invariant subspace, i.e., no invariant subspace different from $\{0\}$ and V. Every representation (π, V) of G with $V \neq \{0\}$ is equivalent to a direct sum of irreducible representations of G.

Let (π, V) and (ρ, W) be two representations of G. Denote by $\mathrm{Hom}_G(\pi, \rho)$ the set of linear maps $T : V \to W$ which are G-equivariant:

$$T\pi(g) = \rho(g)T \quad (g \in G).$$

Then (π, V) and (ρ, W) are said to be *equivalent* if there exists an invertible linear map in $\mathrm{Hom}_G(\pi, \rho)$.

Lemma 6.9 (Schur's Lemma) *Let (π, V) and (ρ, W) be two finite-dimensional irreducible representations of G. Then*

$$\dim_{\mathbf{C}} \mathrm{Hom}_G(\pi, \rho) = \begin{cases} 0 & \text{if } \pi \text{ and } \rho \text{ are not equivalent;} \\ 1 & \text{if } \pi \text{ and } \rho \text{ are equivalent.} \end{cases}$$

Starting from two representations (π, V) and (ρ, W) we can naturally form the representations $\pi \oplus \rho$ in the vector space direct sum $V \oplus W$ and $\pi \otimes \rho$ in the tensor product $V \otimes W$. Moreover, given (π, V) we can form the dual representation π^* in the dual vector space V^* of V.

The *character* of a representation π is the function $\chi_\pi : G \to \mathbf{C}; g \mapsto \mathrm{tr}(\pi(g))$. That characters are useful can be seen in the following theorem which, together with Schur's lemma, shows that the characters of irreducible representations of G form an orthonormal system with respect to the Hermitian scalar product

$$\langle f_1, f_2 \rangle_G = \frac{1}{|G|} \sum_{g \in G} f_1(g) \overline{f_2(g)},$$

defined on the space of functions $G \to \mathbf{C}$.

Theorem 6.10 *Let (π, V) and (ρ, W) be two representations of the finite group G. Then*

$$\langle \chi_\rho, \chi_\pi \rangle_G = \dim_{\mathbf{C}} \mathrm{Hom}_G(\pi, \rho).$$

Remark 6.11 Note that it follows in particular from Theorem 6.10 and Schur's lemma that the representation (π, V) is irreducible if and only if $\langle \chi_\pi, \chi_\pi \rangle_G = 1$. We shall use Theorem 6.10 several times in this form.

The *left regular representation* λ_G of the group in the vector space $\mathbf{C}^G$ of functions $G \to \mathbf{C}$ is given by

$$(\lambda_G(g)f)(x) = f(g^{-1}x).$$

Its character χ_{λ_G} satisfies

$$\chi_{\lambda_G}(g) = \begin{cases} |G| & \text{if } g = 1 \\ 0 & \text{else.} \end{cases}$$

Theorem 6.10 and Schur's lemma then imply that an irreducible representation ρ of G appears $\dim \rho$ times in the decomposition of λ_G into irreducible subrepresentations. Thus

$$\chi_{\lambda_G} = \sum_\rho (\dim \rho) \chi_\rho,$$

where ρ runs through the set of equivalence classes of irreducible representations of G. Theorem 6.10 implies the *degree formula*

$$|G| = \sum_{\rho} (\dim \rho)^2,$$

where ρ runs through the set of equivalence classes of irreducible representations of G.

We shall now use these general facts to prove Theorem 6.8.

6.3.2 Proof of Theorem 6.8

Consider the affine group B on $\mathbf{F}_p = \mathbf{Z}/p\mathbf{Z}$, i.e., the group of transformations

$$z \longmapsto \alpha z + \beta \quad (\alpha \in \mathbf{F}_p^*, \ \beta \in \mathbf{F}_p)$$

of $\mathbf{F}_p$. The group B acts on the space W of complex functions on $\mathbf{F}_p$ by

$$(L(b)f)(z) = f(b^{-1}z).$$

Consider the subspace

$$W_0 = \left\{ f : \mathbf{F}_p \longrightarrow \mathbf{C} \mid \sum_{z \in \mathbf{F}_p} f(z) = 0 \right\}$$

of W. Then B preserves W_0; write L_0 for the representation of B on W_0.

Lemma 6.12 *The linear representation L_0 of B on W_0 is irreducible of dimension $p - 1$.*

Proof In the canonical basis of W, one calculates

$$\frac{1}{|B|} \sum_{b \in B} \chi_{W \otimes W}(b) = 2.$$

Thus

$$\langle \chi_W, \chi_W \rangle_B = \frac{1}{|B|} \sum_{b \in B} \chi_W(b)^2$$

$$= \frac{1}{|B|} \sum_{b \in B} \chi_{W \otimes W}(b)$$

$$= 2.$$

Now observe that the representation W of B decomposes into a direct sum of the trivial representation (in the subspace of constant functions) and the representation W_0. Hence

$$\chi_W = 1 + \chi_{W_0}$$

and

$$
\begin{aligned}
2 &= \langle 1 + \chi_{W_0}, 1 + \chi_{W_0} \rangle_B \\
&= 1 + 2 \langle 1, \chi_{W_0} \rangle_B + \langle \chi_{W_0}, \chi_{W_0} \rangle_B.
\end{aligned}
$$

This in turn forces

$$\langle \chi_{W_0}, \chi_{W_0} \rangle_B = 1$$

and implies (using Theorem 6.10 and Schur's lemma) that the representation W_0 (of dimension $p - 1$) is irreducible.   □

Let B_0 be the stabilizer in $\mathrm{PSL}(2, \mathbf{F}_p)$ of the line generated by the vector $(1, 0)$, i.e., the point at infinity in $\mathbf{P}^1(\mathbf{F}_p)$; the group B_0 can also be viewed as the image in $\mathrm{PSL}(2, \mathbf{F}_p)$ of the subgroup

$$\left\{ \begin{pmatrix} a & b \\ 0 & a^{-1} \end{pmatrix} \;\middle|\; a \in \mathbf{F}_p^*, \, b \in \mathbf{F}_p \right\}$$

of $\mathrm{SL}(2, \mathbf{F}_p)$. The action of B_0 on $\mathbf{F}_p$ is then given by

$$\varphi_A(z) = a^2 z + ab, \quad \text{where } A = \begin{pmatrix} a & b \\ 0 & a^{-1} \end{pmatrix}.$$

We may identify B_0 with an index 2 subgroup of B. Writing

$$\alpha : B \longrightarrow \mathbf{F}_p^*; \quad (z \longmapsto az + b) \longmapsto a,$$

we have $B_0 = \alpha^{-1}(\mathbf{F}_p^{*2})$, where $\mathbf{F}_p^{*2}$ denotes the subgroup of squares in $\mathbf{F}_p^*$.

Proposition 6.13 *Let p be an odd prime. There exists exactly $(p + 3)/2$ irreducible pairwise nonequivalent representations of B_0, consisting of*

- *$(p - 1)/2$ group homomorphisms $B_0 \to \mathbf{C}^*$, factoring through $\alpha_{|B_0}$,*
- *two nonequivalent representations ρ_1 and ρ_2, both of dimension $(p - 1)/2$.*

Proof Since $\mathbf{F}_p^{*2}$ is abelian of order $(p-1)/2$ it admits exactly $(p-1)/2$ irreducible representations, all of dimension 1, given by the homomorphisms $\chi_1, \ldots, \chi_{(p-1)/2}$

of $\mathbf{F}_p^{*2}$ into $\mathbf{C}^*$. Composing them with $\alpha_{|B_0}$ we obtain $(p-1)/2$ homomorphisms $\chi_1 \circ \alpha_{|B_0}, \dots, \chi_{(p-1)/2} \circ \alpha_{|B_0}$ from B_0 into $\mathbf{C}^*$.

Continuing, we restrict the representation L_0 to the group B_0 with the aim of showing that it breaks up into a direct sum of two irreducible nonequivalent representations ρ_1 and ρ_2, both of dimension $(p-1)/2$. We begin by recalling that the homomorphisms

$$e_c : \mathbf{F}_p \longrightarrow \mathbf{C}^*; \quad z \longmapsto e\,(cz/p) \quad (c \in \mathbf{F}_p)$$

form a basis of W. In particular, the e_c, where $c \in \mathbf{F}_p^*$, form a basis of the subspace W_0 of W. Note that if $g \in B$ is given by $g(z) = az + b$, then

$$(L_0(g)e_c)(z) = e_c(g^{-1}z) = e_c\left(\frac{z-b}{a}\right) = e\left(-\frac{cb}{ap}\right)e_{c/a}(z).$$

Denote by W_1 (resp. W_2) the subspace of W_0 generated by the e_c, with $c \in \mathbf{F}_p^{*2}$ (resp. $c \in \mathbf{F}_p^* - \mathbf{F}_p^{*2}$). The above formula implies that the subspaces W_1 and W_2 are left invariant by the restriction of the representation L_0 to the group B_0. Observe moreover that

$$\dim_{\mathbf{C}} W_1 = \dim_{\mathbf{C}} W_2 = \frac{p-1}{2}.$$

We shall show that the representations ρ_1 and ρ_2 of B_0 in W_1 and W_2 respectively are irreducible. For this we first notice that in the decomposition $W_0 = W_1 \oplus W_2$, we have

$$L_0(g) = \begin{cases} \begin{pmatrix} \rho_1(g) & 0 \\ 0 & \rho_2(g) \end{pmatrix} & \text{if } g \in B_0; \\ \begin{pmatrix} 0 & * \\ * & 0 \end{pmatrix} & \text{if } g \in B - B_0. \end{cases}$$

This implies the following identity on characters

$$\chi_{L_0}(g) = \begin{cases} \chi_{\rho_1}(g) + \chi_{\rho_2}(g) & \text{if } g \in B_0; \\ 0 & \text{if } g \in B - B_0. \end{cases}$$

According to Lemma 6.12 the representation is irreducible, we have (Theorem 6.10 and Schur's lemma):

$$1 = \langle \chi_{L_0}, \chi_{L_0} \rangle_B = \frac{1}{|B|} \sum_{g \in B} |\chi_{L_0}(g)|^2$$

$$= \frac{1}{2|B_0|} \sum_{g \in B_0} |\chi_{\rho_1}(g) + \chi_{\rho_2}(g)|^2$$

$$= \frac{1}{2} \langle \chi_{\rho_1} + \chi_{\rho_2}, \chi_{\rho_1} + \chi_{\rho_2} \rangle_{B_0}.$$

which implies (again using Theorem 6.10 and Schur's lemma) that ρ_1 and ρ_2 are irreducible and nonequivalent.

The degree formula

$$|B_0| = \frac{p(p-1)}{2} = \frac{p-1}{2} \cdot 1^2 + 2 \cdot \left(\frac{p-1}{2}\right)^2$$

finally implies that we have obtained all of the representations of B_0. This concludes the proof of Proposition 6.13. ∎

Proof of Theorem 6.8 Let π be a non-trivial representation of $\mathrm{PSL}(2, \mathbf{F}_p)$ on $\mathbf{C}^n$. The restriction $\pi_{|B_0}$ decomposes as a direct sum of irreducible representations of B_0. Since $p \geqslant 5$, the group $\mathrm{PSL}(2, \mathbf{F}_p)$ is simple. The representation $\pi_{|B_0}$ is therefore *faithful*, meaning that for every non-trivial element $b \in B_0$, the transformation $\pi(b)$ is different from the identity. But the $(p-1)/2$ representations of dimension 1 of B_0 are trivial on $\{z \mapsto z+b\} \subset B_0$, so that one of the representations ρ_1, ρ_2 must appear in $\pi_{|B_0}$. We deduce that n is greater than or equal to $(p-1)/2$. ∎

6.4   Lower Bound on the First Non-zero Eigenvalue

We can now finish the proof of the principal result of this chapter.

Proof of Theorem 6.1 We must show that if the first non-zero eigenvalue λ_1 of the Laplacian on $\Gamma(p)\backslash\mathcal{H}$ (for large p) is strictly smaller than $5/36$ then λ_1 is a Laplacian eigenvalue for $\Gamma\backslash\mathcal{H}$. Suppose then that $\lambda_1 = s_1(1-s_1)$ with $s_1 \in (5/6, 1)$.

According to Theorem 6.4, for any $\varepsilon > 0$ we have

$$\mathrm{m}(\lambda_1, \Gamma(p)) = O_\varepsilon\left(p^{6(1-s_1)+\varepsilon}\right).$$

On the other hand, since $\Gamma(p)$ is normal in Γ, we have a group action $\overline{\Gamma}/\overline{\Gamma(p)}$ on $\Gamma(p)\backslash\mathcal{H}$ by deck transformations on the cover $\Gamma(p)\backslash\mathcal{H} \rightarrow \Gamma\backslash\mathcal{H}$. From Lemma 6.5 it follows that for large enough p we have

$$\overline{\Gamma}/\overline{\Gamma(p)} \cong \mathrm{PSL}(2, \mathbf{F}_p).$$

The automorphisms of the cover $\Gamma(p)\backslash\mathcal{H} \rightarrow \Gamma\backslash\mathcal{H}$ then induce a group action $\mathrm{PSL}(2, \mathbf{F}_p)$ on the functions on $\Gamma(p)\backslash\mathcal{H}$. This action preserves the λ_1 eigenspace; we write $V(\lambda_1)$ for the corresponding (finite-dimensional) representation.

Arguing by contradiction, assume that λ_1 is not in the spectrum of $\Gamma\backslash\mathcal{H}$. The representation $V(\lambda_1)$ cannot contain the trivial representation and Theorem 6.8 then states that its dimension is greater than or equal to $(p-1)/2$ (as soon as p is larger than 5). For any positive ε we thus have

$$\frac{p-1}{2} \leqslant \mathrm{m}(\lambda_1, \Gamma(p)) = O_\varepsilon\big(p^{6(1-s_1)+\varepsilon}\big),$$

for large enough p. This contradicts our hypothesis, according to which s_1 is strictly larger than $5/6$.   □

6.5 Comments and References

The proof of Theorem 6.1 described in this chapter is based on an idea of Kazhdan, developed by Sarnak and Xue in [113].

§ 6.1

For the Dirichlet unit theorem the reader can consult for example [75].
The fact that the equation

$$-ax_1^2 - bx_2^2 + abx_3^2 = 0$$

admits a non-trivial solution mod p when a and b are prime to p follows from the Chevalley-Warning theorem (see [118, Chap. I Th. 3]) according to which *every quadratic form in at least 3 variables over the finite field with p elements* $\mathbf{F}_p$ *has a non-trivial zero.*
The extension of the proof of Theorem 6.4 to the case of non-compact surfaces is non-trivial and was the topic of the thesis of Gamburd. We have followed his article [42]. In particular, the reader can find the proof of the lower bound (6.17) in [42, Claim 4.1].

§ 6.3

The proof of Theorem 6.8 is reproduced in the book of Davidoff, Sarnak and Valette [32]. All of the results from the representation theory of finite groups summarized in the first paragraph are proved in detail in [32]. The book of Serre [119] is the standard reference on the subject.
One can find a proof of the simplicity of the group $\mathrm{PSL}(2, \mathbf{F}_p)$ $(p \geqslant 5)$ in [96, Th. 4.1].

Chapter 7
L-Functions and the Selberg Conjecture

Let N be a square-free integer. Recall that

$$\Gamma_0(N) = \left\{ \begin{pmatrix} a & b \\ c & d \end{pmatrix} \in \mathrm{SL}(2, \mathbf{Z}) \ \middle| \ N | c \right\}.$$

The goal of this chapter is to prove, using methods in analytic number theory, the following approximation to the Selberg conjecture.

Theorem 7.1 *We have*

$$\lambda_1(\Gamma_0(N) \backslash \mathcal{H}) \geq \frac{5}{36}.$$

Let $\Gamma = \Gamma_0(N)$ and consider a Maaß cusp form $f \in \mathcal{C}(\Gamma \backslash \mathcal{H})$ such that

$$\Delta f = (1/4 - \nu^2) f \quad (\nu \in [0, 1/2) \cup i\mathbf{R}).$$

We must show that $\mathrm{Re}(\nu) \leq 1/3$. Since

$$\begin{pmatrix} 1 & 1 \\ 0 & 1 \end{pmatrix} \in \Gamma,$$

the function f is 1-periodic: $f(z + 1) = f(z)$. It can therefore be expanded in a Fourier series in the variable $x = \mathrm{Re}(z)$, see (6.15):

$$f(z) = \sum_{r \in \mathbf{Z}, \, r \neq 0} a_r \sqrt{y} \, K_\nu(2\pi |r| y) e(rx). \tag{7.1}$$

(The coefficient a_0 is zero since f is cuspidal.)

© Springer International Publishing Switzerland 2016
N. Bergeron, *The Spectrum of Hyperbolic Surfaces*, Universitext,
DOI 10.1007/978-3-319-27666-3_7

The work of Maaß teaches us to think of the "Fourier coefficients" a_r of f as arithmetical quantities. This point of view is particularly justified by the behavior of the L-functions attached to them. The study of these L-functions is rather similar to that of the Riemann zeta function (see Exercise 1.16), the role of the theta function being played now by the Maaß form. The L-function, once completed, satisfies a functional equation and contains the "spectral information" v. We shall see that the Riemann hypothesis for the Rankin-Selberg L-function implies the Selberg theorem. The method then consists in proving a weak (average) version of the Riemann hypothesis in this special case.

7.1   The L-Function Attached to a Maaß Form

The antiholomorphic involution $\iota : x + iy \mapsto -x + iy$ induces an isometry of $\mathcal{H}$. Since it commutes with the Laplacian one can freely suppose that f is either *even* ($f \circ \iota = f$) or *odd* ($f \circ \iota = -f$). If f is even (resp. odd), its Fourier coefficients satisfy $a_r = a_{-r}$ (resp. $a_r = -a_{-r}$); we shall then write $\varepsilon = 0$ (resp. 1). In either case, one defines the L-function associated with f by

$$L(s,f) = \sum_{n=1}^{+\infty} \frac{a_n}{n^s}. \tag{7.2}$$

The following lemma is due to Hecke.

Lemma 7.2 *The Fourier coefficients of f satisfy*

$$a_n = O(n^{1/2}),$$

for large enough n. In particular, the series (7.2) is absolutely convergent in the half-plane $\mathrm{Re}(s) > 3/2$.

Proof Since f is cuspidal, it is bounded by a certain constant C. For all $y > 0$, we thus have

$$\left| \int_0^1 f(x + iy)e(-nx)\, dx \right| \leq \int_0^1 |f(x + iy)|dx \leq C.$$

But

$$\int_0^1 f(x + iy)e(-nx)\, dx = a_n \sqrt{y}\, K_v(2\pi |n|y)$$

and the lemma is proved by taking $y = 1/n$.   □

On average one can prove the following bound.

Lemma 7.3 *For x sufficiently large we have*

$$\sum_{1 \leqslant m \leqslant x} |a_m|^2 = O_f(x).$$

Proof Since *f* is a cusp form, its Fourier coefficient a_0 is zero and Parseval's identity implies

$$2 \sum_{1 \leqslant m \leqslant x} |a_m|^2 y |K_\nu(2\pi m y)|^2 \leqslant \int_0^1 |f(x + iy)|^2 dx.$$

We deduce from this that for any positive real number *Y* we have

$$2 \sum_{1 \leqslant m \leqslant x} |a_m|^2 \int_Y^{+\infty} |K_\nu(2\pi m y)|^2 \frac{dy}{y} \leqslant \int_Y^{+\infty} \int_0^1 |f(z)|^2 d\mu(z). \tag{7.3}$$

Since *f* is bounded,

$$\int_Y^{+\infty} \int_0^1 |f(z)|^2 d\mu(z) = O(1/Y).$$

Taking $Y = 1/x$ in (7.3) we obtain the stated bound.   □

Remark 7.4 In Sect. 7.6 we prove a non-trivial bound on individual Fourier coefficients of Maaß cusp forms that cannot be deduced from the above lemma. See Theorem 7.42.

The element

$$\sigma = \begin{pmatrix} 0 & -1/\sqrt{N} \\ \sqrt{N} & 0 \end{pmatrix} \in \mathrm{SL}(2, \mathbf{R})$$

normalizes $\Gamma = \Gamma_0(N)$. Define

$$\tilde{f}(z) = f(\sigma z);$$

this function again defines a Maaß cusp form such that $\Delta \tilde{f} = (1/4 - \nu^2)\tilde{f}$.

Proposition 7.5 *Let*

$$\Lambda(f, s) = \pi^{-s} \Gamma\left(\frac{s + \varepsilon + \nu}{2}\right) \Gamma\left(\frac{s + \varepsilon - \nu}{2}\right) L(f, s). \tag{7.4}$$

Then $\Lambda(f,s)$ extends to a holomorphic function on the entire complex plane **C**.
Moreover

$$\Lambda(f,s) = (-1)^\varepsilon N^{1/2-s}\Lambda(\tilde{f},1-s).$$

Proof We first prove Proposition 7.5 in the case where f is even ($\varepsilon = 0$). According
to Lemma 4.27, for Re(s) sufficiently large we have

$$\Lambda(s,f) = 2\int_0^{+\infty} f(iy)y^{s-1/2}\frac{dy}{y}.$$

Given a real number $\alpha > 0$, we break up this last integral so as to obtain

$$\begin{aligned}
\Lambda(s,f) &= 2\int_0^{\alpha} f(iy)y^{s-1/2}\frac{dy}{y} + 2\int_\alpha^{+\infty} f(iy)y^{s-1/2}\frac{dy}{y}\\
&= 2\int_{1/(N\alpha)}^{+\infty} f\left(\frac{i}{Nu}\right)(Nu)^{-s+1/2}\frac{du}{u} + 2\int_\alpha^{+\infty} f(iy)y^{s-1/2}\frac{dy}{y}\\
&= 2\int_{1/(N\alpha)}^{+\infty} \tilde{f}(iu)(Nu)^{-s+1/2}\frac{du}{u} + 2\int_\alpha^{+\infty} f(iy)y^{s-1/2}\frac{dy}{y}\\
&= 2N^{1/2-s}\int_{1/(N\alpha)}^{+\infty} \tilde{f}(iu)u^{-s+1/2}\frac{du}{u} + 2\int_\alpha^{+\infty} f(iy)y^{s-1/2}\frac{dy}{y}.
\end{aligned}$$

The analytic continuation of $\Lambda(s,f)$ comes as an immediate consequence; the
functional equation can be obtained by taking $\alpha = N^{-1/2}$.

If f is odd ($\varepsilon = 1$), note that

$$\Lambda(s,f) = \int_0^{+\infty} g(iy)y^{s+1/2}\frac{dy}{y},$$

where

$$g(z) = \frac{1}{4i\pi}\frac{\partial f}{\partial x}(z) = \sum_{n=1}^{+\infty} a_n n\sqrt{y}\,K_\nu(2i\pi ny)\cos(2\pi nx).$$

We have

$$g\left(i/Nu\right) = (-1)Nu^2\frac{\partial\tilde{f}}{\partial x}(iu),$$

and the functional equation can be deduced as above.   □

According to Maaß, the L-functions associated with Maaß forms, despite the
likely transcendental nature of their Fourier coefficients, should correspond to

"arithmetic" L-functions (associated, for example, with Hecke characters of real quadratic extensions of $\mathbf{Q}$, see § 4.5). We shall approach them in this light, keeping in mind the various properties (some conjectural) of the Riemann zeta function.

The Riemann hypothesis, generalized to $L(s,f)$, implies that the function $s \mapsto L(s,f)$ should never vanish on the half-plane $\mathrm{Re}(s) > 1/2$. But according to Proposition 7.5 the function $s \mapsto \Lambda(s,f)$ is everywhere holomorphic. The Γ-factors which arise in the definition of $\Lambda(s,f)$ should therefore not admit any poles in the half plane $\mathrm{Re}(s) > 1/2$. Assuming f even, and recalling that $\Delta f = (1/4 - \nu^2)f$, these Γ-factors are precisely

$$\Gamma\left(\frac{s-\nu}{2}\right) \quad \text{and} \quad \Gamma\left(\frac{s+\nu}{2}\right).$$

These have poles at ν and $-\nu$, respectively. Thus $|\nu| \leqslant 1/2$ for ν real. It follows that $\lambda_1(\Gamma_0(N)\backslash\mathcal{H})$ must be... greater than or equal to 0!

Admittedly, this is not a very interesting conclusion. It is tempting, though, to consider more general Dirichlet series[1] of the form

$$\sum_{n=1}^{+\infty} \frac{|a_n|^{2k}}{n^s}$$

for $k = 1, 2, \ldots$ to deduce better lower bounds on the first non-zero eigenvalue of the Laplacian. This is the approach we follow in this chapter.

We begin by noting that the "arithmetic" nature of the coefficients a_n is perhaps less surprising if we relate them to the Hecke operators (4.34).

7.2 Hecke Operators and Applications

7.2.1 Hecke Operators

The Hecke operators are associated with "correspondences" on arithmetic hyperbolic surfaces. We define a degree r *correspondence* C on a Riemannian manifold X as a map

$$C : X \longrightarrow (X \times \cdots \times X)/\mathfrak{S}_r$$
$$z \longmapsto \{z_1, \ldots, z_r\},$$

---

[1]This idea, due to Langlands, was used by Deligne in his proof [35] of the Weil conjectures.

where $\mathfrak{S}_r$ is the permutation group on r elements and each function $z \mapsto z_i(z)$ is a local isometry.[2]

Let $\Gamma \subset SL(2, \mathbf{R})$ be a Fuchsian group. Let $\alpha \in GL(2, \mathbf{R})$, having positive determinant, be such that $\Gamma_\alpha = \Gamma \cap \alpha^{-1} \Gamma \alpha$ is of finite index in Γ and $\alpha^{-1} \Gamma \alpha$. We associate with α the correspondence C_α on $\Gamma \backslash \mathcal{H}$ given by

$$C_\alpha : \Gamma z \longmapsto \{\Gamma \alpha \delta_1 z, \dots, \Gamma \alpha \delta_r z\},$$

where the $\delta_1, \dots, \delta_r$ form a set of representatives of the classes $\Gamma_\alpha \backslash \Gamma$ in such a way that

$$\Gamma \backslash \Gamma \alpha \Gamma = \{\Gamma \alpha \delta_1, \dots, \Gamma \alpha \delta_r\}.$$

To such a correspondence there corresponds a *Hecke operator*

$$T_\alpha : L^2(\Gamma \backslash \mathcal{H}) \longrightarrow L^2(\Gamma \backslash \mathcal{H}), \qquad T_\alpha f(z) = \sum_{j=1}^{r} f(\alpha \delta_j z).$$

The operator T_α is well-defined; moreover, since each $\alpha \delta_j$ acts by isometries on $\mathcal{H}$, the operator T_α commutes with the Laplacian. The T_α generate an algebra. When α belongs to Γ, the operator T_α is trivial. But the algebra of Hecke operators is infinite as soon as the *commensurator*

$$\mathrm{Com}(\Gamma) = \{\alpha \in SL(2, \mathbf{R}) \mid \Gamma_\alpha \text{ is of finite index in } \Gamma \text{ and } \alpha^{-1} \Gamma \alpha\}$$

is infinite modulo Γ.

Set $\Gamma = \Gamma_0(N)$ for an integer N greater than or equal to 1 and take $\alpha \in GL(2, \mathbf{Q})$ to have positive determinant. Then the group $\alpha^{-1} \Gamma \alpha$ contains a congruence subgroup $\Gamma(M)$ for some large enough integer M. From this one deduces that Γ_α is of finite index in Γ (and $\alpha^{-1} \Gamma \alpha$). For every integer $n \geq 1$ we consider the operator defined by

$$\sum_{\det \alpha = n} T_\alpha,$$

which clearly lies in the algebra generated by the Hecke operators of $\Gamma \backslash \mathcal{H}$. This operator can also be obtained from the set

$$H_n = \left\{ \begin{pmatrix} a & b \\ c & d \end{pmatrix} \,\middle|\, a, b, c, d \in \mathbf{Z}, \ ad - bc = n, \ N | c, \ (a, N) = 1 \right\}, \qquad (7.5)$$

---

[2]The functions z_j are not individually globally defined but the entire set of such z_j is.

for it is obvious that $\Gamma H_n = H_n \Gamma = H_n$ and, analogously to the quotient $\Gamma \backslash \Gamma \alpha \Gamma$ considered above, the quotient $\Gamma \backslash H_n$ defines a correspondence on $\Gamma \backslash \mathcal{H}$.

We leave as an exercise the verification that

$$H_n = \bigcup_{\substack{ad=n \\ (a,N)=1}} \bigcup_{b=0}^{d-1} \Gamma \begin{pmatrix} a & b \\ 0 & d \end{pmatrix} \quad \text{(disjoint union).} \tag{7.6}$$

The set H_n depends on N but if n and N are relatively prime the relation (7.6) is independent of N. In this case we can speak of "the" correspondence defined by H_n even while varying the level N. For notational convenience, we shall write H_n for the above set as well as the correspondence it defines.

Given an integer n greater than or equal to 1, we denote by T_n^N (or just T_n if the context is sufficiently clear or if $(n, N) = 1$) the operator associated with the modular correspondence determined by H_n and renormalized by $1/\sqrt{n}$. To simplify, we shall write $\frac{1}{\sqrt{n}} H_n$ for this latter correspondence. It suffices in fact to define the Hecke operators by the following formulas, see (4.34), and to study their action on the Fourier coefficients of Maaß forms:

$$(T_n^N f)(z) = \frac{1}{\sqrt{n}} \sum_{\substack{ad=n \\ (a,N)=1}} \sum_{b=0}^{d-1} f\left(\frac{az+b}{d}\right). \tag{7.7}$$

Given two positive integers m and n, the operators T_n^N and T_m^N commute.

We leave as an exercise the verification that

$$T_n^N \circ T_m^N = \sum_{\substack{0<d|(n,m) \\ (d,N)=1}} T_{mn/d^2}^N. \tag{7.8}$$

Hecke introduced these formulas to explain – with the help of (7.8) – the multiplicative properties of the Fourier coefficients of certain modular forms. Observe that it follows from the expression $T_n = (1/\sqrt{n}) H_n$ that each T_n sends $L^2(\Gamma \backslash \mathcal{H})$ to itself and commutes with the Laplacian. The adjoint T_n^* of T_n is the operator associated with the correspondence $1/\sqrt{n}(H_n)^{-1}$. Now one sees from (7.5) that if n and N are relatively prime then $(H_n)^{-1} = n^{-1} H_n$ and thus $T_n^* = T_n$, i.e.,

$$\langle T_n f, g \rangle = \langle f, T_n g \rangle, \quad \text{if } (n, N) = 1.$$

We can therefore choose an orthonormal basis $\{f_j(z)\}$ of the space of cusp forms $\mathcal{C}(\Gamma \backslash \mathcal{H})$ consisting of eigenfunctions of the T_n, $(n, N) = 1$. The other operators are more difficult to manipulate; we shall study them in the next section.

Lemma 7.6 *Let*

$$f(z) = \sum_{r \in \mathbf{Z}} a_r \sqrt{y} K_\nu(2\pi|r|y)e(rx)$$

be a Maaß cusp form ($a_0 = 0$). Then, for all $n \geq 1$,

$$(T_n^N f)(z) = \sum_{r \in \mathbf{Z}} \left(\sum_{\substack{a|(n,r) \\ (a,N)=1 \\ a \geq 1}} a_{rn/a^2} \right) \sqrt{y} K_\nu(2\pi|r|y)e(rx).$$

Proof Let n be an integer ≥ 1. Then

$$(T_n f)(z) = \frac{1}{\sqrt{n}} \sum_{\substack{ad=n \\ (a,N)=1}} \sum_{b=0}^{d-1} f\left(\frac{az+b}{d}\right)$$

$$= \sum_{r \in \mathbf{Z}} \sum_{\substack{ad=n \\ (a,N)=1 \\ d \geq 1}} a_r \sqrt{y} K_\nu(2\pi|r|ay/d)e(rax/d) \left(\frac{1}{d} \sum_{b=0}^{d-1} e(rb/d)\right).$$

The last sum is trivial except when d divides r in which case it equals 1. Note finally that $d/a = n/a^2$, from which the lemma follows directly.   □

Let f be a non-zero Maaß cusp form, even or odd, and an eigenfunction of the Hecke operators: $T_n f = \lambda(n)f$, $(n, N) = 1$. If its first Fourier coefficient a_1 is non-zero, by multiplying f by a non-zero scalar if necessary we can assume without loss of generality that $a_1 = 1$. It then follows from Lemma 7.6 that, for $(n, N) = 1$:

$$a_n = \lambda(n) \quad \text{and} \quad a_{-n} = (-1)^\varepsilon \lambda(n). \tag{7.9}$$

The expected multiplicative properties of the Fourier coefficients (at least when $(n, N) = 1$) can be read off from those of the Hecke operators, see (7.8) and (7.12). But it is possible, in general, that a_1 is zero. If this is the case, if h is a cusp form for a group $\Gamma_0(M) \supset \Gamma_0(N)$ with $M|N$, each Maaß form $f(z) = h(Dz)$, where $DM|N$, lies in $\mathcal{C}(\Gamma_0(N)\backslash\mathcal{H})$ and only the Fourier coefficients a_n, for $n \equiv 0 \pmod{D}$, can be non-zero. If $M < N$, such a Maaß form ($f(z) = h(Dz)$) is called an *oldform*.

In a certain sense the form $h(Dz)$ is linked more to the group $\Gamma_0(M)$ than the group $\Gamma_0(N)$. The idea of Atkin-Lehner theory, which we shall adapt to the context of Maaß forms in the next section, is that by "eliminating" these forms coming from smaller levels, the simultaneous diagonalization by *all* Hecke operators becomes possible; moreover, the common eigenspaces are of dimension 1.

From now on we write $\mathcal{C}_{\text{old}}(\Gamma_0(N)\backslash\mathcal{H})$ for the subspace of $\mathcal{C}(\Gamma_0(N)\backslash\mathcal{H})$ generated by oldforms and $\mathcal{C}_{\text{prim}}(\Gamma_0(N)\backslash\mathcal{H})$ its orthogonal complement. We call the latter the *space of newforms*. The Hecke operators T_n, for $(n, N) = 1$, obviously preserve the subspace $\mathcal{C}_{\text{old}}(\Gamma_0(N)\backslash\mathcal{H})$, since they commute with the operator $f(z) \mapsto f(Dz)$ for all $D|N$. They also preserve the subspace $\mathcal{C}_{\text{prim}}(\Gamma_0(N)\backslash\mathcal{H})$, since they are self-adjoint. There exists then a basis of Maaß forms in $\mathcal{C}_{\text{prim}}(\Gamma_0(N)\backslash\mathcal{H})$ consisting of eigenforms of the T_n, $(n, N) = 1$, even or odd; such a form is called a *newform*.

7.2.2 Atkin-Lehner Theory

The main interest in introducing these newforms is to prove the multiplicativity of the Fourier coefficients, which is based on the following proposition.

Proposition 7.7 *Let*

$$f(z) = \sum_{r \in \mathbf{Z}} a_r \sqrt{y}\, K_\nu(2\pi|r|y)e(rx)$$

be a non-zero Maaß cusp $(a_0 = 0)$ newform in $\mathcal{C}(\Gamma_0(N)\backslash\mathcal{H})$. Then $a_1 \neq 0$.

Proof Suppose the opposite holds. Thus we assume that

$$f(z) = \sum_{r \in \mathbf{Z}} a_r \sqrt{y}\, K_\nu(2\pi|r|y)e(rx)$$

is an even or odd non-zero Maaß cusp form $(a_0 = 0)$ which is a eigenfunction for the Hecke operators T_n, $(n, N) = 1$, and for which $a_1 = 0$. We shall show that f is a linear combination of oldforms.

Since $a_1 = 0$, it follows from Lemma 7.6 that for every integer n relatively prime to N the coefficient a_n is zero. The last of the next four lemmas (each an intermediate step in the proof) implies the existence, for every prime divisor p of N, of forms $f_p \in \mathcal{C}(\Gamma_0(N/p)\backslash\mathcal{H})$ such that

$$f(z) = \sum_{p|N} f_p(pz).$$

This concludes the proof of the proposition. $\qquad\square$

The following lemmas constitute the basic statements of Atkin-Lehner theory.

Lemma 7.8 *Let ℓ be an integer greater than or equal to 1 and f a 1-periodic C^∞-function on $\mathcal{H}$ such that $z \mapsto f(\ell z)$ defines a Maaß form in $\mathcal{C}(\Gamma_0(N)\backslash\mathcal{H})$. Then*

1. *if $N = D\ell$, the function f lies in $\mathcal{C}(\Gamma_0(D)\backslash\mathcal{H})$;*
2. *if $\ell \nmid N$, then f is zero.*

Proof Arguing inductively on the number of prime divisors of ℓ one can reduce to the case when ℓ is a prime number. We begin by showing that f lies in $\mathcal{C}(\Gamma_0(N)\backslash\mathcal{H})$. Put $g(z) = f(\ell z)$,

$$\delta_\ell = \begin{pmatrix} \ell & 0 \\ 0 & 1 \end{pmatrix} \quad \text{and} \quad \Gamma' = \left\{ \begin{pmatrix} a & b \\ c & d \end{pmatrix} \in \Gamma_0(N) \,\bigg|\, \ell|b \right\}.$$

Since $\delta_\ell^{-1}\Gamma'\delta_\ell \subset \Gamma_0(N)$, g is $\delta_\ell^{-1}\Gamma'\delta_\ell$-invariant and f is therefore Γ'-invariant. Now f is moreover invariant by $z \mapsto z + 1$ and so is necessarily invariant under the action of the whole group $\Gamma_0(N)$ (which is generated by Γ' and the matrix $\left(\begin{smallmatrix} 1 & 1 \\ 0 & 1 \end{smallmatrix}\right)$). Since g is cuspidal, the form f indeed lies in $\mathcal{C}(\Gamma_0(N)\backslash\mathcal{H})$.

Finally, the form f is invariant under the action of the groups $\Gamma_0(N)$ and $\delta_\ell\Gamma_0(N)\delta_\ell^{-1}$. According to whether ℓ divides N or not, the group generated by $\Gamma_0(N)$ and the matrix $\left(\begin{smallmatrix} 1 & 0 \\ N/\ell & 1 \end{smallmatrix}\right)$ is discrete or not. In the first case, the generated group is $\Gamma_0(N/\ell)$ and f then lies in $\mathcal{C}(\Gamma_0(D)\backslash\mathcal{H})$. In the second case the generated group is necessarily dense in $SL(2,\mathbf{R})$ and so f is zero. $\square$

Lemma 7.9 *Let*

$$f(z) = \sum_{r\in\mathbf{Z}} a_r \sqrt{y}\, K_\nu(2\pi|r|y)e(rx)$$

be a Maaß form in $\mathcal{C}(\Gamma_0(N)\backslash\mathcal{H})$ and p a prime number. If

$$g(z) = \sum_{(r,p)=1} a_r \sqrt{y}\, K_\nu(2\pi|r|y)e(rx)$$

then for p dividing N the form g is in $\mathcal{C}(\Gamma_0(Np)\backslash\mathcal{H})$ and otherwise g is in $\mathcal{C}(\Gamma_0(Np^2)\backslash\mathcal{H})$.

Proof Replacing N by pN if necessary, we can assume that p divides N. It then follows from Lemma 7.6 that

$$(T_p f)(z) = \sum_{r\in\mathbf{Z}} a_{rp} \sqrt{y}\, K_\nu(2\pi|r|y)e(rx).$$

But since $T_p f$ belongs to $\mathcal{C}(\Gamma_0(N)\backslash\mathcal{H})$, the form

$$(T_p f)(pz) = \sum_{r\in\mathbf{Z}} a_{rp} \sqrt{y}\, K_\nu(2\pi|rp|y)e(rpx)$$

lies in $\mathcal{C}(\Gamma_0(Np)\backslash\mathcal{H})$. From this one deduces that

$$g(z) = f(z) - (T_p f)(pz)$$

lies in $\mathcal{C}(\Gamma_0(Np)\backslash\mathcal{H})$. $\square$

Lemma 7.10 *Let ℓ be a squarefree integer strictly greater than 1 and*

$$f(z) = \sum_{r \in \mathbf{Z}} a_r \sqrt{y} \, K_\nu(2\pi |r| y) e(rx)$$

a Maaß form in $\mathcal{C}(\Gamma_0(N)\backslash\mathcal{H})$. If for every integer r prime to ℓ the coefficient a_r is zero then

$$f(z) = \sum_{p|\ell} g_p(pz),$$

where the sum runs over all prime divisors of ℓ and $g_p \in \mathcal{C}(\Gamma_0(N\ell^2)\backslash\mathcal{H})$ ($g_p \in \mathcal{C}(\Gamma_0(N\ell)\backslash\mathcal{H})$, if $\ell|N$).

Proof The proof is by induction on the number of prime factors of ℓ. Suppose first that ℓ is prime. The function $g(z) = f(z/\ell)$ then satisfies the conditions of Lemma 7.8, so that either g lies in $\mathcal{C}(\Gamma_0(N/\ell)\backslash\mathcal{H})$ or f and g are both zero, according to whether ℓ divides N or not. The last case in degenerate. In the first case, we have a fortiori that g lies in $\mathcal{C}(\Gamma_0(N\ell)\backslash\mathcal{H})$ and

$$f(z) = g(\ell z).$$

Suppose now that ℓ is a product of at least two prime numbers and that the conclusion of the lemma holds for all of the proper divisors of ℓ. Let p be a prime factor of ℓ and set $\ell' = \ell/p$. Define

$$h(z) = \sum_{(r,p)=1} a_r \sqrt{y} \, K_\nu(2\pi |r| y) e(rx).$$

According to Lemma 7.9, h lies in $\mathcal{C}(\Gamma_0(Np^2)\backslash\mathcal{H})$. We write

$$f(z) - h(z) = \sum_{r \in \mathbf{Z}} b_r \sqrt{y} \, K_\nu(2\pi |r| y) e(rx) \in \mathcal{C}(\Gamma_0(Np^2)\backslash\mathcal{H}).$$

If n and p are relatively prime, then the coefficient b_n is zero. The first step in our induction – applied with N replaced by Np^2 – then implies that

$$g_p(z) := f(z/p) - h(z/p)$$

belongs to $\mathcal{C}(\Gamma_0(Np)\backslash\mathcal{H})$. Thus

$$f(z) = g_p(pz) + h(z), \tag{7.10}$$

where h satisfies the conditions of the lemma with Np^2 and ℓ' in place of N and ℓ, respectively. By the induction hypothesis, we can find functions g_q in

$C(\Gamma_0(Np^2\ell'^2)\backslash\mathcal{H})$ for every prime factor ℓ' such that

$$h(z) = \sum_{q|\ell'} g_q(qz).$$

Since $Np^2\ell'^2 = N\ell^2$, if we insert this expression into (7.10) this proves the lemma for ℓ not dividing N. If ℓ does divide N, it follows immediately from the above proof that and Lemma 7.9 that one can take the $g_p \in C(\Gamma_0(N\ell)\backslash\mathcal{H})$, so that Lemma 7.10 is completely proved.   □

Lemma 7.11 *Let ℓ be an integer greater than or equal to 1 and*

$$f(z) = \sum_{r\in\mathbf{Z}} a_r \sqrt{y}\, K_\nu(2\pi|r|y)e(rx)$$

a Maaß form in $C(\Gamma_0(N)\backslash\mathcal{H})$. Suppose a_r is equal to 0 for every integer r relatively prime to ℓ.

1. *If $(\ell, N) = 1$, then f is zero.*
2. *If $(\ell, N) \neq 1$, there exists Maaß forms f_p in $C(\Gamma_0(N/p)\backslash\mathcal{H})$, where p runs through the set of prime factors of (ℓ, N), such that*

$$f(z) = \sum_{p|(\ell,N)} f_p(pz).$$

Proof As the hypothesis of the lemma depends only on the largest squarefree divisor of ℓ, we can assume ℓ to be squarefree. We proceed by induction on the number of prime factors of ℓ. If ℓ is prime, it suffices to apply Lemma 7.8 to $f(z/\ell)$. Suppose then that the result is true for all proper divisors of ℓ. Let p be a prime divisor of ℓ and set $\ell' = \ell/p$.

We write $f = g + h$ where

$$g(z) = \sum_{(r,\ell')=1} a_r \sqrt{y}\, K_\nu(2\pi|r|y)e(rx),$$

and

$$h(z) = \sum_{(r,\ell')\neq 1} a_r \sqrt{y}\, K_\nu(2\pi|r|y)e(rx).$$

According to Lemma 7.9 the form g belongs to $C(\Gamma_0(N\ell'^2)\backslash\mathcal{H})$ and thus so does h.

If p does not divide N then p does not divide $N\ell'^2$ either, and since the r-th Fourier coefficient of g is zero if r is prime to p, Lemma 7.8 implies that g is zero. Hence

$$f(z) = h(z) = \sum_{(r,\ell')\neq 1} a_r \sqrt{y} K_\nu(2\pi|r|y)e(rx)$$

satisfies the hypotheses of the lemma with ℓ replaced by ℓ'. The conclusion therefore holds by the induction hypothesis.

Now assume that p divides N. Let $g_p(z) = g(z/p)$. Lemma 7.8 implies that g_p lies in $\mathcal{C}(\Gamma_0(N\ell'^2/p)\backslash\mathcal{H})$. The quotient

$$\Gamma_0(N\ell'^2)\left(\begin{smallmatrix}1&0\\0&p\end{smallmatrix}\right)\backslash\Gamma_0(N\ell'^2)\left(\begin{smallmatrix}1&0\\0&p\end{smallmatrix}\right)\Gamma_0(N\ell'^2/p)$$

is finite; denote by $\gamma_1,\ldots,\gamma_d$ a system of representatives of $\Gamma_0(N\ell'^2/p)$. The double class

$$T = \Gamma_0(N\ell'^2)\left(\begin{smallmatrix}1&0\\0&p\end{smallmatrix}\right)\Gamma_0(N\ell'^2/p)$$

then operates on the $\Gamma_0(N\ell'^2)$-invariant function g by[3]:

$$(g|T)(z) = \frac{1}{\sqrt{p}}\sum_{i=1}^d g\left(\left(\begin{smallmatrix}1&0\\0&p\end{smallmatrix}\right)\gamma_i z\right) = \sum_{i=1}^d g_p(\gamma_i z).$$

Since g_p lies in $\mathcal{C}(\Gamma_0(N\ell'^2/p)\backslash\mathcal{H})$, we obtain

$$g_p(z) = \frac{\sqrt{p}}{d}(g|T)(z).$$

One can then define a function f_p by $f_p(z) = \frac{\sqrt{p}}{d}(f|T)(z)$. Since p and ℓ' are relatively prime we can replace T by

$$T' = \Gamma_0(N)\left(\begin{smallmatrix}1&0\\0&p\end{smallmatrix}\right)\Gamma_0(N/p)$$

in the expression for f_p. We deduce that f_p lies in $\mathcal{C}(\Gamma_0(N/p)\backslash\mathcal{H})$.

We have, by definition of g_p,

$$f(z) - f_p(pz) = f(z) - f_p(pz) - g(z) + g_p(pz) = h(z) - \frac{\sqrt{p}}{d}(h|T)(pz). \qquad (7.11)$$

---

[3]The factor $p^{-1/2}$ is there to simplify the expression for the action of T on the Fourier expansion, similarly to the appearance of $n^{-1/2}$ in the definition of T_n.

Since the r-th Fourier coefficient of h is zero if r is prime to ℓ', Lemma 7.10 allows us to write

$$h(z) = \sum_{q|\ell'} h_q(qz) \quad (h_q \in C(\Gamma_0(N\ell'^3)\backslash\mathcal{H})).$$

For every prime divisor q of ℓ', since p does not divide ℓ', we have $h|T = h|T_q$, where this time

$$T_q = \Gamma_0(N\ell'^3 q) \left(\begin{smallmatrix}1&0\\0&p\end{smallmatrix}\right) \Gamma_0(N\ell'^3 q/p).$$

It follows that

$$(h|T)(z) = \sum_{q|\ell'}(h_q|T'')(qz),$$

where

$$T'' = \Gamma_0(N\ell'^3) \left(\begin{smallmatrix}1&0\\0&p\end{smallmatrix}\right) \Gamma_0(N\ell'^3/p).$$

This implies in particular that the r-th Fourier coefficient of $h|T$ is zero if r and ℓ' are relatively prime. As this is also the case for h, the expression (7.11) implies that $f - f_p$ satisfies the hypotheses of the theorem with ℓ replaced by ℓ'. The induction hypothesis then implies that

$$f(z) - f_p(pz) = \sum_q f_q(qz) \quad (f_q \in C(\Gamma_0(N/q)\backslash\mathcal{H})),$$

where the sum runs over all prime factors q of ℓ'. This proves Lemma 7.11. $\square$

Thanks to the above, we can always normalize a cuspidal newform f so that $a_1 = 1$. In this case, we will say that f is *normalized*.

Theorem 7.12 (Multiplicity 1) *The space $C_{\mathrm{prim}}(\Gamma_0(N)\backslash\mathcal{H})$ admits a basis of normalized newforms. If $f \in C_{\mathrm{prim}}(\Gamma_0(N)\backslash\mathcal{H})$ and $f' \in C_{\mathrm{prim}}(\Gamma_0(N')\backslash\mathcal{H})$, with $N'|N$, are two newforms of the same parity having the same eigenvalues for the Laplacian and the Hecke operators T_n for all integers $(n, N) = 1$, then $N = N'$ and $f = f'$.*

Proof We begin by noting that if f and f' are two normalized newforms in $C_{\mathrm{prim}}(\Gamma_0(N)\backslash\mathcal{H})$, of the same parity and lying in the same eigenspaces for the Laplacian and the Hecke operators T_n, with $(n, N) = 1$, then Lemma 7.11(2) applied to $f - f'$ with $\ell = N$ implies that $f - f'$ is an oldform and thus zero, so that $f = f'$.

Having fixed some Laplacian eigenvalue λ there exists a finite number of normalized newforms in $C_{\mathrm{prim}}(\Gamma_0(N)\backslash\mathcal{H})$ having λ as their joint Laplacian eigenvalue. Moreover, it follows from the preceding paragraph that these forms are pairwise orthogonal. By induction on the level N we can then show that there exists an orthogonal basis of $C(\Gamma_0(N)\backslash\mathcal{H})$ made up of functions of the form $z \mapsto h(Dz)$, where h is a newform in $C_{\text{prim}}(\Gamma_0(M)\backslash\mathcal{H})$ and DM divides N.

The theorem then follows from the following lemma, which is of interest in and of itself. $\square$

Lemma 7.13 *Let $f \in C_{\text{prim}}(\Gamma_0(N)\backslash\mathcal{H})$ be a normalized newform and let $f' \in C(\Gamma_0(N)\backslash\mathcal{H})$, of the same parity as f, be an eigenfunction of the Laplacian and of the Hecke operators T_n, for all $(n, N) = 1$, with the same eigenvalues as f. Then f' is a multiple of f.*

Proof Expanding f' in an orthogonal basis and renormalizing if necessary, we can assume that $f'(z) = h(Dz)$ where h is a normalized newform in $C_{\text{prim}}(\Gamma_0(M)\backslash\mathcal{H})$, where DM divides N. In particular, the form h lies in $C(\Gamma_0(N)\backslash\mathcal{H})$ and has the same eigenvalues as f' for the Laplacian and the operators T_n, $(n, N) = 1$; it is an oldform if $M < N$. It then follows from Lemma 7.11 that $f - h$ lies in $C_{\text{old}}(\Gamma_0(N)\backslash\mathcal{H})$. Since f is a newform, the form h cannot be an oldform; we thus have $M = N$ (and $D = 1$) and the first part of the proof implies $f = h = f'$. $\square$

The next result we prove is that a newform, remarkably enough, is an eigenfunction for *all* of the Hecke operators, and not just the T_n for $(n, N) = 1$.

Proposition 7.14 *Let f be a non-zero cuspidal Maaß newform in $C_{\text{prim}}(\Gamma_0(N)\backslash\mathcal{H})$. Then f is an eigenfunction for all the Hecke operators.*

Proof Let m be a positive integer (not necessarily prime to N). Then $T_m f$ is of the same parity as f and is an eigenfunction for the Laplacian and the operators T_n, for all $(n, N) = 1$, with the same eigenvalues as f. The proposition then follows from the multiplicity 1 theorem. $\square$

7.2.3   Multiplicative Properties of Fourier Coefficients

It follows from (7.9) and Proposition 7.14 that the Fourier coefficients a_n of a normalized newform f are the eigenvalues of the Hecke operators[4] T_n. Lemma 7.6 and the relation (7.8) can then be used to show

$$a_m a_n = \sum_{d|(m,n)} a_{mn/d^2} \quad \text{if } (n, N) = 1$$

$$a_m a_p = a_{mp} \qquad\qquad \text{if } p|N. \tag{7.12}$$

---

[4]This way of thinking of the Fourier coefficients a_n allows one to generalize the work of Hecke, by dispensing with Fourier theory and hence the necessity of having a surface with cusps.

We return to the function $L(s,f)$ and, taking our cue from Maaß, think of it as an L-function coming from arithmetic (Riemann zeta function, Artin L-function...). The cusp form f plays the role, for this L-function, of the theta series $\theta(x) = \sum_{n\neq0} e^{-\pi n^2 x^2}$, for the Riemann zeta function (see Exercise 1.16). When f is a normalized newform we again have at our disposal an Euler product decomposition, as follows.

Theorem 7.15 *Let*

$$f(z) = \sum_{r\in\mathbf{Z}} a_r \sqrt{y}\, K_\nu(2\pi |r|y) e(rx)$$

be a cuspidal ($a_0 = 0$) *normalized Maaß newform in* $\mathcal{C}_{\mathrm{prim}}(\Gamma_0(N)\backslash\mathcal{H})$. *Then*

$$L(s,f) = \prod_{p|N}(1 - a_p p^{-s})^{-1} \cdot \prod_{p\nmid N}(1 - a_p p^{-s} + p^{-2s})^{-1},$$

in the domain of absolute convergence.

Proof It follows from the multiplicativity (7.12) of the coefficients that

$$L(s,f) = \sum_{n=1}^{+\infty} \frac{a_n}{n^s} = \prod_p \left(\sum_{r=0}^{+\infty} a_{p^r} p^{-rs} \right).$$

It then remains to calculate the series arising from the right-hand side. Equation (7.12) implies, when $p \nmid N$, that

$$a_{p^{r+1}} - a_p a_{p^r} + a_{p^{r-1}} = 0$$

and thus that

$$(1 - a_p X + X^2)\left(\sum_{r=0}^{+\infty} a_{p^r} X^r \right) = 1.$$

In the same way, we find that

$$(1 - a_p X)\left(\sum_{r=0}^{+\infty} a_{p^r} X^r \right) = 1,$$

if $p|N$. Setting $X = p^{-s}$ in the last two equalities we obtain the claimed expression.
□

When f is a cuspidal normalized Maaß newform in $\mathcal{C}_{\text{prim}}(\Gamma_0(N)\backslash\mathcal{H})$, the functional equation of Proposition 7.5 takes on an even more satisfying form due to the following lemma.

Lemma 7.16 *Let f be a cuspidal normalized Maaß newform in $\mathcal{C}_{\text{prim}}(\Gamma_0(N)\backslash\mathcal{H})$. Then $\tilde{f}$ is a multiple of f by a complex number of modulus 1.*

Proof The matrix

$$W_N = \begin{pmatrix} 0 & -1 \\ N & 0 \end{pmatrix}$$

normalizes the group $\Gamma_0(N)$; in particular it sends $\mathcal{C}(\Gamma_0(N)\backslash\mathcal{H})$ into itself. The inverse of the correspondence H_n (see (7.5)) is

$$(H_n)^{-1} = (Nn)^{-1} W_N H_n W_N.$$

Writing $H_n = \bigcup_i \Gamma_0(N)\alpha_i$ as the disjoint union (7.6) we thereby obtain

$$(H_n)^{-1} = \bigcup_i (Nn)^{-1} \Gamma_0(N) W_N \alpha_i W_N$$

and the adjoint of the Hecke operator T_n is given by the formula

$$(T_n)^* = W_N \circ T_n \circ W_N.$$

But if n is prime to N, T_n is self-adjoint. The matrix W_N thus commutes with T_n and Lemma 7.13 implies that $\tilde{f} = f \circ W_N$ is a multiple of f. Since these two Maaß forms have the same L^2 norm, the multiplicative coefficient must have modulus 1. □

We shall need an analog of this lemma for a specific prime divisor of N. Let p be a prime dividing N and α the largest integer such that $p^\alpha | N$. Fix three integers x, y and z such that $p^{2\alpha}x - yNz = p^\alpha$. We write

$$W_p = {}^N W_p = \begin{pmatrix} p^\alpha x & y \\ Nz & p^\alpha \end{pmatrix}.$$

The matrix W_p normalizes $\Gamma_0(N)$ and therefore defines an operator from $\mathcal{C}(\Gamma_0(N)\backslash\mathcal{H})$ into itself.

Lemma 7.17 *Let f be a normalized cuspidal Maaß newform in $\mathcal{C}(\Gamma_0(N)\backslash\mathcal{H})$. Then*

$$f \circ W_p = \eta_p f$$

where $\eta_p \in \mathbf{C}$ is of modulus 1.

Proof This again follows from the fact that the operator W_p commutes with the Hecke operators T_n for $(n, N) = 1$.   □

7.3 Dirichlet Characters and Twisted Maaß Forms

7.3.1 Dirichlet Characters

Let q be a positive integer. Let χ_0 be a character of the finite multiplicative group $(\mathbf{Z}/q\mathbf{Z})^*$, i.e., a group homomorphism with values into the unit circle. A *mod q Dirichlet character* is a q-periodic map $\chi : \mathbf{Z} \to \mathbf{C}$ obtained from a character χ_0 as above by extension to $\mathbf{Z}$:

$$\chi(a) = \begin{cases} 0 & \text{if } (a, q) \neq 1, \\ \chi_0(a \bmod q) & \text{if } (a, q) = 1. \end{cases}$$

Note in particular that χ is multiplicative. If d divides q, the group $\mathbf{Z}/q\mathbf{Z}$ surjects onto $\mathbf{Z}/d\mathbf{Z}$ and therefore $(\mathbf{Z}/q\mathbf{Z})^*$ surjects onto $(\mathbf{Z}/d\mathbf{Z})^*$. The Dirichlet character χ is said to be *primitive* mod q if there is no proper divisor d of q such that χ_0 comes from a character of $(\mathbf{Z}/d\mathbf{Z})^*$.

A character mod a prime number is *primitive* as soon as it is non-trivial, in other words when it is not the character[5] χ_{triv} constantly equal to 1 on $(\mathbf{Z}/q\mathbf{Z})^*$.

To simplify the statements and proofs in the following subsections we shall suppose that q is prime and consider a non-trivial character χ mod q. We write $\tau(\chi) = \sum_{\ell=1}^{q} \chi(\ell) e(\ell/q)$ for the *Gauß sum* associated with χ.

Lemma 7.18 *For all integers $n \in \mathbf{Z}$,*

$$\tau(\overline{\chi}) \cdot \chi(n) = \sum_{\ell=1}^{q} \overline{\chi}(\ell) e(\ell n/q).$$

Proof We distinguish two cases according to whether n is prime to q or not. If n is prime to q, then $|\chi(n)| = 1$ and the change of variable $\ell \mapsto \ell n$ transforms the Gauß sum $\tau(\overline{\chi})$ into

$$\tau(\overline{\chi}) = \sum_{\ell=1}^{q} \overline{\chi}(\ell n) e(\ell n/q) = \overline{\chi}(n) \sum_{\ell=1}^{q} \overline{\chi}(\ell) e(\ell n/q),$$

which, multiplied by $\overline{\chi}(n)^{-1} = \chi(n)$, is the claimed identity.

---

[5]One should be careful here that the character χ_{triv} depends on q.

Suppose now that q divides n. Then $\chi(n) = 0$ and it suffices to show that the expression in the lemma vanishes identically. Write $n = qn_0$. We then have

$$\sum_{\ell=1}^{q} \overline{\chi}(\ell)e\,(\ell n/q) = \sum_{\ell=1}^{q} \overline{\chi}(\ell).$$

This last expression is indeed zero since χ is not the trivial character. $\qquad \square$

Recall that

$$\sum_{\nu} \nu(\ell) = \begin{cases} q-1 & \text{if } \ell \equiv 1 \pmod{q} \\ 0 & \text{else,} \end{cases} \tag{7.13}$$

where the sum runs over all characters ν mod q. Regarding the Gauß sums $\tau(\chi)$ and the *Jacobi sums*

$$J(\chi_1, \chi_2) = \sum_{\ell=1}^{q} \chi_1(\ell)\chi_2(1-\ell),$$

we shall need the following classical lemma.

Lemma 7.19 *We have*

1. $\tau(\chi_{\text{triv}}) = -1$, $J(\chi, \overline{\chi}) = -\chi(-1)$;
2. $\tau(\chi)\tau(\overline{\chi}) = \chi(-1)q$, $|\tau(\chi)| = \sqrt{q}$;
3. $J(\chi_1, \chi_2) = \tau(\chi_1)\tau(\chi_2)/\tau(\chi_1\chi_2)$, *if* $\chi_2 \neq \overline{\chi}_1$.

Proof One verifies immediately the first point. We then calculate

$$\tau(\chi_1)\tau(\chi_2) = \sum_{\ell=1}^{q}\sum_{k=1}^{q} \chi_1(\ell)\chi_2(k)e\,((k+\ell)/q)$$

$$= \sum_{n=1}^{q} e\,(n/q) \sum_{\ell=1}^{q} \chi_1(\ell)\chi_2(n-\ell)$$

$$= \sum_{n\in(\mathbf{Z}/q\mathbf{Z})^*} e\,(n/q)\,\chi_1(n)\chi_2(n) \sum_{\ell=1}^{q} \chi_1(n^{-1}\ell)\chi_2(1-n^{-1}\ell)$$

$$\qquad\qquad\qquad\qquad\qquad\qquad + \sum_{\ell=1}^{q} \chi_1(\ell)\chi_2(-\ell)$$

$$= \tau(\chi_1\chi_2)J(\chi_1, \chi_2) + \chi_1(-1)\sum_{\ell=1}^{q}(\chi_1\chi_2)(\ell).$$

The second sum in the last expression is zero unless $\chi_1\chi_2 = \chi_{\text{triv}}$, in other words when $\chi = \chi_1 = \overline{\chi}_2$. In this last case,

$$\tau(\chi)\tau(\overline{\chi}) = \tau(\chi_{\text{triv}})J(\chi,\overline{\chi}) + \chi(-1)(q-1)$$
$$= \chi(-1)q.$$

Finally, since $\overline{\tau(\chi)} = \chi(-1)\tau(\overline{\chi})$, the lemma is completely proved. $\square$

In the following section we are going to form the product – referred to as the Rankin-Selberg product – of a Maaß form f by the same form f twisted by a Dirichlet character. We must begin by extending the theory of Maaß forms – including the Eisenstein series – to the case of forms *twisted* by a character.

7.3.2 Maaß Forms Twisted by a Character

Let

$$f(z) = \sum_{r\in\mathbf{Z}} a_r \sqrt{y}\, K_\nu(2\pi|r|y)e(rx)$$

be a Maaß form in $\mathcal{C}(\Gamma_0(N)\backslash\mathcal{H})$. Let χ be a primitive Dirichlet character mod a prime q, where $q \nmid N$. We define the *twist* of f by the character χ by the following sum:

$$f_\chi(z) = \sum_{r\in\mathbf{Z}} a_r\chi(r)\sqrt{y}\, K_\nu(2\pi|r|y)e(rx).$$

Lemma 7.20 *For all* $\gamma \in \Gamma_0(Nq^2)$, *we have*

$$f_\chi(\gamma z) = \chi(\gamma)^2 f_\chi(z),$$

where, by abuse of notation, we write

$$\chi\begin{pmatrix} a & b \\ c & d \end{pmatrix} = \chi(d).$$

Proof According to Lemma 7.19, the Gauß sum $\tau(\overline{\chi})$ is non-zero. It then follows directly from Lemma 7.18 that

$$f_\chi(z) = \frac{1}{\tau(\overline{\chi})} \sum_{\ell=1}^{q} \overline{\chi}(\ell) \cdot f\left(z + \ell/q\right). \tag{7.14}$$

If $\gamma = \left(\begin{smallmatrix} * & * \\ * & d \end{smallmatrix}\right) \in \Gamma_0(Nq^2)$ then

$$\begin{pmatrix} 1 & \ell/q \\ 0 & 1 \end{pmatrix} \gamma \in \Gamma_0(N) \cdot \begin{pmatrix} 1 & d^2\ell/q \\ 0 & 1 \end{pmatrix}.$$

In other words, for all $z \in \mathcal{H}$ the points $\gamma(z) + \ell/q$ and $z + d^2\ell/q$ are congruent modulo $\Gamma_0(N)$. We deduce from this that

$$f_\chi(\gamma z) = \tau(\overline{\chi})^{-1} \sum_{\ell=1}^{q} \overline{\chi}(\ell) \cdot f\left(z + d^2\ell/q\right)$$

$$= \tau(\overline{\chi})^{-1} \sum_{\ell=1}^{q} \overline{\chi}(\ell d^2)\chi(d)^2 \cdot f\left(z + d^2\ell/q\right)$$

$$= \chi(d)^2 f_\chi(z). \qquad\qquad \square$$

Recall that $\tilde{f} = f \circ \sigma$ where $\sigma(z) = -1/Nz$.

Lemma 7.21 *We have*

$$f_\chi\left(-1/Nq^2 z\right) = \chi(-N)\frac{\tau(\chi)}{\tau(\overline{\chi})}\tilde{f}_{\overline{\chi}}(z).$$

Proof As in the proof of Lemma 7.20 we have

$$\tau(\overline{\chi})f_\chi\left(\begin{pmatrix} 0 & -1/Nq \\ q & 0 \end{pmatrix} z\right) = \sum_{\ell=1}^{q} \overline{\chi}(\ell)f\left(\begin{pmatrix} q & \ell \\ 0 & q \end{pmatrix}\begin{pmatrix} 0 & -1/Nq \\ q & 0 \end{pmatrix} z\right)$$

$$= \sum_{\ell=1}^{q} \overline{\chi}(\ell)\tilde{f}\left(\begin{pmatrix} 0 & 1 \\ -N & 0 \end{pmatrix}\begin{pmatrix} q & \ell \\ 0 & q \end{pmatrix}\begin{pmatrix} 0 & -1/Nq \\ q & 0 \end{pmatrix} z\right)$$

$$= \sum_{(\ell,q)=1} \overline{\chi}(\ell)\tilde{f}\left(\begin{pmatrix} q & -r \\ -N\ell & s \end{pmatrix}\begin{pmatrix} q & r \\ 0 & q \end{pmatrix} z\right)$$

$$\left(\begin{array}{l} \text{where } r = r(\ell) \text{ and } s = s(\ell) \text{ are such that } qs - rN\ell = 1 \\ \text{and thus } \overline{\chi}(\ell) = \chi(-N)\chi(r). \end{array}\right)$$

$$= \chi(-N) \sum_{(r,q)=1} \chi(r)\tilde{f}\left(\begin{pmatrix} q & -r \\ -N\ell & s \end{pmatrix}\begin{pmatrix} q & r \\ 0 & q \end{pmatrix} z\right)$$

$$= \chi(-N) \sum_{(r,q)=1} \chi(r)\tilde{f}\left(\begin{pmatrix} q & r \\ 0 & q \end{pmatrix} z\right) \ (\text{since } \tilde{f} \in \mathcal{C}(\Gamma_0(N)\backslash\mathcal{H}))$$

$$= \chi(-N)\tau(\chi)\tilde{f}_{\overline{\chi}}. \qquad\qquad \square$$

By averaging (7.14) with the help of (7.13) – and by keeping track of the trivial character χ_{triv} – we obtain the following useful lemma.

Lemma 7.22 *Let q be a prime and let u be an integer not divisible by q. Then*

$$f(z + u/q) = \frac{1}{q-1}\left\{\sum_v v(u)\tau(\bar{v})f_v(z) + (qf_q(z) - f(z))\right\},$$

where the sum runs over all the primitive characters mod q and

$$f_q(z) = \sum_{r\in\mathbf{Z}} a_{rq}\sqrt{y}\,K_v(2\pi|r|qy)e(rqx).$$

7.3.3 Twisted Eisenstein Series

Assume from now on that χ is *even*, i.e., $\chi(-1) = 1$ and that q is prime to N. The character χ extends to an imprimitive character mod Nq^2 that we shall denote χ'; it is given by

$$\chi'(a) = \begin{cases} \chi(a) & \text{if } (a, N) = 1, \\ 0 & \text{else.} \end{cases}$$

Let $z = x + iy \in \mathcal{H}$. Consider the *twisted Eisenstein series*

$$\begin{aligned}
E(z, s, \chi) &= \frac{1}{2} \sum_{(c,d)=1,\, Nq^2|c} \chi(d)\frac{y^s}{|cz + d|^{2s}} \\
&= \frac{1}{2} \sum_{\gamma\in\Gamma_\infty\backslash\Gamma_0(Nq^2)} \chi(\gamma)(\text{Im}(\gamma z))^s.
\end{aligned} \tag{7.15}$$

One passes from the first equality to the second by associating with each pair of relatively prime integers (c, d) the set[6] of matrices in $\Gamma_0(Nq^2)$ whose bottom row is (c, d). Each such matrix represents a unique class in $\Gamma_\infty\backslash\Gamma_0(Nq^2)$.

The proof of Lemma 4.1 implies the following lemma.

Lemma 7.23 *The series (7.15) is absolutely convergent on* $\text{Re}(s) > 1$ *and satisfies*

$$E(\gamma(z), s, \chi) = \chi(\gamma)E(z, s, \chi), \quad \text{for all } \gamma \in \Gamma_0(Nq^2).$$

---

[6]The coefficient $1/2$ again comes from the fact that the element $-I \in \text{SL}(2, \mathbf{Z})$ acts trivially on $\mathcal{H}$.

The fundamental property of these Eisenstein series is again the existence of an analytic continuation and a functional equation. The associated completed Eisenstein series are

$$\pi^{-s}\Gamma(s)L(2s,\chi')E(z,s,\chi)$$

$$= \pi^{-s}\Gamma(s)\frac{1}{2}\sum_{m,n\in\mathbf{Z}}\chi'(n)\frac{y^s}{|mNq^2z+n|^{2s}}$$

$$= \pi^{-s}\Gamma(s)\frac{1}{2}\sum_{d|N}\mu(d)\chi(d)\sum_{m,n\in\mathbf{Z}}\chi(n)\frac{y^s}{|mNq^2z+nd|^{2s}}$$

$$= \pi^{-s}\Gamma(s)\frac{1}{2}\sum_{d|N}\mu(d)\chi(d)d^{-s}\sum_{m,n\in\mathbf{Z}}\chi(n)\frac{(y/d)^s}{|mNq^2z/d+n|^{2s}}$$

$$= \sum_{d|N}\mu(d)\chi(d)d^{-s}E^*(z/d,s,\chi),$$

(7.16)

where μ is the Möbius function ($\mu(n)$ equals 0 if n has a square factor and equals $(-1)^r$ if n is the product of r distinct primes),

$$E^*(z,s,\chi) = \pi^{-s}\Gamma(s)\frac{1}{2}\sum_{m,n\in\mathbf{Z}}\chi(n)\frac{y^s}{|mNq^2z+n|^{2s}}$$ (7.17)

and $L(s,\chi')$ is the ordinary Dirichlet series

$$L(s,\chi') = \sum_{n\geq1}\frac{\chi'(n)}{n^s} = \prod_{p|N}\left(1-\frac{\chi(p)}{p^s}\right)L(s,\chi).$$

Theorem 7.24 *The function $E^*(z,s,\chi)$, defined by (7.17) for* Re$(s) > 1$, *holomorphically continues to all $s \in \mathbf{C}$. It moreover satisfies the functional equation*

$$E^*(z,s,\chi) = \tau(\chi)N^{1-2s}q^{2-5s}E^*(-1/N^2q^3z,1-s,\overline{\chi}),$$ (7.18)

remains bounded on vertical strips $\sigma_1 \leq$ Re$(s) \leq \sigma_2$ ($\sigma_1,\sigma_2 \in \mathbf{R}$) and satisfies

$$E(x+iy,s) - O(y^\sigma) \quad \text{as } y \longrightarrow \infty,$$ (7.19)

where $\sigma =$ Re(s).

Proof The proof is essentially the same as that of Theorem 4.2, only here we consider the twisted theta function

$$\Theta_\chi(t, z) = \sum_{(m,n)\in\mathbf{Z}^2} \chi(n) e^{-\pi|mz+n|^2 t/y},$$

where $t > 0$ and $z \in \mathcal{H}$.

Lemma 7.25 *The theta function* $\Theta_\chi(t, z)$ *satisfies the functional equation*

$$\Theta_\chi(t, z) = \frac{\tau(\chi)}{qt} \Theta_{\overline{\chi}}(1/qt, -q/z).$$

Proof To prove the lemma we will need to make use of a twisted Poisson summation formula. Let us show that for any rapidly decreasing function $f : \mathbf{R}^2 \to \mathbf{C}$ we have

$$\sum_{(m,n)\in\mathbf{Z}^2} \chi(n)f(m, n) = \frac{\tau(\chi)}{q} \sum_{(m,n)\in\mathbf{Z}^2} \overline{\chi}(n)\hat{f}(m, n/q). \tag{7.20}$$

To prove (7.20) note that according to Lemma 7.18,

$$\chi(n) = \frac{\tau(\chi)}{q} \sum_{\ell=1}^{q} \overline{\chi}(\ell) e\,(n\ell/q).$$

The left-hand side of (7.20) is therefore equal to the sum $\sum_{(m,n)\in\mathbf{Z}^2} f_1(m, n)$, where

$$f_1(u_1, u_2) = \frac{\tau(\chi)}{q} \sum_{\ell=1}^{q} \overline{\chi}(\ell) e\,(u_2\ell/q) f(u_1, u_2)$$

is a rapidly decreasing function to which we can apply the usual Poisson summation formula. One finds that

$$\hat{f}_1(\xi_1, \xi_2) = \frac{\tau(\chi)}{q} \sum_{\ell=1}^{q} \overline{\chi}(\ell)\hat{f}\,(\xi_1, \xi_2 + \ell/q).$$

The left-hand side of (7.20) is therefore equal to

$$\frac{\tau(\chi)}{q} \sum_{\ell=1}^{q} \sum_{(m,n)\in\mathbf{Z}^2} \overline{\chi}(\ell)\hat{f}\,(m, n + \ell/q).$$

Since $\chi(\ell) = \chi(qn+\ell)$, using the fact that $qn+\ell$ is a parametrization of the integers as ℓ runs through the set $\{1, \ldots, q\}$, we conclude that this last sum is indeed equal to the right-hand side of (7.20).

Recall now that if

$$f_{z,t}(u_1, u_2) = e^{-\pi|u_1 z + u_2|^2 t/y}$$

then

$$\widehat{f_{z,t}}(\xi_1, \xi_2) = t^{-1} e^{-\pi[(y\xi_2)^2 + (\xi_1 - x\xi_2)^2]/ty}.$$

The twisted Poisson summation formula then implies that

$$\Theta_\chi(t, z) = \frac{\tau(\chi)}{qt} \sum_{(m,n)\in\mathbf{Z}^2} \overline{\chi}(n) e^{-\pi[(ny)^2 + (qm - xn)^2]/tq^2 y}$$

$$= \frac{\tau(\chi)}{qt} \sum_{(m,n)\in\mathbf{Z}^2} \overline{\chi}(n) e^{-\pi|nz - qm|^2/tq^2 y}$$

$$= \frac{\tau(\chi)}{qt} \sum_{(m,n)\in\mathbf{Z}^2} \overline{\chi}(n) e^{-\pi|n - qmz^{-1}|^2 |z|^2/tq^2 y}$$

$$= \frac{\tau(\chi)}{qt} \Theta_{\overline{\chi}}(1/qt, -q/z),$$

where we have used the fact that the imaginary part of $-q/z$ is $qy/|z|^2$. $\qquad\qquad\square$

This time we get

$$E^*(z, s, \chi) = \frac{1}{2} \int_0^{+\infty} \Theta_\chi(Nq^2 t, Nq^2 z) t^s \frac{dt}{t}$$

$$= \frac{1}{2} \int_0^{1/Nq^2} \Theta_\chi(Nq^2 t, Nq^2 z) t^s \frac{dt}{t} + \frac{1}{2} \int_{1/Nq^2}^{+\infty} \Theta_\chi(Nq^2 t, Nq^2 z) t^s \frac{dt}{t}$$

$$= E_0^*(z, s, \chi) + E_1^*(z, s, \chi).$$

Since for every $z \in \mathcal{H}$ the function $t \mapsto \Theta_\chi(Nq^2 t, Nq^2 z)$ is of rapid decay, the function $E_1(z, s)$ is holomorphic on the entire complex s-plane and bounded on vertical strips. Lemma 7.25 and the change of variables $t \mapsto N^{-2} q^{-4} t^{-1}$ implies moreover that if $\mathrm{Re}(s) > 0$,

$$E_1^*(z, s, \chi) = \frac{\tau(\chi)}{2Nq^3} \int_{1/Nq^2}^{+\infty} \Theta_{\overline{\chi}}(1/Nq^3 t, -1/Nqz) t^{s-1} \frac{dt}{t}$$

$$= \frac{\tau(\chi) N^{1-2s}}{2q^{4s-1}} \int_0^{1/Nq^2} \Theta_{\overline{\chi}}(Nqt, -1/Nqz) t^{1-s} \frac{dt}{t}.$$

Finally, changing variables $t \mapsto t/q$ gives

$$E_1^*(z, s, \chi)$$

$$= \frac{\tau(\chi)N^{1-2s}}{2q^{5s-2}} \int_0^{1/Nq^3} \Theta_{\overline{\chi}}(Nq^2t, Nq^2(-1/N^2q^3z))t^{1-s}\frac{dt}{t}. \qquad (7.21)$$

This last expression implies that the function

$$s \longmapsto E_0^*(-1/N^2q^3z, s, \overline{\chi}) = \frac{1}{2}\int_0^{1/Nq^2} \Theta_{\overline{\chi}}(Nq^2t, Nq^2(-1/N^2q^3z))t^s\frac{dt}{t}$$

holomorphically continues[7] to all $s \in \mathbf{C}$. Replacing χ by $\overline{\chi}$ and z by $-1/N^2q^3z$ we find that the function $E_0^*(z, s, \chi)$, and thus $E^*(z, s, \chi)$, admits a holomorphic continuation to all $s \in \mathbf{C}$, and that

$$E^*(z, s, \chi) = \frac{1}{2}\int_0^{+\infty} \Theta_{\chi}(Nq^2t, Nq^2z)t^s\frac{dt}{t}$$

$$= \tau(\chi)N^{1-2s}q^{2-5s}\frac{1}{2}\int_0^{+\infty} \Theta_{\overline{\chi}}(Nq^2t, Nq^2(-1/N^2q^3z))t^{1-s}\frac{dt}{t}$$

$$= \tau(\chi)N^{1-2s}q^{2-5s}E^*(-1/N^2q^3z, 1-s, \overline{\chi}).$$

We deduce from this the stated functional equation and the fact that $E^*(z, s, \chi)$ is bounded in vertical strips.

The asymptotic (7.19) can be proved by calculating the Fourier expansion of $E^*(z, s, \chi)$ associated with the cusp at infinity. Using the expression

$$E^*(z, s, \chi) = \pi^{-s}\Gamma(s)L(2s, \chi)y^s + \sum_{m=1}^{+\infty}\sum_{n\in\mathbf{Z}}\chi(n)\frac{y^s}{|mNq^2z + n|^{2s}}$$

$$= \pi^{-s}\Gamma(s)L(2s, \chi)y^s$$

$$+ \left(\frac{y}{N^2q^4}\right)^s\sum_{m=1}^{+\infty}m^{-2s}\sum_{r=1}^{mNq^2}\chi(r)\sum_{d\in\mathbf{Z}}\left|d + z + \frac{r}{mNq^2}\right|^{-2s},$$

---

[7]The fact that one integrates only from 0 to $1/Nq^3$ in (7.21) is not a problem: the integral is defined for all $s \in \mathbf{C}$ away from 0.

the computation is similar to the one carried out for the classical Eisenstein series in §4.1. This time we use

$$\sum_{d \in \mathbf{Z}} \left| d + z + \frac{r}{mNq^2} \right|^{-2s} = 2\pi \frac{\Gamma(2s-1)}{\Gamma(s)^2} |2y|^{1-2s}$$

$$+ \frac{2\pi^s}{\Gamma(s)} \sum_{n \neq 0} \left| \frac{y}{n} \right|^{1/2-s} K_{s-1/2}(2\pi |n|y) \, e\big(n(x + r/mNq^2)\big).$$

The Fourier expansion of $E^*(z, s, \chi)$ has constant term

$$\pi^{-s} \Gamma(s) L(2s, \chi) y^s$$

and what remains is a sum of terms involving Bessel functions which decay rapidly in y; see Appendix B. The bound (7.19) follows immediately from these considerations.   □

We note that Theorem 7.24 remains true when $\chi = \chi_{\text{triv}}$ as long as we substitute 1 for $\tau(\chi)$.

7.4 Rankin-Selberg L-Functions

We fix two normalized cuspidal ($a_0 = b_0 = 0$) Maaß newforms

$$f(z) = \sum_{r \in \mathbf{Z}} a_r \sqrt{y} \, K_\nu(2\pi |r|y) e(rx) \quad \text{and} \quad g(z) = \sum_{r \in \mathbf{Z}} b_r \sqrt{y} \, K_{\nu'}(2\pi |r|y) e(rx)$$

of the same parity ε in $\mathcal{C}_{\text{prim}}(\Gamma_0(N) \backslash \mathcal{H})$. In particular, f (resp. g) is an eigenfunction for the Laplacian with eigenvalue $1/4 - \nu^2$ (resp. $1/4 - \nu'^2$). We fix furthermore an *even* Dirichlet character χ, i.e., $\chi(-1) = 1$, which we require to be primitive to a prime modulus q which does not divide N; to simplify further we shall assume that N is square-free.[8]

We now consider the *Rankin-Selberg L*-function $L(s, f_\chi \times g)$ defined by

$$L(s, f_\chi \times g) = L(2s, \chi'^2) \sum_{n=1}^{+\infty} \chi(n) \frac{a_n b_n}{n^s}. \tag{7.22}$$

---

[8]The hypotheses that f and g be of the same parity, that χ should be even, and that q should be prime are not necessary. They only serve to simplify the presentation.

It follows from Lemma 7.2 that this series converges absolutely in the half-plane $\mathrm{Re}(s) > 2$. Rankin and Selberg succeeded in showing that this series admits a meromorphic continuation to the complex s-plane and satisfies a certain functional equation. The proof is based on the study of the twisted Eisenstein series treated in the previous section.

Recall that one says a function $\phi : \mathcal{H} \to \mathbf{C}$ is of *rapid decay at infinity* if

$$\phi(x + iy) = O(y^{-A}) \quad \text{for all } A > 0 \text{ as } y \longrightarrow +\infty.$$

Since Bessel functions decay rapidly (see Appendix B), it follows from its Fourier expansion that a Maaß cusp form is of rapid decay at infinity; the same is therefore true of the product function $z \mapsto f_\chi(z)g(z)$ and it follows from Theorem 7.24 that the integral

$$\int_{\Gamma_0(Nq^2)\backslash\mathcal{H}} E(z, s, \chi^2) f_\chi(z) g(z)\, d\mu(z)$$

converges absolutely on $\mathrm{Re}(s) > 1$. It could happen that $\chi^2 = \chi_{\mathrm{triv}}$ but, as we remarked at the end of the previous section, this is not a problem.

The Rankin-Selberg method is based on the following proposition.

Proposition 7.26 *For* $\mathrm{Re}(s) > 1$ *we have*

$$\int_{\Gamma_0(Nq^2)\backslash\mathcal{H}} E(z, s, \chi^2) f_\chi(z) g(z)\, d\mu(z)$$

$$= (-1)^\varepsilon \frac{\pi^{-s}}{4} \Gamma(s)^{-1} \left(\prod \Gamma\left(\frac{s \pm \nu \pm \nu'}{2} \right) \right) \sum_{n=1}^{+\infty} \chi(n) \frac{a_n b_n}{n^s}.$$

(On the right-hand side, the product has four factors, corresponding to all possible combinations of signs.)

Proof Since $\mathrm{Re}(s) > 1$, we may unfold the integral to obtain

$$\int_{\Gamma_0(Nq^2)\backslash\mathcal{H}} E(z, s, \chi^2) f_\chi(z) g(z)\, d\mu(z)$$

$$= \frac{1}{2} \sum_{\gamma \in \Gamma_\infty \backslash \Gamma_0(Nq^2)} \int_{\Gamma_0(Nq^2)\backslash\mathcal{H}} \chi^2(\gamma)(\mathrm{Im}(\gamma z))^s f_\chi(z) g(z)\, d\mu(z)$$

$$= \frac{1}{2} \sum_{\gamma \in \Gamma_\infty \backslash \Gamma_0(Nq^2)} \int_{\Gamma_0(Nq^2)\backslash\mathcal{H}} (\mathrm{Im}(\gamma z))^s f_\chi(\gamma z) g(\gamma z)\, d\mu(z) \qquad (7.23)$$

$$= \int_0^{+\infty} \int_0^1 y^s f_\chi(x + iy) g(x + iy) y^{-1} dx \frac{dy}{y}.$$

(One goes from the second to the third equality by using Lemma 7.20.)

Replacing f_χ and g by their Fourier expansions, we get

$$\int_0^1 f_\chi(x+iy)g(x+iy)\,dx$$

$$= y \sum_{n,m\in\mathbf{Z}} \int_0^1 \chi(n)a_n b_m K_\nu(2\pi|n|y)K_{\nu'}(2\pi|m|y)e((n+m)x)\,dx$$

$$= y \sum_{n\in\mathbf{Z}} \chi(n)a_n b_{-n} K_\nu(2\pi|n|y)K_{\nu'}(2\pi|n|y).$$

From this it is clear that

$$\int_0^{+\infty}\int_0^1 y^s f_\chi(x+iy)g(x+iy)y^{-1}dx\frac{dy}{y}$$

$$= (-1)^\varepsilon \sum_{n\in\mathbf{Z}} \chi(n)a_n b_n \int_0^{+\infty} K_\nu(2\pi|n|y)K_{\nu'}(2\pi|n|y)y^s\frac{dy}{y}. \qquad (7.24)$$

Similarly to the proof of Lemma 4.27, one calculates the Mellin transform of the product of two Bessel functions in the next lemma.

Lemma 7.27 *The integral*

$$\int_0^{+\infty} K_\nu(y)K_{\nu'}(y)y^s\frac{dy}{y} = 2^{s-3}\Gamma(s)^{-1}\prod\Gamma\left(\frac{s\pm\nu\pm\nu'}{2}\right) \qquad (7.25)$$

is absolutely convergent on $\mathrm{Re}(s) > |\mathrm{Re}(\nu)| + |\mathrm{Re}(\nu')|$.

Proof The Mellin inversion formula, since it is nothing other than a rewriting of Fourier inversion, transforms the product of two functions into the convolution product of their Mellin inverses. It then follows from Lemma 4.27 that the left-hand side of (7.25) equals

$$\frac{1}{2i\pi}\int_{\sigma-i\infty}^{\sigma+i\infty} 2^{u-2}\Gamma\left(\frac{u+\nu}{2}\right)\Gamma\left(\frac{u-\nu}{2}\right)$$

$$\times 2^{s-u-2}\Gamma\left(\frac{s-u+\mu}{2}\right)\Gamma\left(\frac{s-u-\mu}{2}\right)du$$

$$= \frac{2^{s-3}}{2i\pi}\int_{\sigma/2-i\infty}^{\sigma/2+i\infty} \Gamma\left(u+\frac{\nu}{2}\right)\Gamma\left(u-\frac{\nu}{2}\right)$$

$$\times \Gamma\left(\frac{s+\mu}{2}-u\right)\Gamma\left(\frac{s-\mu}{2}-u\right)du.$$

The stated formula is then a consequence of the Barnes lemma – see [138, §14.5] – according to which

$$\int_{(\sigma)} \Gamma(s-\alpha)\Gamma(s-\beta)\Gamma(s+\gamma)\Gamma(s+\delta)ds$$

$$= \frac{\Gamma(\alpha+\gamma)\Gamma(\alpha+\delta)\Gamma(\beta+\gamma)\Gamma(\beta+\delta)}{\Gamma(\alpha+\beta+\gamma+\delta)}$$

where the line $\mathrm{Re}(s) = \sigma$ is to the right of the poles of $\Gamma(s-\alpha)$ and $\Gamma(s-\beta)$ and to the left of the poles of $\Gamma(s+\gamma)$ and $\Gamma(s+\delta)$.   □

Here $|\mathrm{Re}(v)|$ and $|\mathrm{Re}(v')| \leqslant 1/2$ and one finds that for $\mathrm{Re}(s) > 1$,

$$\int_0^{+\infty} K_v(2\pi|n|y)K_{v'}(2\pi|n|y)y^s\frac{dy}{y}$$

$$= \frac{1}{8}(\pi|n|)^{-s}\Gamma(s)^{-1}\prod\Gamma\left(\frac{s\pm v\pm v'}{2}\right). \qquad (7.26)$$

The identities (7.23), (7.24) and (7.26) finally imply[9]:

$$\int_{\Gamma_0(Nq^2)\backslash\mathcal{H}} E(z,s,\chi^2)f_\chi(z)g(z)\,d\mu(z)$$

$$= (-1)^\varepsilon\frac{1}{8}\pi^{-s}\Gamma(s)^{-1}\left(\prod\Gamma\left(\frac{s\pm v\pm v'}{2}\right)\right)\sum_{n\in\mathbf{Z}}\chi(n)\frac{a_nb_n}{|n|^s}$$

$$= (-1)^\varepsilon\frac{\pi^{-s}}{4}\Gamma(s)^{-1}\left(\prod\Gamma\left(\frac{s\pm v\pm v'}{2}\right)\right)\sum_{n=1}^{+\infty}\chi(n)\frac{a_nb_n}{n^s}. \qquad □$$

Let $D_0(Nq^2)$ be a fundamental domain for $\Gamma_0(Nq^2)$ inside the strip $|x| \leqslant 1/2$. According to Proposition 7.26, for $\mathrm{Re}(s) > 1$,

$$\frac{\pi^{-2s}}{4}\prod\Gamma\left(\frac{s\pm v\pm v'}{2}\right)L(s,f_\chi\times g)$$

$$= (-1)^\varepsilon\sum_{d|N}\mu(d)\chi^2(d)d^{-s}\int_{D_0(Nq^2)} E^*(z/d,s,\chi^2)f_\chi(z)g(z)\,d\mu(z). \qquad (7.27)$$

---

[9]One uses here the fact that f and g are of the same parity.

For a prime divisor p of N, note that[10] p does not divide Nq^2/p. We fix integers u, v and w such that $p^2u - Nq^2vw = p$ and write

$$W_p = \begin{pmatrix} px & y \\ Nq^2z & p \end{pmatrix}.\tag{7.28}$$

It follows from Lemma 7.17 that

$$f \circ W_p = \xi_p f \quad \text{and} \quad g \circ W_p = \eta_p g,$$

where ξ_p and η_p are two complex numbers of modulus 1.

Theorem 7.28 *Let N be a square-free positive integer. The function*

$$\Lambda(s, f_\chi \times g) = \left(\frac{q}{\pi}\right)^{2s} \left(\prod \Gamma\left(\frac{s \pm v \pm v'}{2}\right)\right)$$

$$\times \left(\prod_{p|N}(1 - \chi^2(p)\xi_p^{-1}\eta_p^{-1}p^{-s})^{-1}\right) L(s, f_\chi \times g),$$

defined for sufficiently large $\text{Re}(s)$*, admits a meromorphic continuation to the entire plane of* $s \in \mathbf{C}$*; it is everywhere holomorphic (since* χ *is primitive) and satisfies a functional equation*

$$\Lambda(s, f_\chi \times g) = w(f, g)\chi^2(N) \left(\frac{\tau(\chi)}{\sqrt{q}}\right)^4 N^{1-2s} \Lambda(1 - s, \tilde{f}_{\bar{\chi}} \times \tilde{g}),$$

where $w(f, g)$ *is a complex number of modulus 1, independently of* χ*.*

Proof The next two lemmas allow one to rewrite $\Lambda(s, f_\chi \times g)$ in a more convenient form.

Lemma 7.29 *We have*

$$f_\chi \circ W_p = \chi(p)\xi_p f_\chi.$$

Proof According to Lemma 7.18 we have

$$f_\chi(z) = \frac{1}{\tau(\bar{\chi})} \sum_{\ell=1}^{q} \bar{\chi}(\ell) \cdot f(z + \ell/q).$$

---

[10]Here we use the fact that N is square-free!

A simple matrix calculation moreover implies that

$$\begin{pmatrix} 1 & \ell/q \\ 0 & 1 \end{pmatrix} W_p \in \Gamma_0(N) W_p \begin{pmatrix} 1 & \ell p/q \\ 0 & 1 \end{pmatrix}$$

for $\ell = 1, \ldots, q$. We see then that

$$f_\chi \circ W_p(z) = \frac{1}{\tau(\overline{\chi})} \sum_{\ell=1}^q \overline{\chi}(\ell) \cdot f\left(W_p(z) + \ell/q\right)$$

$$= \chi(p) \frac{1}{\tau(\overline{\chi})} \sum_{\ell=1}^q \overline{\chi}(\ell p) \cdot (f \circ W_p)(z + \ell p/q)$$

$$= \chi(p) \xi_p \frac{1}{\tau(\overline{\chi})} \sum_{\ell=1}^q \overline{\chi}(\ell) \cdot f(z + \ell/q)$$

$$= \chi(p) \xi_p f_\chi. \qquad \Box$$

Lemma 7.30 *Let d be a divisor of N and p a prime dividing N but not d. Then*

$$\int_{D_0(Nq^2)} E^*(z/dp, s, \chi^2) f_\chi(z) g(z) \, d\mu(z)$$

$$= \overline{\chi}(p) \xi_p^{-1} \eta_p^{-1} \int_{D_0(Nq^2)} E^*(z/d, s, \chi^2) f_\chi(z) g(z) \, d\mu(z).$$

Proof The matrix W_p normalizes the group $\Gamma_0(Nq^2)$ and a simple (but laborious) computation using the definition (7.28) allows us to verify

$$E^*(z/dp, s, \chi^2) = E^*(z/d, s, \chi^2) \circ W_p.$$

The change of variables $z \mapsto W_p(z)$ thus implies the lemma. $\qquad \Box$

It follows from Lemma 7.30 that for any divisor d of N, we have

$$\int_{D_0(Nq^2)} E^*(z/d, s, \chi^2) f_\chi(z) g(z) \, d\mu(z) = \chi(N/d) \prod_{p|(N/d)} \xi_p \eta_p$$

$$\times \int_{D_0(Nq^2)} E^*(z/N, s, \chi^2) f_\chi(z) g(z) \, d\mu(z).$$

On the other hand, we have

$$
\sum_{d|N} \mu(d)\chi(d)^2 d^{-s} \prod_{p|d} \xi_p^{-1}\eta_p^{-1} = \sum_{d|N} \mu(d) \prod_{p|d} \chi^2(p)\xi_p^{-1}\eta_p^{-1}p^{-s}
$$

$$
= \prod_{p|N}(1 - \chi^2(p)\xi_p^{-1}\eta_p^{-1}p^{-s}).
$$

Formula (7.27) can be rewritten in the form

$$
\Lambda(s, f_\chi \times g) = (-1)^\varepsilon 4q^{2s}\chi(N)\left(\prod_{p|N}\xi_p\eta_p\right)
$$

$$
\times \int_{D_0(Nq^2)} E^*(z/N, s, \chi^2)f_\chi(z)g(z)\,d\mu(z). \qquad (7.29)
$$

The first part of Theorem 7.28 (meromorphic continuation and holomorphy) is then
a consequence of Theorem 7.24.

To prove the functional equation of Theorem 7.28 we now hit (7.29) with the
functional equation of the Eisenstein series E^* given by Theorem 7.24. Carrying
this out in detail will occupy the rest of the proof.

Let I denote the integral on the right-hand side of (7.29). The function equation
of E^* implies

$$
I = \tau(\chi^2)N^{1-2s}q^{2-5s}\int_{D_0(Nq^2)} E^*(-1/Nq^3z, 1-s, \overline{\chi}^2)f_\chi(z)g(z)\,d\mu(z)
$$

$$
= \tau(\chi^2)N^{1-2s}q^{2-5s}J \qquad (7.30)
$$

with

$$
J = \int_{D_0(Nq^2)} E^*(z/q, 1-s, \overline{\chi}^2)f_\chi(-1/Nq^2z)g(-1/Nq^2z)\,d\mu(z). \qquad (7.31)
$$

Here we have made the change of variables $z \mapsto -1/Nq^2z$ while using the fact that
$\begin{pmatrix} 0 & -1/(q\sqrt{N}) \\ q\sqrt{N} & 0 \end{pmatrix}$ normalizes $\Gamma_0(Nq^2)$.

The denominator q, in the Eisenstein series of (7.31), is a bit annoying; the
following lemma allows us to get rid of it.

Lemma 7.31 *We have*

$$
q^s E^*(z/q, s, \overline{\chi}^2) = \sum_{u \bmod q} E^*(z, s, \overline{\chi}^2) \circ \begin{pmatrix} 1 & 0 \\ Nq & 1 \end{pmatrix}^u.
$$

Proof Since

$$\mathrm{Im}\left(\left(\begin{smallmatrix}1 & 0\\ Nq & 1\end{smallmatrix}\right)^u z\right) = \frac{\mathrm{Im}(z)}{|Nquz + 1|^2},$$

we have

$$E^*(z, s, \overline{\chi}^2) \circ \left(\begin{matrix}1 & 0\\ Nq & 1\end{matrix}\right)^u$$

$$= \pi^{-s}\Gamma(s)\frac{1}{2}\sum_{m,n\in\mathbf{Z}}\overline{\chi}^2(n)\frac{y^s}{|mNq^2z + nNquz + n|^{2s}}$$

$$= \pi^{-s}\Gamma(s)\frac{1}{2}\sum_{m,n\in\mathbf{Z}}\overline{\chi}^2(n)\frac{y^s}{|(mq + nu)Nq^2(z/q) + n|^{2s}}.$$

The lemma can be deduced directly from a summation on u modulo q. □

It follows from Lemma 7.31 that

$$q^{1-s}J = \sum_{u \bmod q}\int_{D_0(Nq^2)} E^*(z, 1 - s, \overline{\chi}^2) \circ \left(\begin{matrix}1 & 0\\ Nq & 1\end{matrix}\right)^u$$

$$\times f_\chi(-1/Nq^2z)g(-1/Nq^2z)\,d\mu(z). \qquad (7.32)$$

Now $\Gamma_0(Nq^2)$ contains

$$\Gamma_1(Nq^2, q) = \left\{\left(\begin{matrix}a & b\\ c & d\end{matrix}\right) \in \Gamma_0(Nq^2) \,\middle|\, a \equiv d \equiv 1 \;(\mathrm{mod}\ Nq^2) \text{ and } b \equiv 0 \;(\mathrm{mod}\ q)\right\}$$

as a subgroup of finite index equal to[11] $q\phi(Nq^2)$. Let $D_1(Nq^2, q)$ denote a fundamental domain for $\Gamma_1(Nq^2, q)$. We then have

$$J = \frac{q^{s-1}}{q\phi(Nq^2)}\sum_{u \bmod q}\int_{D_1(Nq^2,q)} E^*(z, 1 - s, \overline{\chi}^2) \circ \left(\begin{matrix}1 & 0\\ Nq & 1\end{matrix}\right)^u$$

$$\times f_\chi(-1/Nq^2z)g(-1/Nq^2z)\,d\mu(z). \qquad (7.33)$$

---

[11]Here ϕ denotes the Euler totient function.

Lemma 7.32 *The integral*

$$\int_{D_1(Nq^2,q)} E^*(z, 1 - s, \overline{\chi}^2) \circ \begin{pmatrix} 1 & 0 \\ Nq & 1 \end{pmatrix}^u f_\chi(-1/Nq^2z)g(-1/Nq^2z)\, d\mu(z) \qquad (7.34)$$

is equal to

$$\int_{D_1(Nq^2,q)} E^*(z, 1 - s, \overline{\chi}^2)f_\chi\left(-\frac{1}{Nq^2z} + \frac{u}{q}\right)g\left(-\frac{1}{Nq^2z} + \frac{u}{q}\right) d\mu(z).$$

Proof The element

$$\begin{pmatrix} 1 & 0 \\ Nq & 1 \end{pmatrix}^u \qquad (7.35)$$

normalizes $\Gamma_1(Nq^2, q)$. Now the integral (7.34) runs over a fundamental domain for this group. The change of variable induced by the matrix (7.35) then transforms the integral (7.34) into

$$\int_{D_1(Nq^2,q)} E^*(z, 1 - s, \overline{\chi}^2)f_\chi\left(\begin{pmatrix} 0 & -1 \\ Nq^2 & 0 \end{pmatrix}\begin{pmatrix} 1 & 0 \\ -Nuq & 1 \end{pmatrix}z\right)$$

$$\times g\left(\begin{pmatrix} 0 & -1 \\ Nq^2 & 0 \end{pmatrix}\begin{pmatrix} 1 & 0 \\ -Nuq & 1 \end{pmatrix}z\right) d\mu(z).$$

Finally, this last integral equals

$$\int_{D_1(Nq^2,q)} E^*(z, 1 - s, \overline{\chi}^2)f_\chi\left(\begin{pmatrix} 1 & u/q \\ 0 & 1 \end{pmatrix}\begin{pmatrix} 0 & -1 \\ Nq^2 & 0 \end{pmatrix}z\right)$$

$$\times g\left(\begin{pmatrix} 1 & u/q \\ 0 & 1 \end{pmatrix}\begin{pmatrix} 0 & -1 \\ Nq^2 & 0 \end{pmatrix}z\right) d\mu(z),$$

which is precisely the stated expression. □

Lemma 7.33 *We have*

$$\sum_{u \bmod q} f_\chi(z + u/q)g(z + u/q) = \frac{q}{q - 1}\left[\sum_v f_{\chi v}(z)g_{\overline{v}}(z)\right],$$

where the sum runs over all *characters v modulo q, with the convention that if v is the trivial character, $f_{\chi v} = f_\chi$ and $g_v = g - g_q$ (see Lemma 7.22).*

Proof Lemma 7.22 implies that for $u = 1, \dots, q - 1$,

$$f_\chi(z + u/q)g(z + u/q)$$

$$= \frac{1}{(q-1)^2}\left\{ \sum_{\nu_1, \nu_2} \nu_1(u)\nu_2(u)\tau(\overline{\nu}_1)\tau(\overline{\nu}_2)f_{\chi\nu_1}(z)g_{\nu_2}(z)\right.$$

$$+ \sum_\nu \nu(u)\tau(\overline{\nu})\left[f_{\chi\nu}(z)(qg_q(z) - g(z)) - f_\chi(z)g_\nu(z)\right]$$

$$\left. - f_\chi(z)(qg_q(z) - g(z))\right\},$$

where the sums run over all *primitive* characters modulo q. Summing over u, we find (using Lemmas 7.18 and 7.19):

$$\sum_{u=1}^{q-1} f_\chi(z + u/q)g(z + u/q) = \frac{q}{q-1}\left[\sum_\nu f_{\chi\nu}(z)g_{\overline{\nu}}(z)\right] - f_\chi(z)g(z),$$

where the sum runs over *all* characters ν mod q, with the convention described in the lemma. Since f_χ and g are 1-periodic, we deduce the lemma. $\square$

Starting from the expression (7.33), Lemmas 7.32 and 7.33 imply

$$J = \frac{q^s}{q-1}\sum_\nu \int_{D_0(Nq^2)} f_{\chi\nu}\left(-\frac{1}{Nq^2 z}\right)g_{\overline{\nu}}\left(-\frac{1}{Nq^2 z}\right)$$

$$\times E^*(z, 1 - s, \overline{\chi}^2)\,d\mu(z), \qquad (7.36)$$

the sum again running over all characters ν modulo q.

Lemma 7.34 *The integral*

$$\int_{D_0(Nq^2)} f_{\chi\nu}\left(-\frac{1}{Nq^2 z}\right)g_{\overline{\nu}}\left(-\frac{1}{Nq^2 z}\right)E^*(z, 1 - s, \overline{\chi}^2)\,d\mu(z) \qquad (7.37)$$

equals

$$\chi(-N)\frac{(\tau(\chi\nu)\tau(\overline{\nu}))^2}{q^2}\int_{D_0(Nq^2)} \tilde{f}_{\overline{\chi}}(z)\tilde{g}(z)E^*(z, 1 - s, \overline{\chi}^2)\,d\mu(z).$$

Proof Since the characters $\chi\nu$ and $\overline{\nu}$ are both primitive, Lemma 7.21 implies that the integral (7.37) is equal to

$$\chi(-N)\frac{\tau(\chi\nu)\tau(\overline{\nu})}{\tau(\overline{\chi\nu})\tau(\nu)}\int_{D_0(Nq^2)} \tilde{f}_{\overline{\chi\nu}}(z)\tilde{g}_\nu(z)E^*(z, 1 - s, \overline{\chi}^2)\,d\mu(z).$$

Next, Lemma 7.19 implies

$$\frac{\tau(\chi\nu)\tau(\overline{\nu})}{\tau(\overline{\chi\nu})\tau(\nu)} = \frac{(\tau(\chi\nu)\tau(\overline{\nu}))^2}{q^2}.$$

The proof of Proposition 7.26 finally implies

$$\int_{D_0(Nq^2)} \tilde{f}_{\overline{\chi\nu}}(z)\tilde{g}_{\nu}(z)E^*(z, 1-s, \overline{\chi}^2)\,d\mu(z)$$

$$= \int_{D_0(Nq^2)} \tilde{f}_{\overline{\chi}}(z)\tilde{g}(z)E^*(z, 1-s, \overline{\chi}^2)\,d\mu(z),$$

whence the lemma.

It remains to treat the cases where ν is trivial or equal to $\overline{\chi}$. Suppose for example that $\nu = \overline{\chi}$; then we have

$$f_{\chi\nu}(z) = f(z) - f_q(z) = \frac{1}{\tau(\overline{\chi}_{\text{triv}})}\sum_{\ell=1}^{q}\overline{\chi}_{\text{triv}}(\ell)f\left(z + \ell/q\right).$$

The proof of Lemma 7.21 implies

$$(f - qf_q)\left(-\frac{1}{Nq^2z}\right) = (\tilde{f} - q\tilde{f}_q)(z). \tag{7.38}$$

Since

$$f\left(-\frac{1}{Nqz^2}\right) = \tilde{f}(q^2z),$$

Eq. (7.38) can be rewritten as

$$f_q\left(-\frac{1}{Nq^2z}\right) = \tilde{f}_q(z) - \frac{1}{q}\tilde{f}(z) + \frac{1}{q}\tilde{f}(q^2z). \tag{7.39}$$

We conclude in the same way as before. $\qquad\qquad\qquad\qquad\square$

Lemma 7.35 *We have*

$$\frac{1}{q^2}\sum_{\nu}(\tau(\chi\nu)\tau(\overline{\nu}))^2 = \frac{q-1}{q}\left(\frac{\tau(\chi)}{\sqrt{q}}\right)^4\tau(\overline{\chi}^2).$$

Proof From Lemma 7.19 and the hypothesis that χ is even we have

$$\frac{1}{q^2}\sum_v (\tau(\chi v)\tau(\overline{v}))^2 = \frac{1}{q^2}\sum_v \left[\left(\sum_{k=1}^{q-1}(\chi v)(k)e\left(\frac{k}{q}\right)\right)\left(\sum_{\ell=1}^{q-1}\overline{v}(\ell)e\left(\frac{\ell}{q}\right)\right)\right]^2$$

$$= \frac{1}{q^2}\sum_v \left(\sum_{k=1}^{q-1}\chi(k)\sum_{\ell=1}^{q-1}\overline{v}(\ell k^{-1})e\left(\frac{k(1+\ell k^{-1})}{q}\right)\right)^2$$

$$= \frac{1}{q^2}\sum_{k_1=1}^{q}\chi(k_1)\sum_{k_2=1}^{q}\chi(k_2)$$

$$\times \sum_{\ell_1,\ell_2=1}^{q} e\left(\frac{[k_1(1-\ell_1)+k_2(1-\ell_2)]}{q}\right)\sum_v \overline{v}(\ell_1\ell_2),$$

where the last sum is zero if $\ell_1\ell_2 \equiv 1 \pmod q$ and equals $q-1$ otherwise. Thus

$$\sum_v \frac{\tau(\chi v)\tau(\overline{v})}{\tau(\overline{\chi v})\tau(v)} = \frac{q-1}{q^2}\sum_{k_1=1}^{q}\chi(k_1)\sum_{k_2=1}^{q}\chi(k_2)\sum_{(\ell,q)=1}e\left(\frac{(1-\ell)[k_1-k_2\ell^{-1}]}{q}\right)$$

$$= \frac{q-1}{q^2}\sum_{k_2=1}^{q}\sum_{(\ell,q)=1}\chi(k_2)e\left(-\frac{(1-\ell)k_2\ell^{-1}}{q}\right)$$

$$\times \underbrace{\left(\sum_{k_1=1}^{q}\chi(k_1)e\left(\frac{(1-\ell)k_1}{q}\right)\right)}_{=\overline{\chi}(1-\ell)\tau(\chi)}$$

$$= \frac{q-1}{q^2}\tau(\chi)\sum_{(\ell,q)=1}\overline{\chi}(1-\ell)\underbrace{\sum_{k=1}^{q}\chi(k)e\left(-\frac{(1-\ell)k\ell^{-1}}{q}\right)}_{=\overline{\chi}(-(1-\ell)\ell^{-1})\tau(\chi)=\overline{\chi}(1-\ell)\chi(\ell)\tau(\chi)}$$

$$= \frac{q-1}{q^2}\tau(\chi)^2 J(\chi,\overline{\chi}^2).$$

We end the proof by appealing to Lemma 7.19.   □

Starting from the expression (7.36) we can use Lemmas 7.34 and 7.35 to show

$$J = q^{s-1}\chi(N)\left(\frac{\tau(\chi)}{\sqrt{q}}\right)^4\tau(\overline{\chi}^2)$$

$$\times \int_{D_0(Nq^2)} \tilde{f}_{\overline{\chi}}(z)\tilde{g}(z)E^*(z,1-s,\overline{\chi}^2)\,d\mu(z). \qquad (7.40)$$

An inductive argument using Lemma 7.30 again implies, as for (7.29), that

$$\int_{D_0(Nq^2)} \tilde{f}_{\overline{\chi}}(z)\tilde{g}(z)E^*(z, 1-s, \overline{\chi}^2)\, d\mu(z)$$

$$= \overline{\chi}(N)\left(\prod_{p|N}\overline{\xi}_p\overline{\eta}_p\right)\int_{D_0(Nq^2)} \tilde{f}_{\overline{\chi}}(z)\tilde{g}(z)E^*(z/N, 1-s, \overline{\chi}^2)\, d\mu(z).$$

Finally, the identities (7.29), (7.30) and (7.40) together imply that there is a complex number $w(f,g)$ – which we could write down explicitly if needed – of modulus 1 such that

$$\Lambda(s,f_\chi \times g) = w(f,g)\chi^2(N)\left(\frac{\tau(\chi)}{\sqrt{q}}\right)^4 N^{1-2s}\Lambda(1-s,\tilde{f}_{\overline{\chi}} \times \tilde{g}). \qquad \square$$

In addition to the above properties, Selberg noticed that $L(s,f_\chi \times g)$ admits an Euler product decomposition. To see this, we begin by factoring the Hecke polynomials (for p not dividing N):

$$1 - a_pX + X^2a = (1 - \alpha_1(p)X)(1 - \alpha_2(p)X),$$

and

$$1 - b_pX + X^2 = (1 - \beta_1(p)X)(1 - \beta_2(p)X).$$

Here a_p and b_p are the p Fourier coefficients of f and g and $\alpha_i(p), \alpha_i(p)$ ($i = 1, 2$) lie in **C**.

Theorem 7.36 *We have*

$$L(s,f_\chi \times g) = \prod_{p|N}(1 - \chi(p)a_pb_pp^{-s})^{-1}\prod_{p\nmid N}\prod_{i,j=1}^{2}(1 - \chi(p)\alpha_i(p)\beta_j(p)p^{-s})^{-1}.$$

Proof First let p be a prime divisor of N. Then

$$\sum_{k\geq 0}\chi(p^k)a_{p^k}b_{p^k}X^k = \sum_{k\geq 0}(\chi(p)a_pb_p)^kX^k$$
$$= \frac{1}{1 - \chi(p)a_pb_pX}.$$

Now let p be a prime such that $p \nmid N$. It follows from the proof of Proposition 7.15 that for $k \geq 0$,

$$a_{p^k} = \frac{\alpha_1^{k+1} - \alpha_2^{k+1}}{\alpha_1 - \alpha_2} \quad \text{and} \quad b_{p^k} = \frac{\beta_1^{k+1} - \beta_2^{k+1}}{\beta_1 - \beta_2}.$$

Then

$$\sum_{k \geq 0} \chi(p^k) a_{p^k} b_{p^k} X^k = \frac{1}{(\alpha_1 - \alpha_2)(\beta_1 - \beta_2)} \sum_{i,j=1}^{2} (-1)^{i+j} \frac{\alpha_i \beta_j}{1 - \chi(p)\alpha_i \beta_j X}$$

$$= \frac{1 - \chi^2(p)X^2}{\prod_{i,j=1}^{2}(1 - \chi(p)\alpha_i \beta_j X)}.$$

From the two preceding cases one immediately deduces that $L(s, f_\chi \times g)$ indeed admits the stated Euler product decomposition, valid for $\mathrm{Re}(s)$ sufficiently large. $\quad\square$

Remark 7.37 Given a cuspidal Maaß newform

$$f \in \mathcal{C}(\Gamma_0(N)\backslash \mathcal{H}),$$

the generalized Riemann hypothesis applied to $L(s, f \times \bar{f})$ would imply $L(s, f \times \bar{f}) \neq 0$ for all s in the half-plane $\mathrm{Re}(s) > 1/2$. But if

$$\Delta f = (1/4 - v^2)f$$

the function $\Lambda(s, f \times \bar{f})$ has at most simple poles at $s = 0$ and $s = 1$. Now $2\,|\,\mathrm{Re}(v)|$ is a pole of $\Gamma((s - 2\,|\,\mathrm{Re}(v)|)/2)$ and all other local factors in $\Lambda(s, f \times \bar{f})$ are non-zero for $\mathrm{Re}(s) > 1/2$. We must have $2|\,\mathrm{Re}(v)| \leq 1/2$. It follows that $\lambda_1(\Gamma_0(N)\backslash \mathcal{H})$ must be greater than or equal to $3/16$.

In view of the above remark, the "right method" (proposed by Langlands) to prove the Selberg conjecture would be to use the generalized Rankin-Selberg L-function

$$L(s, \underbrace{f \times \bar{f} \times \cdots \times f \times \bar{f}}_{n \text{ times}}).$$

We can in fact define such a function by its Euler product. We can then naturally complete this Euler product at infinity with powers of π which are polynomial in s and with explicit Γ-functions, in which $\Gamma((s - 2n|\,\mathrm{Re}(v)|)/2)$ would appear. This completed L-function should again satisfy a functional equation. The first pole, starting from the right, of these Γ-factors would then be at $2n|\,\mathrm{Re}(v)|$ and (without having to use the Riemann hypothesis) we could conclude that $|\,\mathrm{Re}(v)| \leq 1/2n$ (since the L-function is non-vanishing to the right of the line $\mathrm{Re}(s) = 1$). Letting n go to infinity, we would then obtain the Selberg conjecture.

Now we do not in fact know how to extend the method of Rankin-Selberg to an arbitrary product of Maaß forms, nor can we prove the Riemann hypothesis for Rankin-Selberg L-functions. What we *can* do is consider a family of Rankin-Selberg L-functions each having the same Γ-factor. This is the reason we have been interested in the behavior of L-functions under twisting by Dirichlet characters. In the following section we shall show that a suitably strong approximation to the Riemann hypothesis is true "on average" over Rankin-Selberg L-functions twisted by Dirichlet characters. This result – due to Luo, Rudnick and Sarnak – suffices to show Theorem 7.1.

7.5 The Luo-Rudnick-Sarnak Theorem

We again fix a normalized cuspidal ($a_0 = 0$) Maaß newform

$$f(z) = \sum_{r \in \mathbf{Z}} a_r \sqrt{y}\, K_v(2\pi|r|y)e(rx),$$

in $\mathcal{C}_{\text{prim}}(\Gamma_0(N)\backslash\mathcal{H})$, where N is squarefree. We again denote the roots of the Hecke polynomial $1 - a_p X + X^2$ by $\alpha_1(p)$ and $\alpha_2(p)$.

Let χ be an even primitive Dirichlet character of prime conductor q not dividing N. We write

$$L_\infty(s, f_\chi \times \bar{f}) = \left(\frac{q}{\pi}\right)^{2s} \Gamma\left(\frac{s - 2i\,\text{Im}(v)}{2}\right)\Gamma\left(\frac{s + 2i\,\text{Im}(v)}{2}\right)$$

$$\times \Gamma\left(\frac{s - 2\,\text{Re}(v)}{2}\right)\Gamma\left(\frac{s + 2\,\text{Re}(v)}{2}\right)\left(\prod_{p|N}(1 - \chi^2(p)|\xi_p|^{-2}p^{-s})^{-1}\right),$$

where ξ_p is the complex number of modulus 1 defined by $f \circ W_p = \xi_p f$. The product of $L(s, f_\chi \times \bar{f})$ by $L_\infty(s, f_\chi \times \bar{f})$ is equal to the completed L-function $\Lambda(s, f_\chi \times \bar{f})$ which enjoys the properties stated in Theorem 7.28. We deduce that $L(s, f_\chi \times \bar{f})$ is holomorphic on the entire complex plane since L_∞ never vanishes. The following theorem, due to Luo, Rudnick and Sarnak, is the key step towards a uniform lower bound on $\lambda_1(\Gamma_0(N))$ via L-functions.

Theorem 7.38 *Fix a real number $\beta > 2/3$. There exists an infinite set of even primitive Dirichlet characters of prime modulus not dividing N such that*

$$L(\beta, f_\chi \times \bar{f}) \neq 0.$$

Proof According to Theorem 7.36, the function $L(s, f_\chi \times \bar{f})$ can be written as an Euler product and it follows from Lemma 7.2 that the factors $(1 - \chi(p)a_p b_p p^s)^{-1}$ and $(1 - \chi(p)\alpha_i(p)\overline{\alpha_j(p)}p^{-s})^{-1}$ are non-zero for $s = \beta \geqslant 1$; the same is thus true for $L(s, f_\chi \times \bar{f})$. We can assume then that $2/3 < \beta < 1$ and Theorem 7.38 is an immediate consequence of the following proposition. □

Proposition 7.39 *Let β be a real number lying strictly between 2/3 and 1. For every $\varepsilon > 0$, there exists a constant $c = c(f, \beta) > 0$ such that*

$$\left| \sum_{\chi(q)} L(\beta, f_\chi \times \bar{f}) \right| \geq cq^{1-\varepsilon},$$

as $q \to +\infty$. Here the sum runs over all primitive even Dirichlet characters χ of prime modulus q.

Proof Fix a primitive even Dirichlet character χ of prime conductor q not dividing N. The only tool that we have at our disposal is the functional equation of $L(s, f_\chi \times \bar{f})$ that we now rewrite in the form

$$L(s, f_\chi \times \bar{f})$$
$$= w(f, \bar{f}) \chi^2(N) \left(\frac{\tau(\chi)}{\sqrt{q}} \right)^4 \left(\frac{q}{\pi} \right)^{2-4s} N^{1-2s} G_\infty(s) L(1 - s, f_\chi \times \bar{f}), \qquad (7.41)$$

where

$$G_\infty(s)$$
$$= \frac{\Gamma\left(\frac{1-s-2i\,\mathrm{Im}(v)}{2} \right) \Gamma\left(\frac{1-s+2i\,\mathrm{Im}(v)}{2} \right) \Gamma\left(\frac{1-s-2\,\mathrm{Re}(v)}{2} \right) \Gamma\left(\frac{1-s+2\,\mathrm{Re}(v)}{2} \right)}{\Gamma\left(\frac{s-2i\,\mathrm{Im}(v)}{2} \right) \Gamma\left(\frac{s+2i\,\mathrm{Im}(v)}{2} \right) \Gamma\left(\frac{s-2\,\mathrm{Re}(v)}{2} \right) \Gamma\left(\frac{s+2\,\mathrm{Re}(v)}{2} \right)}$$
$$\times \prod_{p|N} \frac{(1 - \chi^2(p)|\xi_p|^{-2} p^{-(1-s)})^{-1}}{(1 - \chi^2(p)|\xi_p|^{-2} p^{-s})^{-1}}.$$

The proof of the proposition then consists in deriving from (7.41) a manageable expression of $L(\beta, f_\chi \times \bar{f})$. This is not immediate since β lies outside of the region of absolute convergence. Nevertheless, there exists a standard method in analytic number theory which addresses this problem: the "approximate functional equation". This method is based on the knowledge of the functional equation and shifting integration contours. In this case we obtain the following lemma. Put

$$\beta_0 = 2|\,\mathrm{Re}(v)|. \qquad (7.42)$$

Lemma 7.40 *There exist two real-valued C^∞-functions V_1, V_2 satisfying*

$$V_1(y), \; V_2(y) = O_A(y^{-A}) \quad \text{as } y \to +\infty,$$
$$V_1(y) = 1 + O_A(y^A), \; V_2(y) = O_\varepsilon(1 + y^{1-\beta_0-\beta-\varepsilon}), \quad \text{as } y \to 0, \qquad (7.43)$$

for all $A > 0$ and $\varepsilon > 0$ and such that for all parameters $Y \geqslant 1$ we have

$$L(\beta, f_\chi \times \bar{f}) = \sum_{n=1}^{+\infty} \frac{|a_n|^2}{n^\beta} \chi(n) V_1 \left(\frac{n}{Y}\right)$$

$$- \frac{w(f, \bar{f}) \chi^2(N)}{(q^2 N)^{2\beta-1} \pi^{2-4\beta}} \sum_{n=1}^{+\infty} \frac{|a_n|^2}{n^{1-\beta}} \bar{\chi}(n) \left(\frac{\tau(\chi)}{\sqrt{q}}\right)^4 V_2 \left(\frac{n Y \pi^4}{q^4 N^2}\right). \qquad (7.44)$$

Proof We introduce a C^∞ test function $V : \mathbf{R} \to \mathbf{R}$ which is non-negative, compactly supported in $[A, B] \subset \mathbf{R}_+^*$, and such that

$$\int_0^{+\infty} \frac{V(y)}{y} dy = 1.$$

Its Mellin transform

$$\widetilde{V}(s) = \int_0^{+\infty} V(y) y^s \frac{dy}{y}$$

is a holomorphic function which decays rapidly in vertical strips.
From Lemma 7.3 we deduce that the series

$$L(s, f_\chi \times \bar{f}) = \sum_{n=1}^{+\infty} \chi(n) \frac{|a_n|^2}{n^s}$$

converges absolutely for $\mathrm{Re}(s) > 1$. For any $Y > 1$ we can then consider the integral

$$I(\beta, Y) = \frac{1}{2i\pi} \int_{(2)} \widetilde{V}(s) L(s + \beta, f_\chi \times \bar{f}) Y^s \frac{ds}{s}$$

$$= \sum_{n=1}^{+\infty} \chi(n) \frac{|a_n|^2}{n^\beta} \frac{1}{2i\pi} \int_{(2)} \widetilde{V}(s) \left(\frac{Y}{n}\right)^s \frac{ds}{s}.$$

We let

$$V_1(y) = \frac{1}{2i\pi} \int_{(2)} \widetilde{V}(s) y^{-s} \frac{ds}{s}$$

so that

$$I(\beta, Y) = \sum_{n=1}^{+\infty} \chi(n) \frac{|a_n|^2}{n^\beta} V_1 \left(\frac{n}{Y}\right). \qquad (7.45)$$

We immediate observe that by Mellin inversion,

$$V_1(y) = \int_y^{+\infty} \frac{V(x)}{x} dx,$$

and thus

$$0 \le V_1(y) \le 1 \quad \text{and} \quad V_1(y) = \begin{cases} 1 & \text{if } 0 < y \le A, \\ 0 & \text{if } y \ge B. \end{cases}$$

Since $L(s, f_\chi \times \bar{f})$ is bounded in vertical strips, we can shift the integration contour $\sigma = 2$ to the left to reach $\sigma = -1$. Along the way we pick up a pole at $s = 0$, which gives

$$I(\beta, Y) = L(\beta, f_\chi \times \bar{f}) + \frac{1}{2i\pi} \int_{(-1)} \widetilde{V}(s) L(s + \beta, f_\chi \times \bar{f}) Y^s \frac{ds}{s}. \tag{7.46}$$

The functional equation (7.41) implies that the integral on the right-hand side of (7.46) is equal to

$$w(f, \bar{f}) \chi^2(N) \left(\frac{\tau(\chi)}{\sqrt{q}} \right)^4 \left(\frac{Nq^2}{\pi^2} \right)^{1-2\beta}$$

$$\times \frac{1}{2i\pi} \int_{(-1)} \widetilde{V}(s) \left(\frac{\pi^4 Y}{q^4 N^2} \right)^s G_\infty(s + \beta) L(1 - \beta - s, f_\chi \times \bar{f}) \frac{ds}{s}.$$

Changing variables $s \mapsto -s$ transforms this last expression into

$$w(f, \bar{f}) \chi^2(N) \left(\frac{\tau(\chi)}{\sqrt{q}} \right)^4 \left(\frac{Nq^2}{\pi^2} \right)^{1-2\beta}$$

$$\frac{1}{2i\pi} \int_{(1)} \widetilde{V}(-s) \left(\frac{\pi^4 Y}{q^4 N^2} \right)^{-s} G_\infty(-s + \beta) L(1 - \beta + s, \tilde{f}_{\bar{\chi}} \times \bar{f}) \frac{ds}{s}$$

which is then equal to

$$w(f, \bar{f}) \chi^2(N) \left(\frac{\tau(\chi)}{\sqrt{q}} \right)^4 \left(\frac{Nq^2}{\pi^2} \right)^{1-2\beta} \sum_{n=1}^{+\infty} \bar{\chi}(n) \frac{|a_n|^2}{n^{1-\beta}} V_2\left(\frac{nY\pi^4}{q^4 N^2} \right), \tag{7.47}$$

where we have put, for $y > 0$,

$$V_2(y) = \frac{1}{2i\pi} \int_{(1)} \widetilde{V}(-s) G_\infty(-s + \beta) y^{-s} \frac{ds}{s}.$$

Note that if $A > 1$, by shifting the integration contour to the right of $\sigma = 1$ to $\sigma = A$, we obtain

$$V_2(y) = O_A(y^{-A}) \quad \text{as } y \to +\infty,$$

since G_∞ is of at most polynomial growth in vertical strips whereas $\widetilde{V}(-s)$ decays rapidly. We now show that

$$V_2(y) = O_\varepsilon(1 + y^{1-\beta_0-\beta-\varepsilon}) \quad \text{as } y \to 0.$$

If $\beta_0 + \beta - 1 \geq 0$, this is obvious. Otherwise, we shift contours to the left; as the first pole (of $G_\infty(-s+\beta)$) we cross is at $\beta_0 + \beta - 1$, the conclusion follows at once.

We conclude the proof of the lemma by putting (7.45) equal to (7.46), after having replaced the integral on the right-hand side of (7.46) by (7.47). We get

$$L(\beta, f_\chi \times \bar{f}) = \sum_{n=1}^{+\infty} \frac{|a_n|^2}{n^\beta} \chi(n) V_1\left(\frac{n}{Y}\right)$$

$$- w(f, \bar{f}) \chi^2(N) \left(\frac{\tau(\chi)}{\sqrt{q}}\right)^4 \left(\frac{Nq^2}{\pi^2}\right)^{1-2\beta} \sum_{n=1}^{+\infty} \bar{\chi}(n) \frac{|a_n|^2}{n^{1-\beta}} V_2\left(\frac{nY\pi^4}{q^4 N^2}\right),$$

as claimed. $\qquad\qquad\qquad\qquad\qquad\qquad\qquad\qquad\qquad\qquad\qquad\qquad\qquad$ $\square$

We can now average the approximate functional equation over the set of even Dirichlet characters which are non-trivial modulo a prime q (we assume that q is sufficiently large so that q stays prime to N). The sum

$$\sum_\chi L(\beta, f_\chi \times \bar{f})$$

decomposes into a sum $T_1 + T_2$ of two terms corresponding to the terms in the approximate functional equation. Thus

$$T_1 = \sum_\chi \sum_{n=1}^{+\infty} \frac{|a_n|^2}{n^\beta} \chi(n) V_1\left(\frac{n}{Y}\right) \tag{7.48}$$

and

$$T_2 = \frac{w(f, \bar{f})}{(\pi^2 N)^{1-4\beta}} q^{-4\beta} \sum_{n=1}^{+\infty} \frac{|a_n|^2}{n^{1-\beta}} \left(\sum_\chi \chi^2(N) \bar{\chi}(n) \tau(\chi)^4\right) V_2\left(\frac{nY\pi^4}{q^4 N^2}\right). \tag{7.49}$$

We begin by evaluating the term T_1. It is not difficult to deduce from (7.13) that

$$
\sum_{\substack{\chi \neq \chi_{\text{triv}} \text{ even}}} \chi(n) = \begin{cases} 0 & \text{if } n \equiv 0 \ (\text{mod } q) \\ \dfrac{q-1}{2} - 1 & \text{if } n \equiv \pm 1 \ (\text{mod } q) \\ -1 & \text{else.} \end{cases} \tag{7.50}
$$

We thus obtain

$$
T_1 = \frac{q-1}{2} \sum_{n \equiv \pm 1 \ (\text{mod } q)} \frac{|a_n|^2}{n^\beta} V_1 \left(\frac{n}{Y} \right) - \sum_{q \nmid n} \frac{|a_n|^2}{n^\beta} V_1 \left(n/Y \right). \tag{7.51}
$$

All terms in the first sum are non-negative. Hence one may obtain a lower bound on the first sum by considering only the contribution from $n = 1$ term:

$$
\frac{q-1}{2} \sum_{n \equiv \pm 1 \ (\text{mod } q)} \frac{|a_n|^2}{n^\beta} V_1 \left(n/Y \right) \geq \frac{q-1}{2} V_1 \left(1/Y \right). \tag{7.52}
$$

The right-hand side of (7.52) is equal to $(q-1)/2$ for Y sufficiently large.

On the other hand, one has

$$
\sum_{q \nmid n} \frac{|a_n|^2}{n^\beta} V_1 \left(n/Y \right) \leq \sum_{n=1}^{+\infty} \frac{|a_n|^2}{n^\beta} V_1 \left(n/Y \right).
$$

We deduce from Lemma 7.3 that for Y large enough

$$
\sum_{n=1}^{+\infty} \frac{|a_n|^2}{n^\beta} |V_1 \left(n/Y \right)| = O_f \left(Y^{1-\beta} \right)
$$

and thus that

$$
T_1 \geq \frac{q-1}{2} - O_f \left(Y^{1-\beta} \right). \tag{7.53}
$$

We finally arrive at the contribution from T_2. Since for primitive χ we have $|\tau(\chi)| = \sqrt{q}$, we easily bound the inner sum of (7.49) as

$$
\sum_{\chi \neq \chi_{\text{triv}} \text{ even}} \overline{\chi}(nN'^2) \tau(\chi)^4 = O(q^3). \tag{7.54}
$$

We deduce that

$$|T_2| = O\left(q^{3-4\beta} \sum_{n=1}^{+\infty} \frac{|a_n|^2}{n^{1-\beta}} \left|V_2\left(\frac{nY\pi^4}{q^4N^2}\right)\right|\right)$$

$$= O\left(q^{3-4\beta} \int_1^{+\infty} \left|V_2\left(\frac{rY\pi^4}{q^4N^2}\right)\right| r^\beta \frac{dr}{r}\right)$$

$$= O\left(q^3 Y^{-\beta}\right),$$

whence it follows that

$$|T_2| = O\left(q^3 Y^{-\beta}\right), \tag{7.55}$$

for large enough q and Y.

The upper bounds (7.53) and (7.55) imply the inequality

$$\left|\sum_\chi L(\beta, f_\chi \times \bar{f})\right| \geq \frac{q-1}{2} - O_f(Y^{1-\beta} + q^3 Y^{-\beta}), \tag{7.56}$$

for large enough q and Y. Taking $Y \sim q^3$, we obtain Proposition 7.39 (recall that $2/3 < \beta < 1$). $\qquad\square$

We now show that Theorem 7.38 implies Theorem 7.1.

Proof of Theorem 7.1 We restrict ourselves to cusp forms. Let

$$f(z) = \sum_{r \in \mathbf{Z}} a_r \sqrt{y}\, K_\nu(2\pi|r|y) e(rx) \in \mathcal{C}(\Gamma_0(N)\backslash\mathcal{H})$$

satisfy $\Delta f = (1/4 - \nu^2)f$. Suppose that ν is real and strictly greater than $1/3$; we seek to obtain a contradiction. Let

$$\beta = 2|\nu|.$$

In view of the decomposition of $\mathcal{C}(\Gamma_0(N)\backslash\mathcal{H})$ into an orthogonal sum of spaces of oldforms and newforms, we can safely assume that f is a newform. We can moreover take f to be normalized. According to Theorem 7.38, there then exists a non-trivial even Dirichlet character of prime modulus q not dividing N such that

$$L(\beta, f_\chi \times \bar{f}) \neq 0.$$

According to Theorem 7.28, the function

$$\Lambda(s, f_\chi \times \bar{f}) = L_\infty(s, f_\chi \times \bar{f}) L(s, f_\chi \times \bar{f})$$

is entire whereas β is a pole of $L_\infty(s, f_\chi \times \bar{f})$. Contradiction. $\qquad\square$

7.6   Bounds on Fourier Coefficients

The initial motivation of Rankin and Selberg to prove the existence of an analytic continuation of Rankin-Selberg L-functions was the study of the Fourier coefficients of modular forms. Their method applies equally well to the study of the Fourier coefficients of Maaß forms.

Let

$$f(z) = \sum_{r \in \mathbb{Z}} a_r \sqrt{y} \, K_\nu(2\pi|r|y) e(rx)$$

be a cuspidal ($a_0 = 0$) normalized Maaß newform in $\mathcal{C}_{\mathrm{prim}}(\Gamma_0(N)\backslash\mathcal{H})$. The following conjecture is referred to as the Ramanujan-Petersson conjecture; it should be compared with the "trivial" bound of Hecke (Lemma 7.2).

Conjecture 7.41 *For every $\varepsilon > 0$, the sequence of Fourier coefficients of f satisfies*

$$|a_n| = O_\varepsilon(n^\varepsilon),$$

for n sufficiently large.

Any progress towards this conjecture is significant and rich in arithmetic applications. We shall see such an application in Chap. 9.

In this section we give a proof of the following bound, under the simplifying hypothesis that $N = 1$.

Theorem 7.42 *For any $\varepsilon > 0$, the sequence of Fourier coefficients of f satisfies*

$$|a_n| = O_\varepsilon(n^{3/10+\varepsilon}),$$

for n sufficiently large.

Proof in the case $N = 1$ The theorem is a corollary of the following lemma – due to Landau – whose proof we shall admit.

Lemma 7.43 *Let $L(s) = \sum_n b_n n^{-s}$ be a Dirichlet series with non-negative coefficients b_n which converges on $\mathrm{Re}(s) > 1$. Assume that $L(s)$ admits a meromorphic continuation to all of $\mathbb{C}$ with at most a simple pole at $s = 1$ in the right half-plane $\mathrm{Re}(s) \geqslant 1/2$. Assume moreover that L is of finite order in the half-plane $\mathrm{Re}(s) \geqslant -1$ and satisfies a functional equation of the form*

$$q^s L_\infty(s) L(s) = w q^{1-s} L_\infty(1-s) L(1-s),$$

where w and q are positive constants and

$$L_\infty(s) = \prod_{i=1}^{d} \Gamma(\alpha_i s + \beta_i),$$

with $d \geq 1$ and $\alpha_i \geq 0$, $\beta_i \in \mathbf{C}$ for $i = 1, \ldots, d$. Then for all $\varepsilon > 0$ we have

$$\sum_{n \leq x} b_n = Cx + O_{\varepsilon, L}\left(x^{\frac{2\eta-1}{2\eta+1}+\varepsilon}\right),$$

where the constant $C > 0$ depends only on L and $\eta = \sum_{i=1}^{d} \alpha_i$.

We now apply the lemma to the function

$$L(s, f \times \bar{f}) = \zeta(2s) \sum_{n=1}^{+\infty} \frac{|a_n|^2}{n^s}.$$

Note that Theorem 7.28 applies here as well. This time the above L-function admits two simple poles at $s = 0$ and 1. It satisfies the functional equation

$$\pi^{-2s} L_\infty(s) L(s, f \times \bar{f}) = \pi^{-2(1-s)} L_\infty(1-s) L(1-s, f \times \bar{f})$$

with

$$L_\infty(s) = \Gamma\left(\frac{s}{2}\right)^2 \Gamma\left(\frac{s-2\nu}{2}\right) \Gamma\left(\frac{s+2\nu}{2}\right).$$

In order to appeal to Landau's lemma, we must take into account the factor $\zeta(2s)$ in front of the sum $\sum_{n=1}^{+\infty} |a_n|^2 n^{-s}$.

Observe that

$$\zeta(2s) \sum_{n=1}^{+\infty} |a_n|^2 n^{-s} = \sum_{n=1}^{+\infty} b_n n^{-s},$$

where $(b_n)_{n \in \mathbf{N}}$ is a sequence of positive reals. Here the sequence $(b_n)_{n \in \mathbf{N}}$ is obtained by *Dirichlet convolution* of the characteristic function of the squares ζ_2 with the sequence $(|a_n|^2)_{n \in \mathbf{N}}$:

$$b_n = \sum_{kd=n} \zeta_2(k) |a_d|^2.$$

We then deduce from Landau's lemma that

$$\sum_{n \leq x} b_n = Cx + O_{\varepsilon, L}(x^{3/5+\varepsilon}).$$

Now $\zeta_2(1)$ is non-zero so the function ζ_2 is invertible under Dirichlet convolution; let μ_2 be its inverse. Then we have

$$\frac{1}{\zeta(2s)} = \sum_{n=1}^{+\infty} \mu_2(n) n^{-s}.$$

and

$$\sum_{n\leqslant x} |a_n|^2 = \sum_{n\leqslant x} \sum_{kd=n} \mu_2(d) b_k$$

$$= \sum_{d\leqslant x} \mu_2(d) \sum_{k\leqslant x/d} b_k$$

$$= C \sum_{d\leqslant x} \mu_2(d) \frac{x}{d} + O_{\varepsilon,L}\left(x^{3/5+\varepsilon} \sum_{d\leqslant x} \mu_2(d) d^{-3/5-\varepsilon} \right)$$

$$= \frac{C}{\zeta(2)} x + O_{\varepsilon,L}(x^{3/5+\varepsilon}).$$

Taking $x = n$ and $x = n - 1$ in the last inequality, we obtain $|a_n|^2 = O_{\varepsilon,L}(n^{3/5+\varepsilon})$, whence the theorem. □

7.7   Comments and References

The proof of Theorem 7.1 that we gave in this chapter is due to Luo, Rudnick and Sarnak [84]. In fact their theorem is much more general, since it treats all congruence groups and in fact applies to cuspidal automorphic representations of GL(n).

In particular, it is not necessary to assume that N is squarefree; and we could just as well replace $\Gamma_0(N)$ by $\Gamma(N)$. We can also improve the lower bound from $5/36 (= 1/4 - (1/3)^2)$ to $4/25 (= 1/4 - (1/2 - 1/5)^2)$, as we shall explain below.

The original proof of Selberg gives the lower bound of $3/16$. There are nevertheless several advantages to the above method: (1) it can be adapted to any number field, (2) it extends to the groups GL(n). Finally, this method arises in the recent proof of Kim and Sarnak [68] of the best known approximation to the Selberg conjecture, namely $\frac{1}{4} - \left(\frac{7}{64}\right)^2$. At present the article [11] of Blomer and Brumley contains the more complete results. We have limited ourselves to the case treated here in order to simplify the exposition, but the argument generalizes in a natural way.

§7.1

A deep theorem of Margulis [1, 89] states that if Γ is a Fuchsian group, then $\mathrm{Com}(\Gamma)/\Gamma$ is infinite if and only if Γ comes from a quaternion algebra (possibly defined over a finite extension of $\mathbf{Q}$). In this case the commensurator is dense in G.

§7.2

Equality (7.6) is proved in [92, (4.5.25)]. It allows one to recover the "classical" Hecke operators defined, for example, in [118, Chap. VI §5]. On the other hand, the equality (7.8) is proved in [92, Th. 4.5.13].

Atkin-Lehner theory is developed in [5] in the case of modular forms. We have followed the book of Miyake [92], transposing the statements there into the context of Maaß forms. The proofs are completely identical in the two cases: Lemma 7.8 reproduces [92, Th. 4.6.4], Lemma 7.9 reproduces [92, Lem. 4.6.5], Lemma 7.9 reproduces [92, Lem. 4.6.7] and Lemma 7.11 reproduces [92, Th. 4.6.8].

Theorem 7.12 is due to Atkin and Lehner. It is not necessary to assume that N' divides N nor that f and f' lie in the same Laplacian eigenspace; it suffices to assume that f and f' share the same T_p eigenvalue for all but a finite number of primes p. However, to operate under such weak assumptions seems to require passing to the study of L-functions and the use of non-trivial estimates on the Fourier coefficients of Maaß forms. We refer the reader to [5] for more details in the case of modular forms.

§7.3

Lemmas 7.18 and 7.19 are classical, due to Gauß and Jacobi. For more details one can consult [31, 70].

§7.4

In formula (7.16), the character χ is primitive. We can show that the function $L(s, \chi)$ has a "good functional equation", see [20, 31, 75].

Theorem 7.28 is remarkable: even though we do not know how to realize the Rankin-Selberg L-function as the L-function of an automorphic form, we can nevertheless prove the existence of a meromorphic continuation and functional equation. This important theorem – in the case of modular forms – is due to Rankin [100] and Selberg [114] when $N = 1$. The extension to arbitrary level is more delicate and requires the use of newforms. Rather than aim for maximal generality, we have stated and proved the result which was needed in the proof of Theorem 7.1. The extension to arbitrary level of the theorem of Rankin-Selberg was obtained by Winnie Li [81] in the classical language (and again for modular forms) and by Jacquet [66] in the adelic language. Theorem 7.28 remains true if N has square factors or if χ is odd: one must simply adjust the factor $\prod_{p|N}(1 - \chi^2(p)\xi_p^{-1}\eta_p^{-1}p^{-s})^{-1}$ and the computations are a bit more intricate, see [81]. The right point of view for approaching the general case is the theory of representations of adelic groups, see [27, 66] or [9].

§7.5

The proof of Proposition 7.39 is based on the method of the "approximate functional equation", see [46, 49, 61, 64, 78].

The upper bound (7.54) is not optimal. In Appendix C, Valentin Blomer and Farrell Brumley explain more generally how to relate the sums

$$S_n(r, q) = \sum_{\chi \ (\mathrm{mod}\ q)}^{*} \overline{\chi}(r)\tau(\chi)^n$$

(here q, n are two integers with $q, n \geq 2$, r is an integer relatively prime to q, and the asterisk indicates that the sum is taken over primitive characters) to certain exponential sums called "hyper-Kloosterman sums". A theorem of Deligne furnishes an optimal upper bound on these sums and allows one to replace the bound (7.54) by

$$\sum_{\chi \neq \chi_{\text{triv}} \text{ even}} \overline{\chi}(nN'^2)\tau(\chi)^4 = O(q^{5/2}).$$

In Theorem 7.38 one can then replace $2/3$ by $1 - 2/5$; this is in fact the exact result that Luo, Rudnick and Sarnak prove.

The proof of Deligne's theorem is extremely difficult and passes through mountains of algebraic geometry. It is natural to ask whether it is absolutely necessary to appeal to such a powerful result. Indeed, one can in fact refine the argument of Luo, Rudnick and Sarnak without recourse to the Deligne bounds. In Appendix C Blomer and Brumley give an elementary proof of the upper bound on hyper-Kloosterman sums when q is a *non-trivial* power of a prime. It suffices then to go through the above proof, taking q to be a non-trivial power of a prime to arrive at a proof which avoids the "mountains of algebraic geometry" (however beautiful such peaks may be) we made reference to earlier. In this way, one can prove the theorem of Luo, Rudnick and Sarnak independently of that of Deligne.

§7.6

It is rather the analog of Conjecture 7.41 for modular forms which is classically referred to as the "Ramanujan-Petersson conjecture". This version of the conjecture was moreover proved by Deligne as a consequence of his proof of the Weil conjectures, see [35].

The proof of Theorem 7.42 goes back, in principle, to Rankin and Selberg. The first general versions are due to Serre, Moreno-Shahidi and Murty, and of several other authors; see [21] for a brief summary of the literature.

Lemma 7.43 is an improvement of the famous result known as *Landau's lemma* according to which a Dirichlet series with non-negative coefficients converges absolutely to the right of its first pole; see [118, Chap. VI Prop. 7]. The improvement we present is also due to Landau [74], see [10] as well.

Theorem 7.42 remains true for an arbitrary congruence subgroup. One must work a bit harder, however; see [21].

We shall see in Chap. 9 that Conjecture 7.41 is in fact a *p*-adic analog of the Selberg conjecture. We can moreover give a proof of Theorem 7.42 which is essentially identical to the proof of Theorem 7.1:

Write

$$L^p(s, f_\chi \times f) = L(s, f_\chi \times f) \cdot \prod_{i,j=1}^{2} (1 - \chi(p)\alpha_i(p)\beta_j(p)p^{-s})$$

for the (partial) *L*-function, where we have removed the Euler factor at p. As before, one must show that if β is a real number strictly greater than $1 - \frac{2}{5}$, then there exists an infinite set of primitive Dirichlet characters such that $\chi(p) = 1$ and $L^p(\beta, f_\chi \times f) \neq 0$ (analogously to Theorem 7.38). See Luo, Rudnick and Sarnak in [86].

7.8 Exercises

Exercise 7.44 Show that

$$H_n = \bigcup_{\substack{ad=n \\ (a,N)=1}} \bigcup_{b=0}^{d-1} \Gamma \begin{pmatrix} a & b \\ 0 & d \end{pmatrix} \quad \text{(disjoint union)}.$$

Exercise 7.45 Show that

$$T_n^N \circ T_m^N = \sum_{\substack{0<d\mid(n,m) \\ (d,N)=1}} T_{mn/d^2}^N.$$

Chapter 8
Jacquet-Langlands Correspondence

In this chapter we shall explain how to use the Selberg trace formula to relate the study of the spectrum of a Fuchsian group associated with a quaternion division algebra to that of congruence covers of the modular surface.

8.1 Arithmetic of Quaternion Algebras

In this and the following section we develop the necessary arithmetical tools to obtain a more explicit expression for the hyperbolic and elliptic terms in the trace formula for the unit group of a maximal order of a quaternion division algebra or for the group $\Gamma_0(N)$ (with N squarefree). In later sections, we compare these expressions so as to find a correspondence between them – the Jacquet-Langlands correspondence.

8.1.1 Orders in Quaternion Algebras

Let A be a quaternion algebra over $\mathbf{Q}$ with centre $\mathbf{Q} \subset A$, see § 2.2. An *order* of A is a subring $\mathcal{O}$ of A of the form

$$\mathcal{O} = \{x_1 e_1 + x_2 e_2 + x_3 e_3 + x_4 e_4 \mid x_1, \ldots, x_4 \in \mathbf{Z}\},$$

where (e_1, e_2, e_3, e_4) is a basis of A over $\mathbf{Q}$. In this way, if (e_1, e_2, e_3, e_4) is a basis of A over $\mathbf{Q}$, the set $\mathcal{O}$ of linear combinations with integer coefficients of the e_j is an order in A if and only if this set is a ring; it suffices then to verify that the products $e_i e_j$ lie in $\mathcal{O}$ and that the identity element $1 \in A$ is in $\mathcal{O}$. Two such bases

© Springer International Publishing Switzerland 2016
N. Bergeron, *The Spectrum of Hyperbolic Surfaces*, Universitext,
DOI 10.1007/978-3-319-27666-3_8

(e_1, e_2, e_3, e_4) and (f_1, f_2, f_3, f_4) define the same order if and only if the change of base matrix is in $GL(4, \mathbf{Z})$.

Lemma 8.1 *If $\mathcal{O}$ is an order in A, then $\mathcal{O} \cap \mathbf{Q} = \mathbf{Z}$.*

Proof Since $\mathcal{O}$ is a $\mathbf{Z}$-module containing 1, it contains $\mathbf{Z}$; the inclusion $\mathbf{Z} \subset \mathcal{O} \cap \mathbf{Q}$ is then clear. To prove the opposite inclusion, consider a non-zero element $u = a/b$ of $\mathcal{O} \cap \mathbf{Q}$ with relatively prime a, b. Since $u \in \mathcal{O}$, we have

$$\frac{a}{b} = x_1 e_1 + x_2 e_2 + x_3 e_3 + x_4 e_4$$

for certain $x_1, \ldots, x_4 \in \mathbf{Z}$. As $\mathcal{O}$ is a ring containing u, it also contains all powers $(a/b)^n$, $n \in \mathbb{N}$. Upon multiplying both sides of the preceding equality by $(a/b)^{n-1}$, where $n \in \mathbb{N}^*$, we find

$$\left(\frac{a}{b}\right)^n = \frac{a^{n-1} x_1}{b^{n-1}} e_1 + \frac{a^{n-1} x_2}{b^{n-1}} e_2 + \frac{a^{n-1} x_3}{b^{n-1}} e_3 + \frac{a^{n-1} x_4}{b^{n-1}} e_4 \in \mathcal{O}.$$

From this we deduce that b^{n-1} must divide $a^{n-1} x_i$ for $i = 1, \ldots, 4$ and every integer $n > 0$; thus $b = \pm 1$. This establishes the second inclusion.   □

It follows from Lemma 8.1 that we can assume, changing bases for $\mathcal{O}$ if necessary, that the unit $1 \in A$ is a basis element, say $e_1 = 1$. For $i = 2, 3, 4$, we then have (see (2.1) and (2.2)):

$$e_i^2 = -N_{\text{red}}(e_i) \cdot 1 + \text{tr}(e_i) \cdot e_i;$$

as a consequence, $\text{tr}(e_i)$ and $N_{\text{red}}(e_i)$ are integers and the equality

$$\overline{e_i} = \text{tr}(e_i) - e_i$$

implies that $\overline{e_i} \in \mathcal{O}$. This proves

Lemma 8.2 *Every order $\mathcal{O}$ of A is stable by conjugation: $\overline{\mathcal{O}} = \mathcal{O}$.*

The discriminant of a basis (e_1, e_2, e_3, e_4) of A over $\mathbf{Q}$ is the determinant of the matrix of traces of pairwise products of basis elements:

$$\text{disc}(e_1, e_2, e_3, e_4) = \det((\text{tr}(e_i e_j))_{1 \leq i, j \leq 4}) \in \mathbf{Q}.$$

For example, for the standard basis $(1, i, j, k)$ of $A = D_{a,b}(\mathbf{Q})$, we find

$$\text{disc}(1, i, j, k) = \det \begin{pmatrix} 2 & 0 & 0 & 0 \\ 0 & 2a & 0 & 0 \\ 0 & 0 & 2b & 0 \\ 0 & 0 & 0 & -2ab \end{pmatrix} = -2^4 a^2 b^2.$$

Let (e_1, e_2, e_3, e_4) and (f_1, f_2, f_3, f_4) be two bases and write P for the change of basis matrix from the first to the second. The matrix of the bilinear form $(x, y) \mapsto \mathrm{tr}(xy)$ in the first basis is $(\mathrm{tr}(e_i e_j))_{i,j}$. Its matrix in the basis (f_1, f_2, f_3, f_4) is therefore equal to ${}^t P (\mathrm{tr}(e_i e_j))_{i,j} P$; but it is also $(\mathrm{tr}(f_i f_j))_{i,j}$. From this we deduce

$$\mathrm{disc}(f_1, f_2, f_3, f_4) = \mathrm{disc}(e_1, e_2, e_3, e_4)(\det P)^2. \tag{8.1}$$

If

$$\mathcal{O} = \{x_1 e_1 + x_2 e_2 + x_3 e_3 + x_4 e_4 \mid x_1, \dots, x_4 \in \mathbf{Z}\}$$

is an order of A, then $e_i e_j \in \mathcal{O}$ for $i, j = 1, \dots, 4$ and Lemmas 8.1 and 8.2 imply that

$$\mathrm{tr}(e_i e_j) = e_i e_j + \overline{e_i e_j} \in \mathcal{O} \cap \mathbf{Q} = \mathbf{Z}.$$

Consequently, $\mathrm{disc}(e_1, \dots, e_4)$ is an integer which does not depend on the choice of basis of $\mathcal{O}$; denote it by $\mathrm{disc}(\mathcal{O})$. According to the preceding arguments, if $A = D_{a,b}(\mathbf{Q})$ we have

$$\mathrm{disc}(\mathcal{O}) = \mathrm{disc}(e_1, \dots, e_4) = \mathrm{disc}(1, i, j, k)(\det P)^2 = -2^4 a^2 b^2 (\det P)^2,$$

where P is the change of basis matrix from $(1, i, j, k)$ to $(e_1, \dots, e_4)$. We can then write $\mathrm{disc}(\mathcal{O}) = -r^2$, where $r = 2^2 ab(\det P)$. As $\mathrm{disc}(\mathcal{O}) \in \mathbf{Z}$ and $\det P \in \mathbf{Q}$, we must have $r \in \mathbf{Z}$; we call this integer the *reduced discriminant* of $\mathcal{O}$, denoted $d(\mathcal{O})$. By definition we have

$$\mathrm{disc}(\mathcal{O}) = -d(\mathcal{O})^2.$$

The following proposition follows immediately from (8.1).

Proposition 8.3 *Let $\mathcal{O}$ and $\mathcal{O}'$ be two orders of A. If $\mathcal{O} \supset \mathcal{O}'$, then $d(\mathcal{O})$ divides $d(\mathcal{O}')$. Moreover, if $\mathcal{O} \supset \mathcal{O}'$ and $d(\mathcal{O}) = d(\mathcal{O}')$, then $\mathcal{O} = \mathcal{O}'$.*

As every integer has a finite number of divisors, the preceding proposition shows that an increasing sequence of orders is necessarily stationary. From this it follows that every order is contained in a *maximal order*, i.e., an order which is not strictly contained in any other one.

The preceding results generalize to the case where one replaces the base ring $\mathbf{Z}$ by a principal ideal domain, such as the ring of p-adic integers $\mathbf{Z}_p$, or in the case where the algebra A is replaced by a division algebra of arbitrary finite dimension, in particular by a field. We shall briefly return to the case of quadratic extensions of $\mathbf{Q}$ in the following subsection. Let us now consider a quaternion algebra over the ring of p-adic integers (we refer the reader to Sect. 8.5 at the end of this chapter for references pertaining to these notions). The ring of p-adic integers is principal and its unique (up to multiplication by units) irreducible element is p. Moreover, it contains the ring $\mathbf{Z}$, so that its fraction field $\mathbf{Q}_p$ contains the field $\mathbf{Q}$. Every rational quaternion algebra $A = D_{a,b}(\mathbf{Q})$ embeds naturally in a p-adic quaternion algebra

$$A \longhookrightarrow A_p = D_{a,b}(\mathbf{Q}_p)$$

and if $\mathcal{O} = \{x_1 e_1 + x_2 e_2 + x_3 e_3 + x_4 e_4 \mid x_1, \ldots, x_4 \in \mathbf{Z}\}$ is an order of A, then

$$\mathcal{O}_p = \{x_1 e_1 + x_2 e_2 + x_3 e_3 + x_4 e_4 \mid x_1, \ldots, x_4 \in \mathbf{Z}_p\}$$

is an order of A_p over $\mathbf{Z}_p$. As $(e_1, \ldots, e_4)$ is simultaneously a basis of $\mathcal{O}$ over $\mathbf{Z}$ and a basis of $\mathcal{O}_p$ over $\mathbf{Z}_p$, it is clear that $d(\mathcal{O})$ is a representative of $d(\mathcal{O}_p)$ (this latter element being well-defined up to multiplication by a unit in $\mathbf{Z}_p$).

In what follows, we admit the classification of p-adic quaternion algebras: up to isomorphism, there is a unique quaternion division algebra over $\mathbf{Q}_p$. Moreover, this quaternion division algebra contains a unique maximal order: the set of elements whose norm has positive p-adic valuation; the reduced discriminant of this order is p. From this it follows that if p is a prime not dividing the reduced discriminant $d(\mathcal{O})$, the algebra A_p is a matrix algebra; indeed, this algebra contains the order $\mathcal{O}_p$ whose reduced discriminant is invertible in $\mathbf{Z}_p$, and thus it cannot be a division algebra. In addition, if $\mathcal{O}$ is a maximal order of A, then $\mathcal{O}_p$ is a maximal order of A_p for every prime p. It follows that the discriminant $d(\mathcal{O})$ of a maximal order $\mathcal{O}$ is the product of the primes p such that A_p is a division algebra; as a result, all the maximal orders of A have the same reduced discriminant, that we shall call (by abuse of language) the *discriminant of the algebra* A. We admit that this discriminant is always the product of an *even* number of primes.

8.1.2 Orders in Quadratic Extensions of $\mathbf{Q}$

Recall that a quadratic field extension of $\mathbf{Q}$ is of the form $\mathbf{Q}(\sqrt{D})/\mathbf{Q}$ for D a *fundamental discriminant*, i.e., an integer

$$D = \begin{cases} 4d & \text{if } d \equiv 2, 3 \ (\mathrm{mod}\ 4), \\ d & \text{if } d \equiv 1 \ (\mathrm{mod}\ 4), \end{cases}$$

where $d \neq 0, 1$ is squarefree.

An order in $K = \mathbf{Q}(\sqrt{D})$ is necessarily of the form

$$\Omega(n) = \mathbf{Z} + n\omega\mathbf{Z}, \quad \text{where } \omega = \frac{D + \sqrt{D}}{2} \tag{8.2}$$

and $n = 1, 2, 3, \ldots$ The order $\Omega(1)$ is in particular the unique maximal order of $\mathbf{Q}(\sqrt{D})$. Since $[\Omega(1) : \Omega(n)] = n$, we call $\Omega(n)$ the *order of index n* of $\mathbf{Q}(\sqrt{D})$. The discriminant of $\Omega(1)$ is equal to D, the discriminant of the field K.

Let p be a prime. The algebra $K_p = K \otimes \mathbf{Q}_p$ is of dimension 2 over $\mathbf{Q}_p$. It is therefore either isomorphic to the product $\mathbf{Q}_p \times \mathbf{Q}_p$ or equal to a quadratic field extension of $\mathbf{Q}_p$. In the latter case the ideal generated by (p) in the ring of integers of K_p is of the form $\mathfrak{p}^e$, where $\mathfrak{p}$ is the unique maximal ideal in the ring of integers of K_p and the ramification index e is equal to 1 or 2. We say that K_p is a *ramified* extension if $e = 2$ and *unramified* otherwise. For an even prime p it is convenient to describe K_p with the help of the *Legendre symbol* $\left(\frac{\cdot}{p}\right)$. This is the Dirichlet character of modulus p induced by the character of $(\mathbf{Z}/p\mathbf{Z})^*$ equal to 1 on squares and -1 otherwise. When $p = 2$ we may extend the Legendre symbol to fundamental discriminants by putting

$$\left(\frac{D}{2}\right) = \begin{cases} 1 & \text{if } D \equiv 1 \ (\text{mod } 8) \\ 0 & \text{if } D \equiv 0 \ (\text{mod } 4) \\ -1 & \text{if } D \equiv 5 \ (\text{mod } 8). \end{cases}$$

The algebra K_p is then one of the following:

1. a ramified quadratic field extension $\mathbf{Q}_p$ if $\left(\frac{D}{p}\right) = 0$;

2. the unique unramified quadratic field extension $\mathbf{Q}_p$ if $\left(\frac{D}{p}\right) = -1$;

3. the product $\mathbf{Q}_p \times \mathbf{Q}_p$ if $\left(\frac{D}{p}\right) = 1$.

When p is odd, the group $\mathbf{Q}_p^*/(\mathbf{Q}_p^*)^2$ is of order 4 so that there are two isomorphism classes of quadratic ramified field extensions. When $p = 2$ there are six isomorphism classes of ramified field extensions.

In all cases, the algebra K_p contains a unique maximal order, and its discriminant is invertible precisely when the extension is unramified. On the other hand, when the extension is ramified, the discriminant of the maximal order is equal to p if p is odd, to 8 if $p = 2$ and $d \equiv 2 \ (\text{mod } 4)$ and to 4 if $p = 2$ and $d \equiv 3 \ (\text{mod } 4)$. The discriminant D is therefore equal to the product of the local discriminants.

The various orders of K_p are the subrings

$$\Omega[p^k] = \mathbf{Z}_p + p^k \omega \mathbf{Z}_p, \quad k = 0, 1, 2, \ldots,$$

where ω is as in (8.2) via the embedding $K \hookrightarrow K_p$. We have $\Omega[1] \supset \Omega[p] \supset \Omega[p^2], \ldots, [\Omega[1] : \Omega[p^k]] = p^k$ and

$$\Omega(n)_p = \Omega[p^{v_p(n)}].$$

The unique maximal order is $\Omega[1]$; it is of discriminant D.

To conclude, we recall that $h(D)$ denotes the number of equivalence classes of primitive quadratic forms (positive definite, if $D < 0$) of discriminant D. Since D is a fundamental discriminant, the number $h(D)$ is also equal to the number of equivalence classes – in the narrow sense – of ideals in the fundamental order $\Omega(1)$, see §4.4.1. Given an order Ω of K, we shall denote more generally by $h(\Omega)$ the number of equivalence classes, in the narrow sense, of ideals in the order Ω. For $\Omega = \Omega(n)$, the number $h(\Omega)$ is the same as the number of equivalence classes of binary primitive quadratic forms (positive definite, if $D < 0$) of discriminant Dn^2.

8.1.3   Strong Approximation Theorems

Let A be an algebra over $\mathbf{Q}$ (for example a quaternion algebra or a quadratic extension of $\mathbf{Q}$). The "local-global" principle consists in deducing global properties of the algebra A from "local" properties of the $A_p = A \otimes_{\mathbf{Q}} \mathbf{Q}_p$ for $p \in \{2, 3, 5, \dots\}$ and $A_\infty = A \otimes_{\mathbf{Q}} \mathbf{R}$. It is thus convenient to introduce the ring of adeles of A. We shall not recall the entire theory here; we content ourselves with citing the results we need and send the reader to the book of Miyake [92] for a more detailed description as well as for proofs.

Fix an order $\mathcal{O}$ of A and write $\mathcal{O}_p = \mathcal{O} \otimes_{\mathbf{Z}} \mathbf{Z}_p$ for the associated order of A_p. The adelic points of A consists of

$$(\alpha_p) \in \prod_{p \in \{\infty, 2, 3, 5, \dots\}} A_p$$

where all but a finite number of α_p belong to $\mathcal{O}_p$; we denote this set[1] by $\mathbf{A}_A$. Note that we have $A_\infty = A \otimes_{\mathbf{Q}} \mathbf{R}$.

Remark 8.4 The definition of $\mathbf{A}_A$ depends a priori on the choice of order $\mathcal{O}$. It is not difficult to show that this is in fact not the case and that replacing $\mathcal{O}$ by any other rational lattice in A does not change the set $\mathbf{A}_A$.

The algebra A embeds diagonally in $\mathbf{A}_A$ as

$$\alpha \in A \longmapsto (\alpha, \alpha, \dots) \in \mathbf{A}_A.$$

One can think of the following theorems – which we admit – as certain generalizations of the Chinese Remainder Theorem for the multiplicative group of a quadratic extension of $\mathbf{Q}$ or a quaternion algebra; we call them *Strong Approximation Theorems*.

---

[1] We shall not need to endow it with its natural topology.

Theorem 8.5 *Let K be a quadratic field extension of $\mathbf{Q}$ and Ω an order in K. The group K^* embeds diagonally in $\mathbf{A}_K^*$ and the cardinality of the double quotient*

$$K^* \backslash \mathbf{A}_K^* \Big/ \left(\left\{ \begin{matrix} (\mathbf{R}_+^*)^2 & \text{if } D > 0 \\ \mathbf{C}^* & \text{if } D < 0 \end{matrix} \right\} \times \prod_p \Omega_p^* \right)$$

is finite, equal in fact to $h(\Omega)$.

Theorem 8.6 *Let A be a quaternion algebra over $\mathbf{Q}$ such that $A \otimes_{\mathbf{Q}} \mathbf{R} \cong \mathcal{M}_2(\mathbf{R})$. Let $\mathcal{O}$ be an order in A such that*

$$N_{\mathrm{red}}(\mathcal{O}_p^*) = \mathbf{Z}_p^*$$

for all primes p. Then A^* embeds diagonally in $\mathbf{A}_A^*$ and, relative to the coordinate-wise product $\cdot$ in $\mathbf{A}_A^*$, we have

$$\mathbf{A}_A^* = A^* \cdot \left(\mathrm{GL}^+(2, \mathbf{R}) \times \prod_p \mathcal{O}_p^* \right).$$

8.2 Optimal Embeddings of Quadratic Fields

Henceforth we shall assume that we are in one of the two following cases:

Case I. The quaternion algebra A is equal to $D_{1,1}(\mathbf{Q}) = \mathcal{M}_2(\mathbf{Q})$ and

$$\mathcal{O} = M(N) = \left\{ \begin{pmatrix} a & b \\ c & d \end{pmatrix} \in \mathcal{M}_2(\mathbf{Z}) \;\Big|\; c \equiv 0 \;(\mathrm{mod}\; N) \right\},$$

for a certain squarefree integer $N \geqslant 1$.

Case II. The quaternion algebra A is a division algebra $D_{a,b}(\mathbf{Q})$ ($a, b > 0$) of discriminant $d_A > 1$ and $\mathcal{O}$ is a maximal order of A.

Note that in the first case the order is not maximal if $N = 1$.

Let K be a quadratic field extension of $\mathbf{Q}$ of discriminant D and Ω an order in K. Suppose that there exists an embedding $\iota : K \hookrightarrow A$. The order Ω is said to be *optimally embedded* in $\mathcal{O}$ relative to ι if $\iota(\Omega) = \mathcal{O} \cap \iota(K)$. Two optimal embeddings ι_1 and ι_2 are *conjugated* by γ if $\iota_1(\Omega) = \gamma \iota_2(\Omega) \gamma^{-1}$.

In what follows we write $\mathcal{O}^1 = \mathrm{SL}(1, \mathcal{O})$ for the elements of norm 1 in $\mathcal{O}$.

Example 8.7 The map

$$\iota : \begin{cases} \mathbf{Q}(\sqrt{2}) \longrightarrow \mathcal{M}_2(\mathbf{Q}) \\ a + b\sqrt{2} \longmapsto \begin{pmatrix} a & 2b \\ b & a \end{pmatrix} \end{cases}$$

realizes an optimal embedding of $\mathbf{Z}[\sqrt{2}]$ in the order $\mathcal{M}_2(\mathbf{Z})$.

More generally if D is a fundamental discriminant, let $\iota : \mathbf{Q}(\sqrt{D}) \to \mathcal{M}_2(\mathbf{Q})$ be the map which sends 1 to the identity matrix and ω – defined in (8.2) – to

$$\iota(\omega) = \begin{pmatrix} 0 & \frac{D-D^2}{4} \\ 1 & D \end{pmatrix}.$$

Then ι realizes an optimal embedding of the maximal order $\Omega(1) \subset \mathbf{Q}(\sqrt{D})$ in $\mathcal{M}_2(\mathbf{Z})$.

An optimal embedding does not always exist. The following theorem gives, amongst other things, an existence criterion.

Theorem 8.8 *The number $E(\Omega, \mathcal{O}^1)$ of optimal embeddings of Ω modulo conjugation by elements of $\mathcal{O}^1 = \mathrm{SL}(1, \mathcal{O})$ is given by the formulas*

Case I: $E(\Omega, \mathcal{O}^1) = h(\Omega) \cdot \begin{cases} 2 & \text{if } D < 0 \\ 1 & \text{if } D > 0 \end{cases} \cdot \prod_{p|N} \left(1 + \left(\frac{\Omega}{p}\right)\right)$

Case II: $E(\Omega, \mathcal{O}^1) = h(\Omega) \cdot \begin{cases} 2 & \text{if } D < 0 \\ 1 & \text{if } D > 0 \end{cases} \cdot \prod_{p|d(\mathcal{O})} \left(1 - \left(\frac{\Omega}{p}\right)\right),$

where

$$\left(\frac{\Omega}{p}\right) = \begin{cases} 1 & \text{if } \left(\frac{D}{p}\right) = 1 \text{ or } \Omega_p \text{ is not maximal in } K_p, \\ -1 & \text{if } \left(\frac{D}{p}\right) = -1 \text{ and } \Omega_p \text{ is maximal in } K_p, \\ 0 & \text{if } \left(\frac{D}{p}\right) = 0 \text{ and } \Omega_p \text{ is maximal in } K_p. \end{cases}$$

Proof Given a prime p and an embedding $\iota_p : K_p \hookrightarrow A_p$, the order Ω_p is said to be optimally embedded with respect to ι_p if $\iota_p(\Omega_p) = \mathcal{O}_p \cap \iota_p(K_p)$. Write $E_p(\Omega, \mathcal{O}^*)$ for the number of optimal embeddings of Ω_p modulo conjugation by elements of $\mathcal{O}_p^*$ (the invertible elements of $\mathcal{O}_p$). Finally, let $E_\infty(\Omega, \mathcal{O}^*)$ denote the number of embeddings of $K_\infty = K \otimes_{\mathbf{Q}} \mathbf{R}$ in $A_\infty = A \otimes_{\mathbf{Q}} \mathbf{R}$, counted modulo conjugation by elements in $\mathrm{GL}^+(2, \mathbf{R})$. The group $\mathrm{GL}^+(2, \mathbf{R})$ acts by homographies on the Riemann sphere $\mathbf{C} \cup \{\infty\}$ and preserves both the upper and lower half-planes. The two distinct roots of the minimal polynomial of a primitive element for K are either real or complex conjugate, according to whether D is positive or negative. Only in the first case can the roots be sent to one another by a transformation in $\mathrm{GL}^+(2, \mathbf{R})$. We deduce from this that

$$E_\infty(\Omega, \mathcal{O}^*) = \begin{cases} 2 & \text{if } D < 0 \\ 1 & \text{if } D > 0. \end{cases} \tag{8.3}$$

For almost every prime p, the algebra A_p is unramified and therefore isomorphic to the 2×2 matrix algebra over $\mathbf{Q}_p$. We begin then by calculating $E_p(\Omega, \mathcal{O}^*)$ when $A_p = \mathcal{M}_2(\mathbf{Q}_p)$.

Let p be a prime. Suppose that $A_p = \mathcal{M}_2(\mathbf{Q}_p)$ and

$$\mathcal{O}_p = \left\{ \begin{pmatrix} a & b \\ c & d \end{pmatrix} \in \mathcal{M}_2(\mathbf{Z}_p) \,\middle|\, c \in p^\nu \mathbf{Z}_p \right\}$$

for a certain integer $\nu \geq 0$. Let ι_p be an embedding of K_p in A_p. As an order in K_p, the set Ω_p is a free $\mathbf{Z}_p$-module, and one can then write

$$\Omega_p = \mathbf{Z}_p[\alpha] \quad \text{with } \alpha \in K_p.$$

Let us write

$$f_\alpha(X) = X^2 - tX + n \quad (t, n \in \mathbf{Z}_p)$$

for the minimal polynomial of α.

Lemma 8.9 *The embedding ι_p defines an optimal embedding of Ω_p if and only if there exists*

$$\xi \in E(\alpha, \Omega_p) = \{\xi \in \mathbf{Z}_p \mid f_\alpha(\xi) \equiv 0 \ (\mathrm{mod}\ p^\nu)\}$$

and

$$u_p \in N(\mathcal{O}_p) = \begin{cases} \mathbf{Q}_p^* \mathcal{O}_p^* & \text{if } \nu = 0 \\ \mathbf{Q}_p^* \mathcal{O}_p^* \cup \begin{pmatrix} 0 & 1 \\ p^\nu & 0 \end{pmatrix} \mathbf{Q}_p^* \mathcal{O}_p^* & \text{if } \nu > 0, \end{cases}$$

such that

$$u_p \iota_p(\alpha) u_p^{-1} = \alpha_\xi := \begin{pmatrix} \xi & 1 \\ -f_\alpha(\xi) & t - \xi \end{pmatrix}.$$

Proof We write

$$\iota_p(\alpha) = \begin{pmatrix} a & b \\ c & d \end{pmatrix}.$$

The embedding ι_p defines an optimal embedding of Ω_p if and only if for every integer $r \geq 1$, the set $(\mathbf{Z}_p + \iota_p(\alpha)) \cap p^r \mathcal{O}_p$ is empty. This last property is equivalent to the fact that the elements b, $a - d$ and $p^{-\nu} c$ are relatively prime integers. It is clear that, if $\xi \in E(\alpha, \Omega_p)$,

$$\alpha_\xi \in \mathcal{M}_2(\mathbf{Z}_p) \quad \text{and} \quad \alpha_\xi \notin \mathbf{Z}_p + p\mathcal{M}_2(\mathbf{Z}_p),$$

so that

$$\begin{pmatrix} 0 & 1 \\ p^\nu & 0 \end{pmatrix}^{-1} \alpha_\xi \begin{pmatrix} 0 & 1 \\ p^\nu & 0 \end{pmatrix},$$

lies in $\mathcal{M}_2(\mathbf{Z}_p)$ but not in $\mathbf{Z}_p + p\mathcal{M}_2(\mathbf{Z}_p)$. We easily deduce that the condition of the lemma is satisfied; let us show that it is necessary.

Assume then that b, $a - d$ and $p^{-\nu}c$ are relatively prime. Note that $f_\alpha(a) = -bc$. If b is a unit, put

$$u_p = \begin{pmatrix} 1 & 0 \\ 0 & b \end{pmatrix}.$$

Then $\xi = a$ lies in $E(\alpha, \Omega_p)$ and $u_p \iota_p(\alpha) u_p^{-1} = \alpha_\xi$. We must therefore reduce the argument to the case where b is a unit. If $p^{-\nu}c$ is a unit, we conjugate $\iota_p(\alpha)$ by $\begin{pmatrix} 0 & 1 \\ p^\nu & 0 \end{pmatrix}$. Otherwise we conjugate $\iota_p(\alpha)$ by $\begin{pmatrix} 1 & 1 \\ 0 & 1 \end{pmatrix}$, which replaces b by $-(a+c) + b + d$. The latter element is a unit, since p does not divide $a-d$. This establishes Lemma 8.9.   □

Let $F(\alpha, \Omega_p)$ denote the set

$$\{\xi \in \mathbf{Z}_p \mid f_\alpha(\xi) \equiv 0 \ (\mathrm{mod} \ p^{\nu+1})\},$$

if $t^2 - 4n \equiv 0 \ (\mathrm{mod} \ p)$ and $\nu \geq 1$, and the empty set otherwise.

Lemma 8.10 *The number of optimal embeddings of Ω_p modulo inner automorphisms induced by $\mathcal{O}_p^*$ is equal to the sum of the cardinality of the image of $E(\alpha, \Omega_p)$ in $\mathbf{Z}/p^\nu \mathbf{Z}$ and the cardinality of the image of $F(\alpha, \Omega_p)$ in $\mathbf{Z}/p^\nu \mathbf{Z}$.*

In particular, if $\nu = 0$ this number is equal to 1.

Proof Let

$$w = \begin{pmatrix} 0 & 1 \\ p^\nu & 0 \end{pmatrix}.$$

To count the equivalence classes modulo conjugation by $\mathcal{O}_p^*$ we begin by observing that, according to Lemma 8.9, two optimal embeddings ι_p and κ_p of Ω_p are equivalent modulo $N(\mathcal{O}_p)$ if and only if ι_p is equivalent to κ_p or to $w\kappa_p w^{-1}$ modulo $\mathcal{O}_p^*$. According to Lemma 8.9, it remains then to determine when α_ξ and $\alpha_{\xi'}$ (resp. α_ξ and $w\alpha_{\xi'}w^{-1}$), with $\xi, \xi' \in E(\alpha, \Omega_p)$, are equivalent modulo $\mathcal{O}_p^*$.

We start by noting that if $\xi \equiv \xi' \ (\mathrm{mod} \ p^\nu)$, we have

$$u_p = \begin{pmatrix} 1 & 0 \\ \xi - \xi' & 1 \end{pmatrix} \in \mathcal{O}_p^*$$

and

$$u_p \alpha_\xi u_p^{-1} = \begin{pmatrix} \xi' & 1 \\ * & * \end{pmatrix} = \alpha_{\xi'}.$$

Conversely, suppose that α_ξ is equivalent to $\alpha_{\xi'}$ mod $\mathcal{O}_p^*$. Every element of $\mathcal{O}_p^*$ being upper triangular mod p^ν, the same is true of α_ξ. Thus if $u_p \in \mathcal{O}_p^*$ conjugates $\alpha_{\xi'}$ to α_ξ, the matrix $u_p \alpha_\xi u_p^{-1}$ has the same diagonal entries mod p^ν as α_ξ. We deduce that $\xi \equiv \xi' \pmod{p^\nu}$. Finally,

$$\alpha_\xi \text{ is equivalent to } \alpha_{\xi'} \iff \xi \equiv \xi' \pmod{p^\nu}. \tag{8.4}$$

We now show in the same way that if $\nu \geqslant 1$,

$$\alpha_\xi \text{ is equivalent to } w\alpha_{\xi'}w^{-1}$$

$$\iff \begin{cases} \xi \equiv t - \xi' \pmod{p^\nu} \text{ and } f_\alpha(\xi') \not\equiv 0 \pmod{p^{\nu+1}} & \text{if } p|(t^2 - 4n), \\ \xi \equiv t - \xi' \pmod{p^\nu} & \text{else.} \end{cases} \tag{8.5}$$

If $p^{-\nu}f_\alpha(\xi')$ is a unit, it follows from Lemma 8.9 (and its proof) that the matrix $w\alpha_{\xi'}w^{-1}$ is equivalent to

$$\begin{pmatrix} t - \xi' & 1 \\ -f_\alpha(\xi') & \xi' \end{pmatrix}.$$

In this way, according to (8.4) α_ξ is equivalent to $w\alpha_{\xi'}w^{-1}$ mod $\mathcal{O}_p^*$ if and only if $\xi = t - \xi' \pmod{p^\nu}$. If now $p^{-\nu}f_\alpha(\xi')$ is not a unit, we put

$$u_p = \begin{pmatrix} 1 & b \\ 0 & 1 \end{pmatrix},$$

with $b \in \mathbf{Z}_p$ to be chosen later. Modulo p^ν, we have

$$u_p w\alpha_{\xi'}w^{-1}u_p^{-1} = \begin{pmatrix} t - \xi' & b(2\xi' - t) - p^{-\nu}f_\alpha(\xi') \\ * & * \end{pmatrix}.$$

Thus if $2\xi' - t$ is a unit, which is to say that $t^2 - 4n$ is a unit, we can choose b in such a way that $u_p w\alpha_{\xi'}w^{-1}u_p^{-1}$ is again equivalent to

$$\begin{pmatrix} t - \xi' & 1 \\ -f_\alpha(\xi') & \xi' \end{pmatrix}$$

mod $\mathcal{O}_p^*$. Finally, suppose that $p^{-\nu}f_\alpha(\xi')$ and $t^2 - 4n$ are not units. Observing that $\mathcal{O}_p^*$ is generated modulo p^ν by diagonal matrices and matrices of the form

$$\begin{pmatrix} 1 & b \\ 0 & 1 \end{pmatrix},$$

we deduce that for all $u_p \in \mathcal{O}_p^*$, if $u_p w \alpha_{\xi'} w^{-1} u_p^{-1} = (x_{ij})$, x_{12} is never a unit, and thus $w \alpha_{\xi'} w^{-1}$ cannot be equivalent to α_ξ mod $\mathcal{O}_p^*$.

The first part of the lemma follows from (8.4) and (8.5). Note finally that if $\nu = 0$ it is clear that $E(\alpha, \Omega_p)$ is not empty and that the number of optimal embeddings of Ω_p modulo inner automorphisms induced by $\mathcal{O}_p^*$ is equal to 1.   □

Let us come back to the global count and denote now by

$$\Phi : \iota \longmapsto (\iota_p)_{p \in \{\infty, 2, 3, 5, \dots\}}$$

the map which to any optimal embedding ι of Ω associates the sequence of optimal embeddings of Ω_p, obtained by censoring by $\mathbf{Q}_p$; two sequences $(\iota_p^1)_{p \in \{\infty, 2, 3, 5, \dots\}}$, $(\iota_p^2)_{p \in \{\infty, 2, 3, 5, \dots\}}$ are said to be equivalent, written

$$(\iota_p^1)_{p \in \{\infty, 2, 3, 5, \dots\}} \sim (\iota_p^2)_{p \in \{\infty, 2, 3, 5, \dots\}},$$

if there exists a sequence

$$u = (u_p)_{p \in \{\infty, 2, 3, 5, \dots\}} \in U = GL^+(2, \mathbf{R}) \times \prod_p \mathcal{O}_p^*$$

such that for $p \in \{\infty, 2, 3, 5, \dots\}$,

$$\iota_p^2 = u_p \iota_p^1 u_p^{-1}.$$

Lemma 8.11 *We have*

$$E(\Omega, \mathcal{O}^1) = h(\Omega) \cdot \prod_{p \in \{\infty, 2, 3, 5, \dots\}} E_p(\Omega, \mathcal{O}^*).$$

Proof We fix an embedding $\iota : K \hookrightarrow A$ (which, recall, we have assumed to exist). We begin by showing that the strong approximation theorem (Theorem 8.6) implies that Φ is surjective. Indeed, fix a sequence $(\kappa_p)_{p \in \{\infty, 2, 3, 5, \dots\}}$ of optimal embeddings of Ω_p. Since, according to Lemma 8.10, for almost all p, the embeddings ι_p and κ_p are conjugate by an element of $\mathcal{O}_p^*$, and more generally ι_p and κ_p are conjugate by elements in A_p^*, Theorem 8.6 implies that there exists an invertible element $\gamma \in A^*$ such that

$$\Phi(\gamma \iota \gamma^{-1}) \sim (\kappa_p)_{p \in \{\infty, 2, 3, 5, \dots\}};$$

the embedding $\gamma \iota \gamma^{-1}$ is necessarily an optimal embedding of Ω.

Suppose now that ι is an optimal embedding of Ω. Given an element $\gamma \in A^*$, we have

$$\Phi(\gamma \iota \gamma^{-1}) \sim \Phi(\iota) \iff \gamma \in (U \cdot \mathbf{A}_K^*) \cap A^*$$

(here A^* and $\mathbf{A}_K^*$ are viewed as embedded in $\mathbf{A}_A^*$). Thus

$$E(\Omega, \mathcal{O}^1) = C \cdot \prod_{p \in \{\infty, 2, 3, 5, \dots\}} E_p(\Omega, \mathcal{O}^*),$$

where the constant C is given by the formula

$$C = \left| \mathcal{O}^1 \backslash (U \cdot \mathbf{A}_K^*) \cap A^* / K^* \right|.$$

Given $t_1, t_2 \in K^*$, Theorem 8.6 implies that there exists two elements $u_1, u_2 \in U$ such that

$$u_1 t_1 \text{ and } u_2 t_2 \in (U \cdot \mathbf{A}_K^*) \cap A^*.$$

It then follows from $\mathcal{O}^1 = U \cap A^*$ that

$$\mathcal{O}^1 u_1 t_1 \mathbf{A}_K^* = \mathcal{O}^1 u_2 t_2 \mathbf{A}_K^*$$

if and only if

$$t_1 t_2^{-1} \in (\mathbf{A}_K^* \cap U) \cdot (\mathbf{A}_K^* \cap A^*) = (\mathbf{A}_K^* \cap U) \cdot K^*.$$

Therefore

$$C = |\mathbf{A}_K^* / (\mathbf{A}_K^* \cap U) \cdot K^*|$$
$$= |K^* \backslash \mathbf{A}_K^* / (\mathbf{A}_K^* \cap U)|.$$

But

$$\mathbf{A}_K^* \cap U = \left\{ \begin{matrix} \mathbf{R}_+^* & \text{if } D > 0 \\ \mathbf{C}^* & \text{if } D < 0 \end{matrix} \right\} \times \prod_p \mathcal{O}_p^*,$$

and the lemma follows from Theorem 8.5. □

The following lemmas allow us to complete the proof of Theorem 8.8.

Lemma 8.12 *Let p be a prime such that K_p is a quadratic field extension of $\mathbf{Q}_p$, i.e., such that $(D/p) \neq 1$. Let A_p be a division algebra and $\mathcal{O}_p$ a maximal order of A_p. If Ω_p is a maximal order of K_p, then*

$$
E_p(\Omega, \mathcal{O}^*) = \begin{cases} 2 & \text{if } K_p/\mathbf{Q}_p \text{ is unramified} \\ 1 & \text{if } K_p/\mathbf{Q}_p \text{ is (totally) ramified.} \end{cases}
$$

If Ω_p is not maximal, $E_p(\Omega, \mathcal{O}^*) = 0$.

Proof Suppose that $K_p = \mathbf{Q}_p[\alpha]$ is contained in A_p. The set of optimal embeddings of Ω_p is in bijection with a subset of the conjugation class of α in A_p^*, equal to

$$
C(\alpha, \Omega_p) = \{h\alpha h^{-1} \mid h \in A_p^*, \ \mathbf{Q}_p[h\alpha h^{-1}] \cap \mathcal{O}_p = h\Omega_p h^{-1}\}.
$$

Note that the unique maximal order $\mathcal{O}_p$ of A_p coincides with the set of elements of A_p whose reduced norm and trace both belong to $\mathbf{Z}_p$. The conjugacy class of an element α in A_p^* thus meets $\mathcal{O}_p$ if and only if α itself belongs to $\mathcal{O}_p$. Furthermore, if α is a non-scalar element of $\mathcal{O}_p$, the intersection $\mathbf{Q}_p[\alpha] \cap \mathcal{O}_p$ is the unique maximal order of $\mathbf{Q}_p[\alpha]$. Hence one sees that if Ω_p is not maximal $C(\alpha, \Omega_p)$ is empty; there is therefore no optimal embedding of Ω_p.

Suppose now that Ω_p is maximal. There then exists an element in $C(\alpha, \Omega_p)$, which is in fact unique up to conjugation by an element in A_p^*. Note that there exists an element $\pi_p \in A_p$ of reduced norm p such that

$$
A_p = \bigcup_{n=0}^{\infty} \pi_p^{-n} \mathcal{O}_p.
$$

According to whether π_p belongs to $\mathbf{Q}_p[\alpha]$ or not, we deduce

$$
A_p = \begin{cases} \mathbf{Q}_p[\alpha]\mathcal{O}_p, & \text{if } p \text{ ramifies in } \mathbf{Q}_p[\alpha], \\ \mathbf{Q}_p[\alpha]\mathcal{O}_p \cup \pi_p \mathbf{Q}_p[\alpha]\mathcal{O}_p, & \text{else.} \end{cases}
$$

The lemma then follows immediately.   □

Lemma 8.13 *Let p be a prime. Let $A_p = M_2(\mathbf{Q}_p)$ and*

$$
\mathcal{O}_p = \left\{ \begin{pmatrix} a & b \\ c & d \end{pmatrix} \in M_2(\mathbf{Z}_p) \;\middle|\; c \in p\mathbf{Z}_p \right\}.
$$

Then

$$E_p(\Omega, \mathcal{O}^*) = \begin{cases} 2 & \text{if } \Omega_p \text{ is not maximal} \\ 1 + \left(\dfrac{D}{p}\right) & \text{else.} \end{cases}$$

Proof We use Lemma 8.10. Here $v = 1$ and, just as at the end of the proof of Lemma 8.10, we can assume that $r = 0$. Let ℓ denote the index of the order Ω. Since $t^2 - 4n$ is the discriminant $\ell^2 D$ of the order Ω_p, Lemma 8.10 then implies that

$$E_p(\Omega, \mathcal{O}^*) = |E/p| + |F/p|, \tag{8.6}$$

where

$$E = \{\xi \in \mathbf{Z}_p \mid \xi^2 - t\xi + n \equiv 0 \ (\mathrm{mod}\ p)\},$$

$$F = \begin{cases} \{\xi \in \mathbf{Z}_p \mid \xi^2 - t\xi + n \equiv 0 \ (\mathrm{mod}\ p^2)\}, & \text{if } p|\ell^2 D, \\ \varnothing & \text{if } p \nmid \ell^2 D. \end{cases}$$

and E/p (resp. F/p) denotes the image of E (resp. F) in $\mathbf{Z}/p\mathbf{Z}$.

We must therefore calculate (8.6). Assume first of all that $p \neq 2$. Since $t^2 - 4n = \ell^2 D$, the element ξ belongs to E if and only if

$$(2\xi - t)^2 \equiv \ell^2 D \ (\mathrm{mod}\ p). \tag{8.7}$$

We make the temporary assumption that either Ω_p is not maximal (so that $p|\ell$) or $p|D$. Equation (8.7) is then equivalent to $\xi \equiv t/2 \ (\mathrm{mod}\ p)$, which admits a unique solution modulo p. Using the fact that p^2 does not divide the fundamental discriminant D, we similarly obtain

$$|F/p| = \begin{cases} 1 & \text{if } \Omega_p \text{ is not maximal,} \\ 0 & \text{if } \Omega_p \text{ is maximal and } p|D. \end{cases}$$

If we now assume that Ω_p is maximal and that p does not divide D, Eq. (8.7) then admits two solutions modulo p if $\left(\dfrac{D}{p}\right) = 1$ and no solution if $\left(\dfrac{D}{p}\right) = -1$. Moreover, the set F is empty since $p \nmid \ell^2 D$. Finally, we have shown that for $p \neq 2$,

$$|E/p| + |F/p| = \begin{cases} 2 & \text{if } \Omega_p \text{ is not maximal,} \\ 1 & \text{if } \Omega_p \text{ is maximal and } p|D, \\ 2 & \text{if } \Omega_p \text{ is maximal and } \left(\dfrac{D}{p}\right) = 1, \\ 0 & \text{if } \Omega_p \text{ is maximal and } \left(\dfrac{D}{p}\right) = -1. \end{cases} \tag{8.8}$$

By slightly modifying the above calculations, we can show that (8.8) remains valid
for $p = 2$. The lemma then follows immediately from (8.6).   □

Finally, Theorem 8.8 is a consequence of Lemmas 8.11, 8.10, 8.12 and 8.13 as
well as identity (8.3).   □

8.3   The Trace Formula

In this section we reformulate the Selberg trace formula in the case of the groups $\mathcal{O}^1$,
where $\mathcal{O}$ is a maximal order in a quaternion algebra $A = D_{a,b}(\mathbf{Q})$ with $a, b > 0$ of
discriminant $d_A > 1$ (case II) and the congruence subgroup $\Gamma_0(N)$ of $SL(2, \mathbf{Z})$ for
squarefree $N \geqslant 1$ (case I).

We begin with case II. Let $\gamma \in \mathcal{O}$, $\gamma \notin \mathbf{Q}$, such that $\operatorname{tr}(\gamma) = t$ and $N_{red}(\gamma) = n$.
The subalgebra $\mathbf{Q}[\gamma] \subset A$ is commutative and of dimension 2 over $\mathbf{Q}$; it is a
quadratic field extension K of $\mathbf{Q}$, completely determined by t and n. The inclusion
$\mathbf{Q}[\gamma] \subset A$ induces an embedding of K into A. We admit that two such embeddings
are conjugate by an element of A^*. The set of embeddings of K into A is therefore
in bijection with the conjugacy class $\{\gamma\}_{A^*}$ of γ in A.

Consider now the conjugacy classes $\{\delta\}_{\mathcal{O}^1}$ relative to $\mathcal{O}^1$. For $t, n \in \mathbf{Z}$, we denote
by $E(t, n, \mathcal{O}^1)$ the number of conjugacy classes $\{\delta\}_{\mathcal{O}^1}$ with $\delta \in \mathcal{O}$, $\operatorname{tr}(\delta) = t$ and
$N_{red}(\delta) = n$.

Lemma 8.14 *Let $\gamma \in \mathcal{O}$, $\gamma \notin \mathbf{Q}$, such that $\operatorname{tr}(\gamma) = t$ and $N_{red}(\gamma) = n$. Then*

$$E(t, n, \mathcal{O}^1) = \sum_{\Omega \supset \mathbf{Z}[\gamma]} E(\Omega, \mathcal{O}^1),$$

where the sum runs over the set of orders in $\mathbf{Q}[\gamma]$ which contain γ.

Proof If $\alpha \in A^*$, then $\mathbf{Q}[\gamma] \cap \alpha^{-1}\mathcal{O}\alpha$ is an order of $\mathbf{Q}[\gamma]$. We thus have

$$\{\gamma\}_{A^*} = \bigcup_{\Omega} C(\gamma, \Omega), \tag{8.9}$$

where the union is taken over the set of orders of $\mathbf{Q}[\gamma]$ and

$$C(\gamma, \Omega) = \left\{ \alpha\gamma\alpha^{-1} \mid \alpha \in A^*, \ \mathbf{Q}[\gamma] \cap \alpha^{-1}\mathcal{O}\alpha = \Omega \right\}.$$

Note moreover that the union (8.9) is disjoint. To see this, suppose that $\alpha\gamma\alpha^{-1} = \beta\gamma\beta^{-1}$ ($\alpha, \beta \in A^*$). Then $\alpha^{-1}\beta$ commutes with γ and $\mathbf{Q}[\gamma, \alpha^{-1}\beta]$ is an algebra of
dimension 2 or 4 over $\mathbf{Q}$ (it contains $\mathbf{Q}[\gamma]$, which itself is of dimension 2 over $\mathbf{Q}$).

From the commutativity of $\mathbf{Q}[\gamma, \alpha^{-1}\beta]$ we have $\alpha^{-1}\beta \in \mathbf{Q}[\gamma]$ and thus

$$
\begin{aligned}
\mathbf{Q}[\gamma] \cap \beta^{-1}\mathcal{O}\beta &= (\alpha^{-1}\beta) \left(\mathbf{Q}[\gamma] \cap \beta^{-1}\mathcal{O}\beta\right) (\alpha^{-1}\beta)^{-1} \\
&= \mathbf{Q}[\gamma] \cap \alpha^{-1}\mathcal{O}\alpha.
\end{aligned}
$$

It follows from the definition that the set $C(\gamma, \Omega)$ meets $\mathcal{O}$ if and only if γ is contained in the order Ω. But the elements of $C(\gamma, \Omega)$ correspond exactly to the embeddings of K into A relative to which the order Ω is optimally embedded in $\mathcal{O}$. Considering these embeddings modulo conjugation by $\mathcal{O}^1$, we obtain the claimed expression. $\qquad\square$

The isomorphism between A_∞ and $\mathcal{M}_2(\mathbf{R})$ induces an embedding $\mathcal{O} \to \mathcal{M}_2(\mathbf{R})$. According to Theorem 2.3, the image $\mathcal{O}^1 \subset SL(2, \mathbf{R})$ is a co-compact Fuchsian subgroup; we denote it by $\Gamma_\mathcal{O}$. An element $\gamma \in \Gamma_\mathcal{O}$ of trace $t > 2$ is hyperbolic; write $E'(t, 1, \mathcal{O}^1)$ for the number of conjugacy classes in $\Gamma_\mathcal{O}$ of primitive elements of trace t. Note that if γ is hyperbolic, then for $r \geq 1$ we have $\{\gamma\}_{\mathcal{O}^1} = \{\beta\}_{\mathcal{O}^1}$ if and only if $\{\gamma^r\}_{\mathcal{O}^1} = \{\beta^r\}_{\mathcal{O}^1}$. We deduce that

$$
E(t, 1, \mathcal{O}^1) = \sum_{s \leqslant t}{}' E'(s, 1, \mathcal{O}^1), \tag{8.10}
$$

where the sum runs over the set of s such that if $\operatorname{tr}(\gamma) = s$, there exists an integer $r \geq 1$ such that $\operatorname{tr}(\gamma^r) = t$.

We can now reformulate for the group $\Gamma_\mathcal{O}$ the Selberg trace formula proved in Chap. 5.

In what follows, $h : \mathbf{C} \to \mathbf{C}$ is an even function, holomorphic in the vertical strip $|\operatorname{Im} r| < \frac{1}{2} + \varepsilon$ for some $\varepsilon > 0$, and satisfying the following decay property

$$
|h(r)| = O\left((1 + \operatorname{Re}(r))^{-2-\delta}\right)
$$

for a constant $\delta > 0$.

Proposition 8.15 *Denote by $\lambda_k = r_k^2 + 1/4$, $k \in \mathbf{N}$, the set of eigenvalues of the hyperbolic Laplacian in $L^2(\Gamma_\mathcal{O} \backslash \mathcal{H})$, counted with multiplicities. Then*

$$
\begin{aligned}
\sum_{k=0}^{+\infty} h(r_k) = {}& \frac{\operatorname{area}(\Gamma_\mathcal{O})}{4\pi} \int_{-\infty}^{+\infty} r h(r) \tanh(\pi r)\, dr \\
&+ \sum_{t \in \{0,1\}} \frac{E'(t, 1, \mathcal{O}^1)}{2m_t} \sum_{k=1}^{m_t-1} \frac{1}{\sin(k\pi/m_t)} \int_{-\infty}^{+\infty} h(r) \frac{e^{-2k\pi r/m_t}}{1 + e^{-2\pi r}}\, dr \\
&+ \sum_{t=3}^{+\infty} E'(t, 1, \mathcal{O}^1) \operatorname{arccosh}(t/2) \sum_{k=1}^{+\infty} \frac{g(2k \operatorname{arccosh}(t/2))}{\sinh(k \operatorname{arccosh}(t/2))},
\end{aligned}
$$

where area($\Gamma_{\mathcal{O}}$) is the area of a fundamental domain for the action of $\Gamma_{\mathcal{O}}$ on $\mathcal{H}$, m_t for $t \in \{0, 1\}$ is the order of the primitive element of trace t, and

$$g(u) = \frac{1}{2\pi} \int_{-\infty}^{+\infty} h(r) e^{-iru} dr$$

is the Fourier transform of h.

Proof We have just rewritten the trace formula (Theorem 5.8) by replacing the sums over representatives of primitive hyperbolic and elliptic conjugacy classes in $\Gamma_{\mathcal{O}}$ by sums over the traces $t \in \mathbf{N}$, $t \neq 2$, of these representatives. The number of primitive conjugacy classes of trace t is $E'(t, 1, \mathcal{O}^1)$.   □

Note that the area area($\Gamma_{\mathcal{O}}$) can be explicitly calculated.

Lemma 8.16 *We have*

$$\text{area}(\Gamma_{\mathcal{O}}) = \frac{\pi}{3} \prod_{p | d_A} (p - 1).$$

Idea of the proof Fix a **Z**-basis (e_1, e_2, e_3, e_4) of A. Every element of $A_\infty = A \otimes \mathbf{R}$ can be written as

$$x = x_1 e_1 + x_2 e_2 + x_3 e_3 + x_4 e_4 \quad (x_i \in \mathbf{R})$$

and the group $\mathcal{O}^1$ acts by left-multiplication. We obtain the lemma by calculating in two different ways the integral

$$\int_F dx_1 \cdots dx_4, \tag{8.11}$$

where F is a fundamental domain for $\mathcal{O}^1$ in $\{x \in A_\infty \mid |N_{\text{red}}(x)| \leq 1\}$.

The algebra A_∞ is isomorphic to the algebra of 2×2 matrices; let σ be an isomorphism. Denote by E_{ij} $(j, k = 1, 2)$ the canonical basis of $\mathcal{M}_2(\mathbf{R})$ and x_{ij} the corresponding coordinates of an element $x \in A_\infty$, so that

$$\sigma(x) = \begin{pmatrix} x_{11} & x_{12} \\ x_{21} & x_{22} \end{pmatrix}.$$

We then have $N_{\text{red}}(x) = \det(\sigma(x)) = x_{11} x_{22} - x_{12} x_{21}$. By definition of the discriminant we have

$$d_A dx_1 \cdots dx_4 = dx_{11} dx_{12} dx_{21} dx_{22}.$$

A change of variables then implies that the integral (8.11) is equal to

$$\int_F dx_1 \cdots dx_4 = \frac{1}{d_A} \int_0^1 t dt \times \int_{F_2} \frac{dx_{11} dx_{22} dx_{12}}{|x_{12}|}$$

$$= \frac{2}{d_A} \int_{F_2} \frac{dx_{11} dx_{22} dx_{12}}{|x_{12}|},$$

where F_2 is a fundamental domain of $\Gamma_\mathcal{O}$ in the group $SL^\pm(2, \mathbf{R})$ of 2×2 matrices of determinant ± 1. The quotient $SL^\pm(2, \mathbf{R})/O(2)$ can be identified with the Poincaré upper half-plane $\mathcal{H}$ via the map

$$g \longmapsto z = g(i)$$

where the isotropy group of i is $O(2)$. Write dk for the Haar measure normalised in such a way as to give $O(2)$ total volume 1. We then have

$$\frac{dx_{11}dx_{22}dx_{12}}{|x_{12}|} = \pi d\mu(z)dk.$$

Let $\mathcal{F}_\mathcal{O}$ be a fundamental domain for the action of $\Gamma_\mathcal{O}$ on $\mathcal{H}$. Since an element of $\Gamma_\mathcal{O}$ acts trivially on $\mathcal{H}$ if and only if it is equal to ± 1, we find that

$$\int_{\mathcal{F}_\mathcal{O}} d\mu(z) = \frac{2}{\pi} \int_{F_2} \frac{dx_{11}dx_{22}dx_{12}}{|x_{12}|}.$$

In other words

$$\text{area}(\Gamma_\mathcal{O}) = \frac{2d_A}{\pi} \int_F dx_1 \cdots dx_4. \tag{8.12}$$

Furthermore, one can show that the integral (8.11) is equal to the residue at $s = 1$ of the zeta function $\zeta_A(s)$ of the algebra A. This function has an Euler product expansion and one has the identity

$$\zeta_A(s) = \zeta(2s)\zeta(2s - 1) \prod_{p|d_A}(1 - p^{1-2s}).$$

Its residue at $s = 1$ is therefore equal to

$$\frac{\zeta(2)}{d_A} \prod_{p|d_A}(p - 1). \tag{8.13}$$

Finally, the lemma follows from (8.12), (8.13) and the well-known fact that $\zeta(2) = \pi^2/6$. $\square$

Now consider case I. The identity (8.10) remains valid when one replaces $\mathcal{O}^1$ by $\Gamma_0(N)$. Using the fact that N is squarefree, the Selberg trace formula (Theorem 5.21) can be rewritten in the following way.

Proposition 8.17 Let $\mu_k = r_k^2 + 1/4$, $k \in \mathbb{N}$, denote the set of eigenvalues of the hyperbolic Laplacian in $L^2(\Gamma_0(N)\backslash\mathcal{H})$, counted with multiplicity. Then

$$\sum_{k=0}^{+\infty} h(r_k) = \frac{\text{area}(\Gamma_0(N))}{4\pi} \int_{-\infty}^{+\infty} rh(r) \tanh(\pi r)\, dr$$

$$+ \sum_{t\in\{0,1\}} \frac{E'(t, 1, \Gamma_0(N))}{2m_t} \sum_{k=1}^{m_t-1} \frac{1}{\sin(k\pi/m_t)} \int_{-\infty}^{+\infty} h(r)\frac{e^{-2k\pi r/m_t}}{1 + e^{-2\pi r}}\, dr$$

$$+ \sum_{t=3}^{+\infty} E'(t, 1, \Gamma_0(N)) \arccos(t/2) \sum_{k=1}^{+\infty} \frac{g(2k \arccos(t/2))}{\sinh(k \arccos(t/2))}$$

$$+ 2^{\omega(N)} \left\{ g(0) \log(\pi/2) - \frac{1}{2\pi} \int_{-\infty}^{+\infty} h(r) \left[\frac{\Gamma'}{\Gamma} \left(\frac{1}{2} + ir \right) + \frac{\Gamma'}{\Gamma}(1 + ir) \right] dr \right.$$

$$\left. + 2 \sum_{n=1}^{+\infty} \frac{\Lambda(n)}{n} g(2 \log n) - \sum_{p|N, \, p \, prime} \sum_{k=0}^{+\infty} \frac{\log p}{p^k} g(2k \log p) \right\},$$

where area($\Gamma_0(N)$) is the area of a fundamental domain for the action of $\Gamma_0(N)$ on $\mathcal{H}$, m_t for $t \in \{0, 1\}$ denotes the order of the primitive element of tracet, $\omega(N)$ is equal to the number of divisors of N and

$$g(u) = \frac{1}{2\pi} \int_{-\infty}^{+\infty} h(r) e^{-iru} dr$$

is the Fourier transform of h.

8.4   Jacquet-Langlands Correspondence and Applications

The Jacquet-Langlands correspondence links the spectrum of the Laplacian of co-compact congruence hyperbolic surfaces to the spectrum of the Laplacian of congruence covers of the modular surface. The proof proceeds by a comparison of terms on the geometric side of the trace formula. We shall therefore consider the trace formula for different groups, and to avoid all confusion we shall write

$$\sum_{f_k \in S} h(r_k)$$

for the sum over an orthonormal basis of eigenfunctions in a subspace S, where $\lambda_k = r_k^2 + 1/4$ is the Laplacian eigenvalue corresponding to f_k and h is a test function such as those arising in the trace formula of the preceding section.

Theorem 8.18 *Let $\mathcal{O}$ be a maximal order in a quaternion algebra $A = D_{a,b}(\mathbf{Q})$ with $a, b > 1$ of discriminant $d_A > 1$. Then the set of non-zero eigenvalues of the hyperbolic Laplacian in $L^2(\Gamma_{\mathcal{O}} \backslash \mathcal{H})$, counted with multiplicity, coincides with the set of eigenvalues associated with primitive Maaß forms for the group $\Gamma_0(d_A)$.*

Proof We begin by showing that

$$\sum_{f_k \in L^2(\Gamma_{\mathcal{O}} \backslash \mathcal{H})} h(r_k) = \sum_{f_k \in \mathbf{C} \oplus C_{\mathrm{prim}}(\Gamma_0(d_A) \backslash \mathcal{H})} h(r_k) \tag{8.14}$$

for every test function h. The two following lemmas will allow us to reduce the proof of (8.14) to a comparison of the right-hand sides of the trace formulas in Propositions 8.15 and 8.17. Define

$$\beta(n) = \sum_{k|n} \mu(k)\mu(n/k), \qquad (8.15)$$

where $\mu(n)$ is the MÖbius function. In particular, if n is the product of r distinct primes, then $\beta(n) = (-2)^r$. In terms of Dirichlet convolutions the function β is equal to $\mu * \mu$.

Lemma 8.19 *We have*

$$\sum_{f_k \in C_{\mathrm{prim}}(\Gamma_0(d_A)\backslash\mathcal{H})} h(r_k) = \sum_{m|d_A} \beta\,(d_A/m) \sum_{f_k \in C(\Gamma_0(m)\backslash\mathcal{H})} h(r_k).$$

Proof Let m be a divisor of d_A. To a form $f \in C_{\mathrm{prim}}(\Gamma_0(m)\backslash\mathcal{H})$ there corresponds exactly $\tau\,(d_A/m)$ oldforms in $C(\Gamma_0(d_A)\backslash\mathcal{H})$, where $\tau(n)$ is the number of positive divisors of n. Write $\delta(m,\lambda)$ for the dimension of the eigenspace of $C(\Gamma_0(m)\backslash\mathcal{H})$ corresponding to the eigenvalue λ, and write $\delta'(m,\lambda)$ for the dimension of the subspace of those which are primitive. Since $C(\Gamma_0(d_A)\backslash\mathcal{H})$ is a direct sum of spaces of oldforms and newforms, we have

$$\delta(d_A, \lambda) = \sum_{m|d_A} \tau\,(d_A/m)\,\delta'(m, \lambda).$$

Inverting this formula yields

$$\delta'(d_A, \lambda) = \sum_{m|d_A} \beta\,(d_A/m)\,\delta(m, \lambda).$$

The lemma now follows directly. □

Lemma 8.20 *For a squarefree integer $d > 1$ we have*

$$\sum_{m|d} \beta\,(d/m) = (-1)^{\omega(d)},$$

where $\omega(d)$ denotes the number of prime divisors of d.

Proof This follows immediately from the binomial theorem since

$$\sum_{m|d} \beta\,(d/m) = \sum_{i=0}^{\omega(d)} (-2)^{\omega(d)-i}\binom{\omega(d)}{i} = (1-2)^{\omega(d)} = (-1)^{\omega(d)}. \qquad □$$

Remark 8.21 The function μ is the inverse under convolution of the function 1, constantly equal to 1 on **N**. Lemma 8.20 is the formula

$$\beta * 1 = \mu$$

which is the same as saying $\beta * 1 = \mu * (\mu * 1) = \mu$.

In the same way the function β is the inverse under convolution of $1 * 1 = d$, where d is the divisor function: $d(n) = 2^{\omega(n)}$. We deduce that

$$\beta * d = \delta, \tag{8.16}$$

where δ – the unit for convolution product – is the function which is everywhere zero except at 1, where it equals 1.

Lemmas 8.19 and 8.20 imply

$$\sum_{m|d_A} \beta \, (d_A/m) \sum_{f_k \in \mathcal{C} \oplus \mathcal{C}(\Gamma_0(m)\backslash\mathcal{H})} h(r_k)$$

$$= h(r_0) + \sum_{m|d_A} \beta \, (d_A/m) \sum_{f_k \in \mathcal{C}(\Gamma_0(m)\backslash\mathcal{H})} h(r_k)$$

$$= \sum_{f_k \in \mathcal{C} \oplus \mathcal{C}_{\mathrm{prim}}(\Gamma_0(d_A)\backslash\mathcal{H})} h(r_k).$$

We are thus led to establish the following identity

$$\sum_{f_k \in L^2(\Gamma_{\mathcal{O}}\backslash\mathcal{H})} h(r_k) = \sum_{m|d_A} \beta \, (d_A/m) \sum_{f_k \in \mathcal{C} \oplus \mathcal{C}(\Gamma_0(m)\backslash\mathcal{H})} h(r_k), \tag{8.17}$$

since it is equivalent to (8.14). To this end we compare the right-hand sides of the trace formulas given in Propositions 8.15 and 8.17. We do so with the help of the following lemmas, which treat the contributions of each of the conjugacy classes (identity, elliptic, hyperbolic and parabolic) appearing in these formulas.

Lemma 8.22 *We have*

$$\mathrm{area}(\Gamma_{\mathcal{O}}) = \sum_{m|d_A} \beta \, (d_A/m) \, \mathrm{area}(\Gamma_0(m)).$$

Proof According to Lemma 8.16 and Exercise 2.18, we must show that

$$d_A \prod_{p|d_A} \left(1 - \frac{1}{p}\right) = \sum_{m|d_A} \beta \, (d_A/m) \, m \prod_{p|m} \left(1 + \frac{1}{p}\right).$$

This is equivalent to the identity

$$d_A \prod_{p|d_A} \left(1 + \frac{1}{p}\right) = \sum_{m|d_A} \tau \, (d_A/m) \, m \prod_{p|m} \left(1 - \frac{1}{p}\right),$$

which is clearly

$$d_A \prod_{p|d_A} \left(1 + \frac{1}{p}\right) = \sum_{m|d_A} \sum_{d'|m} d' \prod_{p|d'} \left(1 - \frac{1}{p}\right). \tag{8.18}$$

Since both sides of this equation are multiplicative arithmetic functions in the integer d_A, it suffices to verify (8.18) for d_A prime. This is a trivial verification. $\square$

In the elliptic and hyperbolic contributions to the trace formulas, only the numbers $E'(t, 1, \Gamma)$ depend on the group Γ. The following lemma identifies these terms.

Lemma 8.23 *For every trace t, we have*

$$E'(t, 1, \mathcal{O}^1) = \sum_{m|d_A} \beta \, (d_A/m) \, E'(t, 1, \Gamma_0(m)).$$

Proof The elliptic and hyperbolic cases can be treated in the same way. We may therefore assume that $t > 2$. Since $E(3, 1, \mathcal{O}^1) = E'(3, 1, \mathcal{O}^1)$, a simple induction on t starting with (8.10) shows that it suffices to verify

$$E(t, 1, \mathcal{O}^1) = \sum_{m|d_A} \beta \, (d_A/m) \, E(t, 1, \Gamma_0(m)). \tag{8.19}$$

Let us show more generally that if Ω is an order in a quadratic field extension of $\mathbf{Q}$, then

$$E(\Omega, \mathcal{O}^1) = \sum_{m|d_A} \beta \, (d_A/m) \, E(\Omega, \Gamma_0(m)). \tag{8.20}$$

According to Lemma 8.14 the expression (8.20) indeed implies (8.19); moreover, from Theorem 8.8 we see that the above expression is equivalent to

$$\prod_{p|d_A} \left(1 - \left(\frac{\Omega}{p}\right)\right) = \sum_{m|d_A} \beta \, (d_A/m) \prod_{p|m} \left(1 + \left(\frac{\Omega}{p}\right)\right). \tag{8.21}$$

The discriminant d_A is the product of an even number, say $2r$, of primes. Let us put

$$k = \left| \left\{ p \mid p|d_A, \ \left(\frac{\Omega}{p}\right) = 1 \right\} \right| \text{ and } e = \left| \left\{ p \mid p|d_A, \ \left(\frac{\Omega}{p}\right) = 0 \right\} \right|.$$

It is clear that

$$\prod_{p \mid d_A}\left(1 - \left(\frac{\Omega}{p}\right)\right) = \begin{cases} 0 & \text{if } k > 0, \\ 2^{2r-e} & \text{otherwise.} \end{cases}$$

Furthermore, all of the non-zero terms of the right-hand side of (8.21) correspond to the various possible products of the $k + e$ primes such that $\left(\frac{\Omega}{p}\right) \neq -1$, which then leads to the expression

$$\sum_{m \mid d_A} \beta\,(d_A/m) \prod_{p \mid m}\left(1 + \left(\frac{\Omega}{p}\right)\right) = \sum_{j=0}^{e}\sum_{i=0}^{k}(-2)^{2r-i-j} \cdot 2^i \cdot 1^j \binom{k}{i}\binom{e}{j}$$

$$= 2^{2r-e}\sum_{j=0}^{e}(-1)^j 2^{e-j}\binom{e}{j}\sum_{i=0}^{k}(-1)^i\binom{k}{i}.$$

The expression (8.21) now follows from the binomial theorem, and more precisely from

$$\sum_{j=0}^{e}(-1)^j 2^{e-j}\binom{e}{j} = (2-1)^e = 1$$

and

$$\sum_{i=0}^{k}(-1)^i\binom{k}{i} = \begin{cases} (1-1)^k = 0 & \text{if } k > 0, \\ 1 & \text{if } k = 0. \end{cases}$$

$\square$

The parabolic terms only arise in the context of the groups $\Gamma_0(m)$; their contribution to (8.17) is of the form

$$\sum_{m \mid d_A} \beta\,(d_A/m)\,2^{\omega(m)}\left(C + \sum_{p \mid m} f(p)\right),$$

where C is a constant (independent of m) and f depends only on p. The following lemma shows that this contribution vanishes.

Lemma 8.24 *If $d_A > 1$, we have*

$$\sum_{m \mid d_A} \beta\,(d_A/m)\,2^{\omega(m)} = 0.$$

Moreover, if f is an arbitrary function and if d_A has at least two distinct prime divisors, then

$$\sum_{m|d_A} \beta\,(d_A/m)\,2^{\omega(m)} \sum_{p|m} f(p) = 0,$$

where the outer sum runs over all the divisors of d_A and the inner sum runs over all prime divisors of m.

Proof The first part follows from (8.16) or from the following computation:

$$\sum_{m|d_A} \beta\,(d_A/m)\,2^{\omega(m)} = \sum_{i=0}^{\omega(d_A)} (-2)^{\omega(d_A)-i}\cdot 2^i \binom{\omega(d_A)}{i} = (2-2)^{\omega(d_A)} = 0.$$

We deduce from this that

$$\sum_{m|d_A} \beta\,(d_A/m)\,2^{\omega(m)} \sum_{p|m} f(p) = \sum_{p|d_A} f(p) \sum_{m|d_A/p} \beta\,(d_A/mp)\,2^{\omega(mp)}$$

$$= 2\cdot \sum_{p|d_A} f(p) \sum_{m|d_A/p} \beta\left(\frac{d_A/p}{m}\right) 2^{\omega(m)} = 0. \qquad \square$$

We have thus proved formula (8.14); let us now show how to deduce the theorem from this. For this we fix $r_* \in i[-1/2, 0] \cup [0, +\infty[$ and write $\lambda_* = r_*^2 + 1/4$. Let $\delta > 0$ be such that for every eigenvalue $\lambda \neq \lambda_*$ of the hyperbolic Laplacian in $L^2(\Gamma_\mathcal{O}\backslash\mathcal{H})$ or in $\mathcal{C}_{\mathrm{prim}}(\Gamma_0(d_A)\backslash\mathcal{H})$, we have $|\lambda - \lambda^*| > \delta$. Given a constant $C > 0$, we put

$$h_C(r) = \exp\left[-C(r^2 - r_*^2)^2\right]. \qquad (8.22)$$

The function h_C is a test function to which formula (8.14) applies; it satisfies $0 < h_C(r) \leqslant 1$ for $r \in i\mathbf{R} \cup \mathbf{R}$ and

$$(|\lambda - \lambda_*| > \delta \text{ and } C > 2) \Longrightarrow h_C(r) < e^{-C\delta^2/2} \cdot h_1(r). \qquad (8.23)$$

Now recall that each of the sums which arise in (8.14) is absolutely convergent for every function $h = h_C$; in particular this is the case for $h = h_1$. Property (8.23) then implies, by letting C tend to infinity, that the multiplicity with which λ_* appears in the spectrum of the hyperbolic Laplacian in $L^2(\Gamma_\mathcal{O}\backslash\mathcal{H})$ is equal to the multiplicity with which λ_* appears in the spectrum of the hyperbolic Laplacian in $\mathcal{C}_{\mathrm{prim}}(\Gamma_0(d_A)\backslash\mathcal{H})$. This completes the proof of the theorem. $\qquad \square$

As a corollary of Theorems 7.1 and 8.18, we get the following result

Theorem 8.25 *Let $\mathcal{O}$ be a maximal order in a quaternion algebra $A = D_{a,b}(\mathbf{Q})$ with $a, b > 0$ of discriminant $d_A > 1$. Then*

$$\lambda_1(\Gamma_{\mathcal{O}} \backslash \mathcal{H}) \geqslant \frac{5}{36}.$$

Remark 8.26 The theorem remains true when $\Gamma_{\mathcal{O}}$ is replaced by a congruence subgroup. In this case as well, we expect to be able to replace $5/36$ by $1/4$ and this lower bound would then be optimal: Theorem 4.31 and the Jacquet-Langlands correspondence – extended to non-maximal subgroups – allows one to construct a quaternion algebra A of discriminant $d_A > 1$ and a congruence subgroup $\Gamma \subset \Gamma_{\mathcal{O}}$ such that

$$\lambda_1(\Gamma \backslash \mathcal{H}) = \frac{1}{4}.$$

8.5 Commentary and References

For more on the Jacquet-Langlands correspondence, the reader can consult [43, 55, 65]. The first two of these references are written in the adelic language but the proof that we have given here is nevertheless based on the same approach: a comparison of two trace formulas.

§8.1

The reader is encouraged to look at [92, 132] for a more comprehensive study of the arithmetic of quaternion algebras.

Here are a few (non-trivial) examples – taken from Hejhal [55] – of maximal orders ($\left(\frac{\cdot}{\cdot}\right)$ denotes the Legendre symbol).

Example 8.27 Let a and b be two positive relatively prime integers. Assume that a and b are squarefree and such that

$$\begin{cases} ab > 1 \\ a \equiv 1 \ (\mathrm{mod} \ 4), \ b \ \mathrm{odd} \\ \left(\frac{b}{p}\right) = -1 \ \text{for every prime } p \text{ dividing } a \\ \left(\frac{a}{p}\right) = -1 \ \text{for every prime } p \text{ dividing } b. \end{cases} \tag{8.24}$$

Then the quaternion algebra $A = D_{a,b}(\mathbf{Q})$ is a division algebra and

$$\mathcal{O} = \mathbf{Z} \cdot 1 + \mathbf{Z} \cdot \frac{1+i}{2} + \mathbf{Z} \cdot j + \mathbf{Z} \cdot \frac{j+k}{2}$$

is a maximal order. Its reduced discriminant is equal to ab.

The book [47] is an excellent introduction to p-adic numbers. The classification of p-adic quaternion algebras is not hard to prove for odd primes p, see [73, Ch. 6, Th. 2.10]. The reader can consult [101, §12] for a detailed study of its (unique) maximal order. The passage from $\mathcal{O}$ – the maximal order of A – to $\mathcal{O}_p$ can be achieved through a "local-global" principle (see [136, p. 211]). Finally, the fact that the discriminant of A is always the product of an *even* number of primes is proved in [136, XIII §3, Th. 2].

For more details on orders in quadratic field extensions of $\mathbf{Q}$ and for more on the case when $p = 2$, see [137, §6.5].

The strong approximation theorems – Theorems 8.5 and 8.6 – are proved in [92, p. 208].

§8.2

The results of this paragraph are special cases of results of due to Vignéras, see [132], whose proofs we have followed.

§8.3

The fact (which we have admitted) that two embeddings of K into A induced by the inclusions $\mathbf{Q}[\gamma] \subset A$ are conjugate by an element of A^* follows, for instance, from [136, IX, Prop. 3 – Cor. 2].

The computation of the area $\mathrm{area}(\varGamma_{\mathcal{O}})$ is due to Shimizu [120]; see also [88, §11.1].

The fact that the integral (8.11) is equal to the residue at $s = 1$ of the zeta function $\zeta_A(s)$ of the algebra A is shown in [133, Chap. III, §2, Th. 2.2].

§8.4

In order to give an idea – in our classical ($\equiv$ non adelic) case – of the proof, we have contented ourselves here with proving Theorem 8.18 which realizes the Jacquet-Langlands correspondence only for maximal arithmetic subgroups. In doing so we have followed two articles by Bolte and Johansson [12, 13].

Chapter 9
Arithmetic Quantum Unique Ergodicity

Let $S = \Gamma \backslash \mathcal{H}$ be a compact hyperbolic surface (or more generally, of finite area). In this last chapter we shall study the following conjecture.

Conjecture 9.1 (Quantum unique ergodicity) *Let ϕ_i be a sequence of eigenfunctions of the Laplacian Δ on S, whose L^2-norm is equal to 1 and with associated eigenvalue $\lambda_i \rightarrow +\infty$. Then the probability measure $d\mu_i(z) = |\phi_i(z)|^2 d\mu(z)$ converges (in the weak-$*$ topology) towards the normalized measure $\mathrm{area}(S)^{-1} d\mu$.*

Recall that a sequence of measures μ_i converges towards a measure μ in the *weak-$*$ topology* if for every continuous function f of compact support, the sequence of integrals $\int f d\mu_i$ converges towards the integral $\int f d\mu$. In the following, we shall call a *quantum limit*[1] any (weak-$*$) limit ν of a sequence of measures $d\mu_i$ as in the conjecture.

The goal of this chapter is to describe in detail a portion of the recent work of Lindenstrauss which proves an arithmetic version of this conjecture. We begin by briefly explaining the link between Conjecture 9.1 and quantum mechanics on hyperbolic surfaces.

9.1 Quantization of the Geodesic Flow

Let us now explain how, in the context of the hyperbolic plane, *one can think of the equation $\Delta \phi = \lambda \phi$ as the quantization of the Hamiltonian equation generated by the geodesic flow.* We first describe the classical system and then its quantization, from the viewpoint of Schrödinger.

---

[1]The appearance of the word "quantum" here is explained in the first subsection.

© Springer International Publishing Switzerland 2016
N. Bergeron, *The Spectrum of Hyperbolic Surfaces*, Universitext,
DOI 10.1007/978-3-319-27666-3_9

9.1.1 The Classical System: The Geodesic Flow

The hyperbolic plane $\mathcal{H}$ is a homogeneous space $SL(2, \mathbf{R})/SO(2)$. The group $G = SL(2, \mathbf{R})$ acts on $\mathcal{H}$ by isometries and thus – via the differential – on the unit tangent bundle $\mathbf{S}^1\mathcal{H}$; this action is transitive and the stabilizer of $(i, \uparrow)$ (with $\uparrow = \frac{d}{dt}|_0 e^t i$) is $\{\pm I\}$. The map $g \mapsto g \cdot (i, \uparrow)$ then allows us to identify the unit tangent bundle $\mathbf{S}^1\mathcal{H}$ with the group $\mathrm{PSL}(2, \mathbf{R})$. The curve $t \mapsto e^t i = \begin{pmatrix} e^{t/2} & 0 \\ 0 & e^{-t/2} \end{pmatrix} \cdot i$ is the geodesic γ in $\mathcal{H}$ with initial condition $(i, \uparrow)$. For all $g \in G$,

$$g \cdot \gamma(t) = g \begin{pmatrix} e^{t/2} & 0 \\ 0 & e^{-t/2} \end{pmatrix} \cdot i;$$

this is the geodesic in $\mathcal{H}$ with initial condition $g \cdot (i, \uparrow)$. The action of the geodesic flow on $\mathbf{S}^1\mathcal{H}$ then corresponds to the action of the diagonal group

$$A = \left\{ \begin{pmatrix} e^{t/2} & 0 \\ 0 & e^{-t/2} \end{pmatrix} \;\middle|\; t \in \mathbf{R} \right\}$$

on $\mathrm{PSL}(2, \mathbf{R})$. The Iwasawa decomposition (see § 2.1) of G is

$$\begin{pmatrix} a & b \\ c & d \end{pmatrix} = \begin{pmatrix} 1 & x \\ 0 & 1 \end{pmatrix} \begin{pmatrix} y^{1/2} & 0 \\ 0 & y^{-1/2} \end{pmatrix} \begin{pmatrix} \cos\theta & \sin\theta \\ -\sin\theta & \cos\theta \end{pmatrix} \in G = NAK;$$

for any g this decomposition is unique. This in particular gives a coordinate system (x, y) on $\mathcal{H} = G/K$ which is identical with the usual coordinates $z = x + iy$ for the upper half-plane model. In these coordinates $\mathrm{PSL}(2, \mathbf{R})$ is identified with the unit tangent bundle $\mathbf{S}^1(\mathcal{H})$ by the formulae

$$x + iy = y^{1/2}e^{i\theta}(ai + b), \quad y^{-1/2}e^{i\theta} = d - ic$$

with $z = x + iy \in \mathcal{H}$ and 2θ equal to the argument in the fiber. We may also express the Haar measure (the unique up to positive scalars bi-invariant measure on $G = SL(2, \mathbf{R})$) in these coordinates:

$$dg = y^{-2}dx\,dy\,d\theta.$$

We can view the geodesic flow as a *Hamiltonian flow*, see for example [3] for an introduction to Hamiltonian mechanics. The phase space for the geodesic flow is more naturally the product $\mathcal{H} \times \mathbf{R}^2$ (position coordinates, tangent vectors). A point of $\mathcal{H} \times \mathbf{R}^2$ is therefore the joint data of a position in $\mathcal{H}$ along with a tangent vector above it, which we think of as being associated with a "particle". The evolution of this system in time defines a path γ in $\mathcal{H} \times \mathbf{R}^2$. The energy is the map $E : (z, v) \in \mathcal{H} \times \mathbf{R}^2 \mapsto y^{-2}(v_x^2 + v_y^2) \in \mathbf{R}$, where $z = x + iy$ and $v = (v_x, v_y)$ in the basis $(\frac{\partial}{\partial x}, \frac{\partial}{\partial y})$; it remains constant along the path γ.

In the Hamiltonian reformulation of Newtonian mechanics one replaces the tangent bundle by the cotangent bundle and fixes a C^∞-function H on phase space called the *Hamiltonian*, which plays the role of the energy. We naturally identify – via the scalar product defined by the Riemannian metric – the tangent bundle and the cotangent bundle. The new phase space is then the product $\mathcal{H} \times \mathbf{R}^2$ (position coordinates and corresponding momenta). The cotangent bundle is equipped with a natural symplectic form given by

$$\omega = dp_x \wedge dx + dp_y \wedge dy,$$

where $z = x + iy$ and $p = (p_x, p_y)$ in the basis (dx, dy) dual to $(\frac{\partial}{\partial x}, \frac{\partial}{\partial y})$. In these coordinates the Hamiltonian

$$H : \mathcal{H} \times \mathbf{R}^2 \longrightarrow \mathbf{R}$$

associated with the energy E is given by

$$H(z, p) = \frac{1}{2} y^2 (p_x^2 + p_y^2).$$

The evolution of this system again defines a path in $\mathcal{H} \times \mathbf{R}^2$ that we continue to denote by γ. This path is an integral curve of the *Hamiltonian equation*

$$\frac{d\gamma}{dt} = \xi_H(\gamma(t)), \tag{9.1}$$

where ξ_H is the symplectic gradient of H, defined as

$$dH = \omega(\cdot, \xi_H).$$

The flow associated with Eq. (9.1) is the Hamiltonian flow. In our case,

$$\xi_H(z, p) = (y^2 p_x, y^2 p_y, 0, -y(p_x^2 + p_y^2)).$$

The equation of motion for the "particle" is then expressed in the Hamilton equations

$$\dot{x} = \frac{\partial H}{\partial p_x}, \quad \dot{y} = \frac{\partial H}{\partial p_y}, \quad \dot{p}_x = -\frac{\partial H}{\partial x}, \quad \dot{p}_y = -\frac{\partial H}{\partial y},$$

which in our case is

$$\frac{dx}{dt} = y^2 p_x, \quad \frac{dy}{dt} = y^2 p_y, \quad \frac{dp_x}{dt} = 0, \quad \frac{dp_y}{dt} = -y(p_x^2 + p_y^2).$$

The set of *observables* of this system are the functions in $C^\infty(\mathcal{H} \times \mathbf{R}^2)$. The Hamiltonian H is an example of an observable, which in addition remains constant along integral curves of (9.1). The evolution of any observable $f \in C^\infty(\mathcal{H} \times \mathbf{R}^2)$ along an integral curve γ of (9.1) is more generally described by the differential equation

$$\frac{df \circ \gamma}{dt} = \{f, H\} \circ \gamma, \tag{9.2}$$

where $\{f, H\} = \omega(\xi_f, \xi_H)$ is the *Poisson bracket* of f and H for the symplectic structure ω.

Lemma 9.2 *For every solution* $\gamma(t) = (z(t), p(t))$ *of (9.1), the curve* $t \mapsto z(t)$ *is a geodesic in the Poincaré upper half-plane* $\mathcal{H}$.

Proof The function

$$C : \begin{cases} \mathcal{H} \times \mathbf{R}^2 \longrightarrow \mathbf{R} \\ (z, p) \longmapsto x + y\dfrac{p_y}{p_x} \end{cases}$$

is C^∞ and satisfies $\{C, H\} = 0$; it is therefore constant along integral curves of (9.1). Denote by c its value along γ. Then

$$c = x(t) + y(t)\frac{p_y(t)}{p_x(t)},$$

so that

$$(x(t) - c)\frac{dx}{dt} + y(t)\frac{dy}{dt} = 0.$$

There exists then a real number $a > 0$ such that for all t, we have

$$(x(t) - c)^2 + y(t)^2 = a^2. \qquad \square$$

The proof of Lemma 9.2 implies that the Hamiltonian system under consideration here is almost *completely integrable*. A completely integrable system on $\mathcal{H}$ is one for which there exist two independent functions[2] (H and C) constant along integral curves and whose Poisson bracket is identically vanishing. Note, however, that C is not everywhere defined.

When the phase space is *compact* a completely integrable Hamiltonian system is very stable: starting from two nearby initial conditions two trajectories diverge

---

[2]I.e., the corresponding set of gradient vectors is linearly independent at almost every point of $\mathcal{H}$.

at most linearly in t. The trajectories are constrained to lie on level surfaces of independent functionals; when the phase space is compact, these surfaces are tori.

Here the phase space is not compact and the level surfaces $\{H = E, \ C = c\}$ are not tori: the geodesics on $\mathcal{H}$ are not closed. In fact, the dynamics in this case are very unstable. If Γ is a Fuchsian group of the first kind, the above Hamiltonian flow passes to the quotient by Γ. The geodesic flow on the surface $\Gamma \backslash \mathcal{H}$ is again a Hamiltonian flow, but it is not completely integrable. Indeed, the function C is not Γ-invariant and does not pass to the quotient. We can in fact show that the flow is mixing and in particular *ergodic*, i.e., every measurable subset of $\Gamma \backslash G$ invariant by the geodesic flow has either zero of full measure. One can think of Conjecture 9.1 as a quantum analog of the results we have just summarized.

9.1.2   Quantum Systems

In this subsection we provide a rough description of quantum systems via the Heisenberg formalism. We shall not seek to be rigorous or precise; in particular we shall use the term "operator" without being precise as to which type we are considering.

In a quantum system, the phase space is replaced by a complex Hilbert space $\mathbf{H}$, the *space of quantum states*. A *state* of the system is then a line in $\mathbf{H}$ (or a non-zero vector in it). The evolution of the system is a path in the projectivization $\mathbf{PH}$ of the Hilbert space. A physical law of the system which dictates this evolution is a unitary flow $U_t = e^{tA}$, i.e., a 1-parameter subgroup of unitary operators. Whereas the Hamiltonian flow is generated by a function H, the unitary flow is generated by an operator A. The orbits in $\mathbf{PH}$ of the unitary flow U_t are the possible outcomes of the system and if $t \mapsto v(t) \in \mathbf{H}$ is a path of non-zero vectors, the quantum analog of Eq. (9.1) is then

$$\frac{dv}{dt} = Av(t). \tag{9.3}$$

The set of observables of a quantum system is made of of operators B on $\mathbf{H}$. If the system is in state $L \in \mathbf{PH}$, the result of the observation corresponding to B is an expectation

$$\langle Bv, v \rangle, \quad \text{where } v \text{ is a unitary vector in } L.$$

Along the orbit of the 1-parameter subgroup $\{e^{tA}\}$ the expectation becomes

$$\langle e^{-tA} B e^{tA} v, v \rangle.$$

The quantum analog of Eq. (9.2) is then

$$\frac{d\langle Bv, v\rangle}{dt} = \langle [B, A]v(t), v(t)\rangle, \tag{9.4}$$

where $[B, A] = BA - AB$ is the usual Lie bracket on operators on **H**. In particular, the observable B is conserved along a trajectory following the law $\{U_t\}$ if and only if B commutes with (the energy) A.

In the following paragraph we present – in the context of the hyperbolic plane – Schrödinger's approach for associating a quantum system with a classical system. This construction depends on a parameter called the level of quantization.

9.1.3 Quantum Mechanics on the Poincaré Upper Half-Plane

It seems reasonable (at least naively) to privilege the observable "particle position" and thus to choose as our Hilbert space

$$\mathbf{H} = L^2(\mathcal{H}, d\mu(z)).$$

A unit vector $v \in \mathbf{H}$ then gives rise to a function $|v|^2$ on $\mathcal{H}$ which is positive and of integral 1, in other words a probability density on $\mathcal{H}$. Following Born, we can think of this density as giving the probability of finding a particle in the state v in a particular position. The expectation of the x- (resp. y-) coordinate in state v is then

$$\int_{\mathcal{H}} x|v|^2 d\mu(z) \quad \left(\text{resp.} \int_{\mathcal{H}} y|v|^2 d\mu(z)\right).$$

In view of this, it is natural to introduce the operators given by multiplication by x and by y. Taking a cue from the classical counterpart, we also introduce the derivation operators $\frac{\hbar}{i}\frac{\partial}{\partial x}$ and $\frac{\hbar}{i}\frac{\partial}{\partial y}$ and the associated quantum Hamiltonian

$$\widehat{H} = -\frac{1}{2}\hbar^2 y^2\left(\left(\frac{\partial}{\partial x}\right)^2 + \left(\frac{\partial}{\partial y}\right)^2\right).$$

We then have $\widehat{H} = \frac{1}{2}\hbar^2\Delta$, where Δ is the hyperbolic Laplacian on $\mathcal{H}$ and $\hbar$ a physical constant called the Planck constant (10^{-34} kg $\cdot$ m^2/s) which fixes the quantization level. The operator $\widehat{H}$ is self-adjoint, the 1-parameter subgroup $\{e^{-\frac{i}{\hbar}t\widehat{H}}\}$ is a group of unitary operators on **H** and the Eq. (9.3) becomes the *Schrödinger equation*:

$$i\frac{d}{dt}\psi_t = \frac{\hbar}{2}\Delta\psi_t,$$

where the unknown is a function $t \in \mathbf{R} \mapsto \psi_t \in \mathbf{H}$.

Note that in complex coordinates $z = x + iy$ and $w = y(p_x + ip_y)$, we have[3]:

$$\{z, w\} = 0 = -[z, y\partial/\partial\bar{z}], \qquad \{z, \bar{w}\} = 2y = -2[z, y\partial/\partial z],$$
$$\{\bar{z}, w\} = 2y = -2[\bar{z}, y\partial/\partial\bar{z}], \quad \{\bar{z}, \bar{w}\} = 0 = -[\bar{z}, y\partial/\partial z]. \tag{9.5}$$

One is led to ask if there exists a Lie algebra homomorphism between a suitable space $\text{Fn}(\mathcal{H} \times \mathbf{R}^2)$ of C^∞ functions on phase space $\mathcal{H} \times \mathbf{R}^2$ and operators on **H**. This is a bit too much to ask. Nevertheless, on the space $\text{Fn}(\mathcal{H} \times \mathbf{R}^2)$ of C^∞ functions having at most polynomial growth, there does exist a map

$$\text{Op} : f \in \text{Fn}(\mathcal{H} \times \mathbf{R}^2) \longmapsto \text{Op}(f)$$

into the space of linear unbounded operators on **H**, such that

$$\frac{1}{i\hbar}[\text{Op}(f), \text{Op}(g)] = \text{Op}(\{f, g\}) + O(\hbar) \tag{9.6}$$

for $\hbar \to 0$. The construction of this map was proposed by Weyl; one calls it the *Weyl calculus*. When f and g are polynomials of degree at most 2 in the variables z, $\bar{z}$, w and $\bar{w}$, the error term in (9.6) is zero. Moreover, if $f(z, w) = f(z)$ the operator $\text{Op}(f)$ is multiplication by f, and if $f(z, w) = w^n$, resp. $f(z, w) = \bar{w}^n$, the operator $\text{Op}(f)$ is the differential operator $\left(\frac{\hbar y}{i}\right)^n \frac{d^n}{d\bar{z}^n}$, resp. $\left(\frac{\hbar y}{i}\right)^n \frac{d^n}{dz^n}$.

To finish, we remark that if $\phi \in$ **H** is a state of the quantum system, we can form the "measure"

$$f \longmapsto \langle \text{Op}(f)\phi, \phi \rangle. \tag{9.7}$$

The question which shall occupy us in this chapter is the behavior of this "measure" in the semiclassical limit $\hbar \to 0$. We have recalled that the geodesic flow on the compact surface $\Gamma \backslash \mathcal{H}$ is ergodic. The quantum states in the associated quantum system are vectors $\phi \in L^2(\Gamma \backslash \mathcal{H}, d\mu(z))$. To any such quantum state, there corresponds, by (9.7), a "measure" on the unit tangent bundle. Understanding the behavior in the semiclassical limit of the measures associated with eigenstates $\widehat{H}\phi = \lambda\phi$ brings us precisely to the question of quantum unique ergodicity.

In the following paragraph, we construct these measures on the unit tangent bundle "by hand". The construction is motivated by what we have just described but is completely independent.

---

[3]The coordinates w and $\bar{w}$ are natural here since we then have $H = \frac{1}{2}w\bar{w}$.

9.2 Microlocal Lift

The Iwasawa decomposition of $G = \mathrm{SL}(2, \mathbf{R})$

$$\begin{pmatrix} a & b \\ c & d \end{pmatrix} = \begin{pmatrix} 1 & x \\ 0 & 1 \end{pmatrix} \begin{pmatrix} y^{1/2} & 0 \\ 0 & y^{-1/2} \end{pmatrix} \begin{pmatrix} \cos\theta & \sin\theta \\ -\sin\theta & \cos\theta \end{pmatrix} \in G = NAK$$

induces a system of coordinates (x, y) on $\mathcal{H} = G/K$ identical to the usual coordinates $z = x + iy$ for the upper half-plane model of the hyperbolic plane. We recalled in the preceding subsection that the group $\mathrm{PSL}(2, \mathbf{R})$ is identified with the unit tangent bundle $\mathbf{S}^1(\mathcal{H})$ of $\mathcal{H}$, given in the above coordinates by

$$x + iy = y^{1/2}e^{i\theta}(ai + b), \quad y^{-1/2}e^{i\theta} = d - ic$$

with $z = x + iy \in \mathcal{H}$ and 2θ the argument in the fiber.

A function ϕ on $\mathcal{H}$ can be naturally viewed as a right K-invariant function on G. Given a cocompact discrete subgroup $\Gamma \subset G$, we write for every $n \in \mathbf{Z}$

$$A_n = \{f \in C^\infty(\Gamma\backslash G) \mid f(gk(-\theta)) = e^{2in\theta}f(g)\}$$

where

$$k(\theta) = \begin{pmatrix} \cos\theta & -\sin\theta \\ \sin\theta & \cos\theta \end{pmatrix}.$$

In particular, the space A_0 is identified with the space $C^\infty(\Gamma\backslash\mathcal{H})$ of right K-invariant functions. We say that a function $f \in C^\infty(\Gamma\backslash G)$ is K-finite if it belongs to the space generated by the subspaces A_n (so that it has only a finite number of non-zero A_n components).

By definition the Lie algebra $\mathfrak{g}$ of traceless matrices in $\mathcal{M}_2(\mathbf{R})$ operates on G by left-invariant first order differential operators on functions on G as

$$(X \cdot f)(g) = \frac{d}{dt}f(ge^{tX})_{|t=0} \quad (X \in \mathfrak{g}, f \in C^\infty(G), g \in G).$$

In this way, to the matrices

$$H = \begin{pmatrix} 1 & 0 \\ 0 & -1 \end{pmatrix}, \ V = \begin{pmatrix} 0 & 1 \\ 1 & 0 \end{pmatrix}, \ W = \begin{pmatrix} 0 & 1 \\ -1 & 0 \end{pmatrix}, \ X = \begin{pmatrix} 0 & 1 \\ 0 & 0 \end{pmatrix},$$

and

$$E^\pm = H \pm iV,$$

there correspond differential operators that we shall denote by the same symbol. In
particular, the operators

$$E^{\pm} = \pm 2iye^{\pm 2i\theta} \left(\frac{\partial}{\partial x} \mp i\frac{\partial}{\partial y} \right) \mp ie^{\pm 2i\theta} \frac{\partial}{\partial \theta}$$

send A_n to $A_{n\pm 1}$. Moreover

$$H = -2y\sin 2\theta \frac{\partial}{\partial x} + 2y\cos 2\theta \frac{\partial}{\partial y} + \sin 2\theta \frac{\partial}{\partial \theta},$$

which operates by

$$H \cdot f = \frac{d}{dt} f \left(g \left(\begin{smallmatrix} e^t & 0 \\ 0 & e^{-t} \end{smallmatrix} \right) \right)_{|t=0},$$

and generates the geodesic flow on $S = \Gamma\backslash\mathcal{H}$, corresponds to the action (on the
right) of $\left(\begin{smallmatrix} e^t & 0 \\ 0 & e^{-t} \end{smallmatrix} \right)$ on the unit tangent bundle $S^1(S) = \Gamma\backslash G$. Note that $W = \partial/\partial\theta$ is
the infinitesimal generator for the action of K by rotations on the fibers of $S^1(S)$; if
$f \in A_n$,

$$W \cdot f = 2inf. \tag{9.8}$$

The *Casimir operator*

$$\Omega = E^- E^+ - W^2 - 2iW$$

which commutes with all elements in the Lie algebra, coincides with the operator
-4Δ on A_0. Finally, observe that the action of G on $L^2(\Gamma\backslash G)$ is unitary with respect
to the scalar product $\langle \, , \rangle$ defined by

$$\langle f_1, f_2 \rangle = \int_{\Gamma\backslash G} f_1(g)\overline{f_2(g)}dg \quad (f_1, f_2 \in L^2(\Gamma\backslash G)).$$

Given $X \in \mathfrak{g}$ and $f_1, f_2 \in C^\infty(\Gamma\backslash G)$ we therefore have

$$\langle X \cdot f_1, f_2 \rangle = \frac{d}{dt}\langle f_1(\cdot e^{tX}), f_2 \rangle_{|t=0} = \frac{d}{dt}\langle f_1, f_2(\cdot e^{-tX}) \rangle_{|t=0}$$

$$= -\frac{d}{dt}\langle f_1, f_2(\cdot e^{tX}) \rangle_{|t=0} = -\langle f_1, X \cdot f_2 \rangle. \tag{9.9}$$

9.2.1 The Microlocal Lift

Given a non-zero eigenfunction of the Laplacian $\phi \in C^\infty(\Gamma \backslash \mathcal{H})$ with eigenvalue $\lambda = 1/4 + r^2$, $r \in \mathbf{R}_+$, we write

$$\phi_0(g) = \phi(gK)$$

and define ϕ_n by recurrence via the rule

$$(2ir + 2n + 1)\phi_{n+1} = E^+ \phi_n \quad (2ir - 2n + 1)\phi_{n-1} = E^- \phi_n.$$

Since Ω commutes with E^+ and E^- and in view of $\Omega \phi_0 = -(4r^2 + 1)\phi_0$, we have

$$\Omega \phi_n = -(4r^2 + 1)\phi_n. \tag{9.10}$$

Moreover, from the fact that $\phi_n \in A_n$ ($n \in \mathbf{Z}$), Eqs. (9.8) and (9.9) imply – arguing by induction on n – that each ϕ_n has the same L^2-norm as ϕ_0.

We associate with the function ϕ the linear map I_ϕ on the space of K-finite functions in $C^\infty(\Gamma \backslash G)$ defined by

$$I_\phi(f) = \lim_{N \to +\infty} \langle f \sum_{n=-N}^{N} \phi_n, \phi_0 \rangle. \tag{9.11}$$

Note that by the K-finiteness assumption on f in (9.11), there exists an integer N_0 such that

$$f \in \bigoplus_{n=-N_0}^{N_0} A_n$$

where the direct sum in orthogonal. The expression

$$\langle f \sum_{n=-N}^{N} \phi_n, \phi_0 \rangle$$

is therefore constant for $N \geqslant N_0$ and $I_\phi(f)$ is well-defined. We shall write

$$\Phi_\infty = \sum_{n \in \mathbf{Z}} \phi_n$$

which then defines a linear form on the space of K-finite functions such that $I_\phi(f) = \langle f, \phi_0 \overline{\Phi}_\infty \rangle$. Observe finally that

$$I_\phi(f) = \int_S f(z) |\phi(z)|^2 d\mu(z) \quad \text{for } f \in A_0. \tag{9.12}$$

Theorem 9.3 (Microlocal lift) *Every quantum limit ν (a probability measure on S) admits a lift $\widetilde{\nu}$ to the unit tangent bundle $\mathbf{S}^1(S)$ which is invariant under the action of the geodesic flow.*

Proof Assume that ϕ is of L^2-norm 1. Recalling that the eigenvalue of ϕ is $1/4 + r^2$ ($r \geqslant 0$), we write N for the integer part of $r^{1/2}$ and put

$$\Psi_N = \frac{1}{\sqrt{2N+1}} \sum_{n=-N}^{N} \phi_n.$$

Denote by ν_ϕ the measure $|\Psi_N|^2 d \, \mathrm{vol}_{\Gamma \backslash G}$ on $\Gamma \backslash G$. We shall need the following result.

Lemma 9.4

1. *For every K-finite function $f \in C^\infty(\Gamma \backslash G)$, we have*

$$\left| I_\phi(f) - \int_{\Gamma \backslash G} f d\nu_\phi \right| = O_f(r^{-1/2}),$$

where the notation O_f means that the implied constant in the O depends only on f.

2. *Every weak-$*$ limit of the $I_\phi(\cdot)$ is a probability measure on $\Gamma \backslash G$.*

Proof Assume we have

$$f \in \bigoplus_{n=-N_0}^{N_0} A_n.$$

By definition,

$$\langle f\Psi_N, \Psi_N \rangle = \frac{1}{2N+1} \sum_{n,m=-N}^{N} \langle f\phi_n, \phi_m \rangle.$$

Then observe that for $-N \leqslant n, m \leqslant N$,

$$\langle f\phi_n, \phi_m \rangle = \frac{1}{(2ir - 2n - 1)(-2ir - 2m - 1)} \langle fE^- \phi_{n+1}, E^- \phi_{m+1} \rangle$$

$$= \frac{1}{(2ir - 2n - 1)(-2ir - 2m - 1)} (\langle E^-(f\phi_{n+1}), E^- \phi_{m+1} \rangle$$

$$- \langle E^-(f)\phi_{n+1}, E^- \phi_{m+1} \rangle)$$

$$= -\frac{1}{(2ir - 2n - 1)(-2ir - 2m - 1)} (\langle f\phi_{n+1}, E^+ E^- \phi_{m+1} \rangle + O_f(r))$$

(from (9.9) and the fact that each ϕ_n is of norm 1)

$$= -\frac{(-2ir + 2m + 1)(-2ir - 2m - 1)}{(2ir - 2n - 1)(-2ir - 2m - 1)} \langle f\phi_{n+1}, \phi_{m+1}\rangle + O_f(r^{-1})$$

$$= \langle f\phi_{n+1}, \phi_{m+1}\rangle + O_f(|n - m|r^{-1}) + O_f(r^{-1}).$$

Since

$$\langle f\phi_n, \phi_m\rangle = 0 \quad \text{if } |n - m| \geqslant N_0,$$

we finally obtain

$$\langle f\phi_n, \phi_m\rangle = \langle f\phi_{n+1}, \phi_{m+1}\rangle + O_f(r^{-1}). \tag{9.13}$$

Iterating the expression (9.13) at most N times we get

$$\langle f\phi_n, \phi_m\rangle = \langle f\phi_{n-m}, \phi_0\rangle + O_f(r^{-1/2})$$

so that

$$\langle f\Psi_N, \Psi_N\rangle = \sum_{k=-N_0}^{N_0} \frac{2N - 2|k| + 1}{2N + 1} \langle f\phi_k, \phi_0\rangle + O_f(r^{-1/2}).$$

Now using that for $|k| \leqslant N_0$,

$$\frac{2N - 2|k| + 1}{2N + 1} = 1 + O_f(N^{-1})$$

we conclude that

$$\langle f\Psi_N, \Psi_N\rangle = I_\phi(f) + O_f(r^{-1/2}). \tag{9.14}$$

The first part of Lemma 9.4 is then proved.

To prove the second part, note that if the limit $\lim_{r\to+\infty} I_\phi(f)$ exists for a dense subset of $C^\infty(\Gamma\backslash G)$ consisting entirely of K-finite functions, then this limit is given by

$$I(f) := \lim_{r\to+\infty} I_\phi(f) = \lim_{r\to+\infty} \langle f\Psi_N, \Psi_N\rangle.$$

There exists, therefore, a measure $\widetilde{\nu}$ such that

$$I(f) = \int_{\Gamma\backslash G} f d\widetilde{\nu}.$$

Finally, the measure $\widetilde{\nu}$ is a probability measure since $I_\phi(1)$ is always equal to 1. Lemma 9.4 is therefore completely proved.   □

It follows in particular from Lemma 9.4 that every quantum limit admits a lift $\widetilde{\nu}$ to $\Gamma\backslash G$ (and thus to the unit tangent bundle $\mathbf{S}^1(S)$) which we are now going to show is invariant under the action of the geodesic flow. Since the differential operator H generates the geodesic flow, this is the same as to show that for every K-finite function $f \in C^\infty(\Gamma\backslash G)$,

$$\int_{\Gamma\backslash G} (H \cdot f)d\widetilde{\nu} = \lim_{r \to +\infty} I_\phi(H \cdot f) = 0.$$

We conclude the proof of Theorem 9.3 with the help of the following lemma.   □

Lemma 9.5 *We have*

$$I_\phi((-4irH + H^2 + 4X^2) \cdot f) = 0.$$

In particular, $I_\phi(H \cdot f) = O_f(r^{-1})$.

Proof We begin by calculating $H \cdot \Phi_\infty$. Recall that $H = \frac{1}{2}(E^+ + E^-)$. We have

$$E^+ \cdot \Phi_\infty = \sum_n E^+ \cdot \phi_n = \sum_n (2ir + 1 + 2n)\phi_{n+1}$$

$$= \sum_n (2ir - 1 - iW)\phi_n,$$

according to (9.8). We find

$$E^+ \cdot \Phi_\infty = (2ir - 1 - iW)\Phi_\infty \quad \text{and} \quad E^- \cdot \Phi_\infty = (2ir - 1 + iW)\Phi_\infty, \quad (9.15)$$

which is to say

$$H \cdot \Phi_\infty = (2ir - 1)\Phi_\infty \quad \text{and} \quad X \cdot \Phi_\infty = 0 \quad (9.16)$$

(recall that $X = \frac{1}{2}(W + V) = \frac{1}{2}W + \frac{1}{4i}(E^+ - E^-)$). From (9.9) it follows that the lemma is a consequence of

$$(H^2 + 4X^2 - 4irH) \cdot \overline{\phi}\Phi_\infty = [H^2 \cdot \overline{\phi} + 2(2ir - 1)H \cdot \overline{\phi} + (2ir - 1)^2\overline{\phi}$$

$$+ 4X^2 \cdot \overline{\phi} - 4irH \cdot \overline{\phi} - 4ir(2ir - 1)\overline{\phi}]\Phi_\infty$$

$$= [(H^2 - 2H + 4X^2 + (1 + 4r^2)) \cdot \overline{\phi}]\Phi_\infty$$

$$= [(\Omega + (1 + 4r^2)) \cdot \overline{\phi}]\Phi_\infty$$

$$= 0.$$

We have used (9.16) several times above, as well as the fact that we can rewrite Ω in the form

$$\Omega = H^2 + 4X^2 - 4XW - 2H \tag{9.17}$$

and that $W \cdot \overline{\phi} = 0$. $\square$

The linear map I_ϕ is naturally associated with the map $f \mapsto \mathrm{Op}(f)$ to which we made reference at the end of the preceding section. The relation between the two is seen from the formula

$$I_\phi(f) = \langle \mathrm{Op}(f)\phi, \phi \rangle.$$

This is the underlying motivation for the construction of I_ϕ.

9.2.2   Quantum Ergodicity[4]

Each operator $\mathrm{Op}(f)$ has a kernel that one can explicate through the formalism of Zelditch [144]. By applying the Selberg trace formula to this kernel, one can prove the following theorem, see [144].

Theorem 9.6 *Let $\{\phi_k\}$ be an orthonormal basis of Laplacian eigenfunctions on S with*

$$\Delta\phi_k = \lambda_k\phi_k, \quad \lambda_k = 1/4 + r_k^2.$$

Then for every K-finite function $f \in C^\infty(\Gamma\backslash G)$,

$$\lim_{L\to+\infty} \frac{1}{N(L)} \sum_{k \| r_k - L | \leqslant 1/2} I_{\phi_k}(f) = \frac{1}{\mathrm{vol}(\Gamma\backslash G)} \int_{\Gamma\backslash G} f(g)\,dg,$$

where $N(L) = |\{k \mid |r_k - L| \leqslant 1/2\}|$.

We shall study the variance of the linear form I_{ϕ_k}, which one defines as

$$\mathrm{Var}_L(f) = \frac{1}{N(L)} \sum_{k \| r_k - L | \leqslant 1/2} \left| I_{\phi_k}(f) - \frac{1}{\mathrm{vol}(\Gamma\backslash G)} \int_{\Gamma\backslash G} f(g)\,dg \right|^2.$$

The preceding result and the ergodicity[5] of the geodesic flow on S imply the following theorem, see [144] for more details.

---

[4]This paragraph is not needed for what comes after it. We shall freely use the ergodicity of the geodesic flow as well as the L^2 Birkhoff theorem, see [50].

[5]Note that this is the first time we are invoking the ergodicity of the geodesic flow.

Theorem 9.7 *For every K-finite function $f \in C^\infty(\Gamma \backslash G)$, we have*

$$\lim_{L \to +\infty} \operatorname{Var}_L(f) = 0.$$

Proof Let $L \geq 1/2$ be a real number, k a positive integer such that $|r_k - L| \leq 1/2$ and f a K-finite function in $C^\infty(\Gamma \backslash G)$ which we shall assume satisfies

$$\int_{\Gamma \backslash G} f(g)\, dg = 0.$$

Given a real number $T > 0$, write

$$M_T(f) = \frac{1}{T} \int_0^T f\left(\cdot \left(\begin{smallmatrix} e^{t/2} & 0 \\ 0 & e^{-t/2} \end{smallmatrix} \right) \right) dt.$$

According to Lemma 9.5, the distribution I_{ϕ_k} is, up to an error of $O(1/L)$, invariant under the action of the geodesic flow. We thus have

$$I_{\phi_k}(M_T(f)) = I_{\phi_k}(f) + O_{f,T}(1/L),$$

for all real $T > 0$. We deduce that

$$\frac{1}{N(L)} \sum_{k \mid |r_k - L| \leq 1/2} |I_{\phi_k}(f)|^2 = \frac{1}{N(L)} \sum_{k \mid |r_k - L| \leq 1/2} |I_{\phi_k}(M_T(f))|^2 + O_{f,T}(1/L). \qquad (9.18)$$

It follows now from Lemma 9.4 that

$$|I_{\phi_k}(M_T(f))|^2 \leq |\langle M_T(f)\Psi_{k,N}, \Psi_{k,N}\rangle|^2 + O_{f,T}(L^{-1/2}),$$

where N is as in the proof of Lemma 9.4 and $\Psi_{k,N}$ is associated with ϕ_k in the same way that Ψ_N is associated with ϕ. But the Cauchy-Schwarz inequality implies that

$$|\langle M_T(f)\Psi_{k,N}, \Psi_{k,N}\rangle|^2 \leq \langle |M_T(f)|^2 \Psi_{k,N}, \Psi_{k,N}\rangle$$
$$= I_{\phi_k}(|M_T(f)|^2) + O_{f,T}(L^{-1/2}).$$

This gives

$$|I_{\phi_k}(M_T(f))|^2 \leq I_{\phi_k}(|M_T(f)|^2) + O_{f,T}(L^{-1/2}). \qquad (9.19)$$

Equations (9.18) and (9.19) imply

$$\frac{1}{N(L)} \sum_{k \mid |r_k - L| \leq 1/2} |I_{\phi_k}(f)|^2 \leq \frac{1}{N(L)} \sum_{k \mid |r_k - L| \leq 1/2} I_{\phi_k}(|M_T(f)|^2) + O_{f,T}(L^{-1/2}).$$

Passing to the limit $L \to +\infty$ with the help of Theorem 9.6 we find

$$\limsup_{L \to +\infty} \frac{1}{N(L)} \sum_{k \mid |r_k - L| \leq 1/2} |I_{\phi_k}(f)|^2 \leq \frac{1}{\operatorname{vol}(\Gamma \backslash G)} \int_{\Gamma \backslash G} |M_T(f)|^2 dg.$$

We conclude by recalling that the ergodicity of the geodesic flow implies, via the L^2 Birkhoff theorem, that as $T \to +\infty$,

$$M_T(f) \longrightarrow \frac{1}{\text{vol}(\Gamma \backslash G)} \int_{\Gamma \backslash G} f(g) \, dg = 0$$

in $L^2(\Gamma \backslash G)$. $\square$

We define the *density* $D(\Sigma)$ of a subset Σ of the spectrum of S by

$$D(\Sigma) = \lim_{L \to +\infty} \frac{|\{k \mid \lambda_k \in \Sigma \text{ and } |r_k - L| \leq 1/2\}|}{N(L)}.$$

Through the use of a diagonal argument, see [144], we can deduce from Theorem 9.7 the following corollary, announced by Šnirel'man [125], proved in the case of hyperbolic surfaces by Zelditch [144] and vastly generalized by Colin de Verdière [29].

Corollary 9.8 (Quantum Ergodicity) *There exists a density 1 subsequence* $(\lambda_{k_i})_{i \in \mathbb{N}}$ *of the Laplacian spectrum on S such that the measures* $d\mu_{k_i}$ *converge (in the weak-∗ sense) towards the normalized measure* $\text{area}(S)^{-1} d\mu$.

Quantum Unique Ergodicity then asks whether one can avoid having to extract a density 1 subsequence.

9.3 First Links with Ergodic Theory

We recalled in the first subsection how to identify the unit tangent bundle $\mathbf{S}^1 \mathcal{H}$ of the hyperbolic plane $\mathcal{H}$ with the group $\text{PSL}(2, \mathbf{R})$ and the action of the geodesic flow on $\mathbf{S}^1 \mathcal{H}$ with the action by right-translations of the diagonal group

$$A = \left\{ \begin{pmatrix} e^{t/2} & 0 \\ 0 & e^{-t/2} \end{pmatrix} \, \middle| \, t \in \mathbf{R} \right\}$$

on $\text{PSL}(2, \mathbf{R})$.

Let $\Gamma < G$ be a discrete cocompact subgroup (or of the first kind). Similarly to the above, the action of the geodesic flow on $S = \Gamma \backslash \mathcal{H}$ corresponds to the right action of the diagonal group A on $\Gamma \backslash G$, and it follows from the preceding section that to every quantum limit ν there corresponds (by the microlocal lift) a probability measure $\tilde{\nu}$ on $\Gamma \backslash G$ which is A-invariant.

This result is not very restrictive, for there are many such measure on $\Gamma \backslash G$ that one may associate with various A-orbits. For example, the measure supported by a closed geodesic on S is obviously A-invariant. The A-orbits can in fact be quite wild, as is shown in Exercise 9.39.

Nevertheless, we expect the situation to be much simpler in the case when the diagonal subgroup $SL(n, \mathbf{R})$ acts on the space of lattices $SL(n, \mathbf{Z})\backslash SL(n, \mathbf{R})$ as soon as $n \geq 3$. This is a situation where we have two transversal flows which commute with one another.[6] The hope in this direction is explicitly formulated by Margulis. In the case of the space of lattices, Margulis's conjecture is stated in the following way.

Conjecture 9.9 (Margulis) *If $n \geq 3$, every relatively compact orbit of the diagonal subgroup $A \subset SL(n, \mathbf{R})$ in $X_n = SL(n, \mathbf{Z})\backslash SL(n, \mathbf{R})$ is compact.*

Margulis's conjecture is in fact more general since one expects that every A-orbit is homogeneous and that the A-invariant measures are precisely the Haar measures of these homogeneous orbits. We now state this more general version of Margulis's conjecture, but in the simplest case in which it is interesting.

We temporarily consider the group $G = SL(2, \mathbf{R}) \times SL(2, \mathbf{R})$ and Γ a discrete subgroup of finite covolume in G such that the kernel of the projection of Γ onto each factor $SL(2, \mathbf{R})$ is finite. Denote by $B = A_1 \times A_2$ the product of the diagonal subgroup of each of the $SL(2, \mathbf{R})$ factors. We are again in a situation with two transverse dynamical flows which commute with each other. In this case, Margulis's conjecture implies[7]:

Conjecture 9.10 (Margulis) *Every B-ergodic probability measure on $\Gamma\backslash G$ is either the Haar measure or the Haar measure of a compact B-orbit.*

Now, Rudnick and Sarnak show – in the article [108] in which they propose the arithmetic quantum unique ergodicity conjecture – that an "arithmetic" quantum limit measure on $\Gamma\backslash SL(2, \mathbf{R})$ cannot contain an ergodic component supported by a compact A-orbit. The Margulis conjecture then gives hope that an ergodic approach to the arithmetic quantum unique ergodicity conjecture can be found. We can now state a theorem going in the direction of Margulis's conjecture.

Theorem 9.11 (Einsiedler, Katok, Lindenstrauss) *Let G be the group $SL(2, \mathbf{R}) \times SL(2, \mathbf{R})$ and Γ a discrete subgroup of finite covolume in G such that the kernel of the projection of Γ onto each $SL(2, \mathbf{R})$ factor is finite. Let μ be a B-invariant probability measure on $\Gamma\backslash G$. Suppose that each ergodic component of μ relative to the A_1-action has positive entropy. Then the measure μ is $(SL(2, \mathbf{R}) \times \{I\})$-invariant.*

There are then two major obstacles to proving the arithmetic quantum unique ergodicity conjecture by ergodic methods.

---

[6]This type of property rigidifies the situation.

[7]We use here the term *ergodic* without entering into the details. See later sections or [50] for a definition.

- One must find a substitute for the second dynamical flow, which we have seen is indispensable.
- One must verify the positive entropy hypothesis, which is difficult to do in practice.

The underlying arithmetic plays a fundamental role in overcoming these obstacles. We begin by describing how the arithmetic structure arises in a simpler example.

9.4   Multiplication by 2 and 3 on the Circle

In this section we shall be interested in the measures on the circle S^1 which are simultaneously invariant by the action of multiplication by 2 and 3, and are ergodic.

According to a conjecture of Furstenberg, the analog of Margulis's conjecture in this setting, a measure which is invariant under multiplication by 2 and 3 is either atomic, or equal to the Lebesgue measure. Just as for Margulis's conjecture, the general case seems largely inaccessible, but several proofs are available which are valid under the hypothesis that the action has non-zero entropy. Theorem 9.11 is a profound elaboration of these ideas. We present here a proof due to Host. We shall consider multiplication by 2 as being analogous to the A_1-action in Theorem 9.11. We begin by explaining how the choice of an odd prime p (for example $p = 3$) allows one to obtain a second dynamical system which commutes with multiplication by 2.

9.4.1   The Circle as a Foliated Space

Fix an odd prime p. Let

$$T_p = \{x \in S^1 \mid \exists a \in \mathbf{Z}, \, n \in \mathbf{N} \text{ such that } x = a/p^n\}.$$

The set T_p is identified with the quotient $\mathbf{Q}_p/\mathbf{Z}_p$; it is in particular a group under addition. We equip it with the quotient topology induced by the p-adic distance: $d(x, y) = p^n$, where p^n is the smallest denominator of the fraction $x - y$. The space T_p receives an action by the (locally compact additive) group $\mathbf{Q}_p$. The compact subsets of T_p are the bounded subsets with respect to the p-adic distance. Every compact subset of T_p is thus included in one of the

$$T_p^N = \{[a/p^n] \mid n \leqslant N, \, a \in \mathbf{Z}\}.$$

Here $[x]$ denotes the image of an element $x \in \mathbf{R}$ in the circle $S^1 = \mathbf{Z}\backslash\mathbf{R}$. Two points of T_p whose projections to S^1 are close to each other are far from each other in the topology on T_p.

The circle $\mathbf{S}^1$ is then "foliated" by the sets $x + T_p$, $x \in \mathbf{S}^1$, and this "foliation" is preserved by the rotations of the circle. Another way to think of this "foliation" is as a principal bundle without a base:

Lemma 9.12 *The additive group* $\mathbf{Z}[p^{-1}]$ *embedded diagonally in* $\mathbf{R} \times \mathbf{Q}_p$ *is a discrete subgroup; it acts freely on* $\mathbf{R} \times T_p$ *and the quotient*

$$\mathbf{Z}[1/p] \backslash (\mathbf{R} \times T_p)$$

is homeomorphic to the circle $\mathbf{Z} \backslash \mathbf{R} = \mathbf{S}^1$.

Proof Let $(x, [a/p^n]) \in \mathbf{R} \times T_p$. We have $(x, [a/p^n]) = (x - a/p^n, 0) + (a/p^n, [a/p^n])$. In $\mathbf{Z}[1/p] \backslash (\mathbf{R} \times T_p)$,

$$[x, [a/p^n]] = [x - a/p^n, 0].$$

It is thus clear that the map $[x, 0] \mapsto [x] \in \mathbf{Z} \backslash \mathbf{R}$ induces the homeomorphism of the lemma.   □

This "foliation" furnishes the transverse dynamics to the multiplication by 2 action that we seek. We begin by showing that one can "disintegrate" the measure along this "foliation".

9.4.2 Conditional Measures

All measures we consider will implicitly be assumed to be Radon, so that they are defined on the σ-algebra $\mathcal{B}$ of Borel subsets of $\mathbf{S}^1$ and give finite mass to compact subsets. In particular, the measures they assign to the entire space $\mathbf{S}^1$ will be of finite mass. Moreover, we shall fix once and for all a probability measure μ on $\mathbf{S}^1$.

Given a partition F of $\mathbf{S}^1$, we denote by $\mathcal{F}$ the sub-σ-algebra of $\mathcal{B}$ generated by F. An element of $\mathcal{B}$ is in $\mathcal{F}$ if it is the (arbitrary) union of elements in F. Motivated by foliation, we shall call the elements of F the *leaves*; this is clearly an abuse of language. The partition F is said to be *measurable* if there exists a countable number of measurable subsets of $\mathbf{S}^1$ which generate the σ-algebra $\mathcal{F}$.

The partition into points is an example of a measurable partition; the associated σ-algebra is, for example, generated by the open intervals with rational endpoints. We denote by F_p^N, where p is an odd prime and N a positive integer, the partition whose leaves are the sets $x + T_p^N$ (recall that $T_p^N = \{0, 1/p^N, \ldots, (p^N - 1)/p^N\}$). The elements of the associated σ-algebra $\mathcal{F}_p^N$ are the Borel subsets invariant by translation by $1/p^N$. The partition F_p^N is measurable: the σ-algebra $\mathcal{F}_p^N$ is, for example, generated by the open subsets $f_p^{-N}(]\alpha, \beta[)$ where $f_p : x \mapsto px$ and α and β are rational.

Let S^1/F be the space of leaves, i.e., the space obtained by identifying two points if they lie on the same element of F, equipped with the measure $\widetilde{\mu}$ given by $\widetilde{\mu}(A) = \mu(\pi^{-1}A)$ where π is the canonical projection from S^1 to S^1/F. In what follows, we shall identify a point $\xi \in S^1/F$ with its corresponding leaf $\pi^{-1}(\xi)$; this is a convenient abuse of language. We admit the following theorem.

Theorem 9.13 *If the partition F is measurable, then the space of leaves $(S^1/F, \widetilde{\mu})$ is a Lebesgue space, i.e., isomorphic as a measurable space to the union of a compact interval $[0, a]$ equipped with the Lebesgue measure and a countable number of points of positive measure. Moreover, for every leaf $\xi \in F$, there exists a probability measure μ_ξ carried by ξ, and satisfying the property that*

$$\int_{S^1} f(x) \, d\mu(x) = \int_{S^1/F} \left(\int_\xi f(y) \, d\mu_\xi(y) \right) d\widetilde{\mu}(\xi)$$

for every $f \in L^1(\mu)$.

For example, when F is the partition into points (which is generated by open intervals with rational endpoints), then μ_ξ is the Dirac mass at ξ, regardless of the measure μ. There is a slight abuse of notation going on here, since the theorem says nothing about uniqueness. In fact uniqueness only holds for $\widetilde{\mu}$-almost every leaf, for modifying μ_ξ on a measure zero set of leaves does not change the claims of the theorem.

Let x be a point of S^1 and $\xi = x + T_p^N$ the leaf of F_p^N containing it. The measure μ_ξ then canonically defines a measure μ_x^N on T_p^N, upon identifying T_p^N and ξ via the map $\omega \mapsto x + \omega$. For $y \in \xi$, one should think of $\mu_\xi(\{y\})$ as the μ-probability of choosing y knowing that y lies on ξ. By construction, if $y = x + a/p^N$, then μ_y^N is obtained from μ_x^N by a translation of $-a/p^n$ (corresponding to the change of origin). Moreover,

$$\int_\xi f \, d\mu_\xi = \sum_{\omega \in T_p^N} f(x + \omega) \mu_x^N(\{\omega\}). \tag{9.20}$$

On the other hand, we cannot make this construction with the partition whose leaves are the sets $x + T_p$ (corresponding to a sub-σ-algebra $\mathcal{F}$ of $\mathcal{B}$) since it is not measurable. For example, in the case where μ is the Lebesgue measure, every $\mathcal{F}$-measurable set is of measure 0 or 1 (we can deduce this fact from the ergodicity of rational translations), and the measures μ_x^∞ should either be zero or equal to the Lebesgue measure on S^1; they are therefore not measures on T_p. Nevertheless, upon renormalizing and passing to the (direct) limit over N, we obtain the following proposition.

Proposition 9.14 *For $x \in S^1$, there exists a measure μ_x^∞ on T_p, with $\mu_x^\infty(\{0\}) = 1$, such that for μ-almost all $x \in S^1$ and for all $N \in \mathbb{N}$, $\mu_x^N = (\mu_x^\infty)_{|T_p^N}/\mu_x^\infty(T_p^N)$.*

Proof Let $X_N = \{x \mid \mu_x^N(\{0\}) = 0\}$. Then according to (9.20), for every leaf ξ of F_p^N passing through a point $x \in \mathbf{S}^1$, we have

$$\int_\xi 1_{X_N} d\mu_\xi = \sum_{x+\omega \in X_N \cap T_p^N} \mu_x^N(\{\omega\}) = \sum_{y \in X_N \cap T_p^N} \mu_y^N(\{0\}) = 0.$$

Integrating over the space of leaves, we get $\mu(X_N) = 0$. Thus for μ-almost every x, $\mu_x^N(\{0\})$ is positive for all N. Fix an x satisfying this property.

Since $\mathcal{F}_p^N$ is less fine than $\mathcal{F}_p^{N-1}$, μ_x^N is a linear combination of the measures μ_{x+k/p^N}^{N-1} on $T_p^{N-1}+k/p^N$, for $k = 0, \ldots, N-1$. The sum of the coefficients is 1, since each of the measures are probability measures. Moreover, the coefficient of μ_x^{N-1} is non-zero since $\mu_x^N(\{0\})$ is positive. Thus μ_x^{N-1} is proportional to the restriction of μ_x^N to T_p^{N-1}. In particular, if $\nu_x^N = \mu_x^N/\mu_x^N(\{0\})$, we have $(\nu_x^N)_{|T_p^{N-1}} = \nu_x^{N-1}$ since these two measures are proportional and give mass 1 to $\{0\}$.

Finally, we define μ_x^∞ by $\mu_x^\infty(\{\omega\}) = \nu_x^N(\{\omega\})$ for any N sufficiently large so that $\omega \in T_p^N$. By construction it satisfies the required properties. □

Note that if $y = x + \omega$ with $\omega \in T_p^N$, then $\mu_x^N = (+\omega)_*(\mu_y^N)$, where $+\omega$ is translation by ω on T_p. In particular, by passing to the limit, we find that μ_x^∞ is proportional to $(+\omega)_*(\mu_y^\infty)$.

9.4.3 Recurrence

We say that a measure μ on the circle $\mathbf{S}^1$ is T_p-*recurrent* if it satisfies one of the following four equivalent conditions:

1. For every B with $\mu(B) > 0$, there exists $\omega \in T_p - \{0\}$ such that $\mu(B \cap (B + \omega)) > 0$.
2. For every B with $\mu(B) > 0$ and for every compact K of T_p, there exists $\omega \in T_p - K$ such that $\mu(B \cap (B + \omega)) > 0$.
3. For every B with $\mu(B) > 0$ and for almost every point x of B, there exists $\omega \in T_p - \{0\}$ such that $x + \omega \in B$.
4. For every B with $\mu(B) > 0$, for almost every point x of B, and for every compact K of T_p, there exists $\omega \in T_p - K$ with $x + \omega \in B$.

The implications $(2) \Rightarrow (1)$ and $(4) \Rightarrow (3)$ are trivial.

Let us show that $(3) \Rightarrow (4)$. Let B satisfy $\mu(B) > 0$. Let $N \in \mathbb{N}$ and put $B_i = B \cap [i/p^N, (i+1)/p^N)$ for $0 \leq i < p^N$. If i is such that $\mu(B_i) > 0$, then for almost every $x \in B_i$ there exists a/p^n such that $x + a/p^n \in B_i$, and necessarily $n > N$. Thus, almost every $x \in B$ satisfies $x + a/p^n \in B$ for a certain rational number who denominator is $> p^N$, which finishes the proof.

The proof that $(1) \Rightarrow (2)$ is analogous.

We now show that (3) $\Rightarrow$ (1). Write $B_\omega = \{x \in B \mid x + \omega \in B\}$ for $\omega \in T_p - \{0\}$. If $\mu(B) > 0$, the sets B_ω cover almost all B by (3), so that one of them is of non-zero measure, which then shows that $\mu(B \cap (B + \omega)) > 0$.

Finally, we prove (1) $\Rightarrow$ (3) by contradiction. Otherwise there would exist $B' \subset B$ of non-zero measure such that, for all $\omega \in T_p - \{0\}$, $\mu(B \cap (B' + \omega)) = 0$. Then the set B' contradicts (1).

Proposition 9.15 *The probability measure μ is T_p-recurrent if and only if, for almost all $x \in \mathbf{S}^1$, $\mu_x^\infty(T_p) = \infty$.*

Proof We first assume that $\mu_x^\infty(T_p) = \infty$ for almost all x. We show that μ is recurrent by using the first characterization of recurrence. Suppose on the contrary that there exists B with $\mu(B) > 0$ such that, for all $\omega \in T_p - \{0\}$, $\mu(B \cap (B + \omega)) = 0$. Replacing B by $B - \bigcup(B \cap (B + \omega))$ if necessary (which does not change its measure), we can assume that $B \cap (B + \omega) = \varnothing$ for all $\omega \in T_p - \{0\}$. Thus the sets $B + \omega$ are pairwise disjoint so that $\sum \mu(B + \omega) \leqslant 1$. Fix an $\omega \in T_p - \{0\}$, an integer N such that $\omega \in T_p^N$ and denote by $\xi = x + T_p^N$ the leaf of F_p^N passing through a point $x \in B$. Since the sets $B + \omega$ are pairwise disjoint, it follows from (9.20) that

$$\int_\xi 1_{B+\omega} d\mu_\xi = \mu_x^N(\{\omega\})$$

and that

$$\int_\xi 1_B(y) \mu_y^\infty(\{\omega\}) \, d\mu_\xi(y) = \mu_x^\infty(\{\omega\}) \mu_x^N(\{0\}) = \mu_x^N(\{\omega\}).$$

Integrating over the space of leaves we then find that

$$\mu(B + \omega) = \int_B \mu_x^\infty(\{\omega\}) \, d\mu(x). \tag{9.21}$$

Thus, $\sum \mu(B + \omega) = \int_B \mu_x^\infty(T_p) = \infty$, which is absurd.

Now assume that $\mu_x^\infty(T) < \infty$ on a set of non-zero measure. We argue by contradiction to show that μ is not recurrent, this time using the fourth characterization of recurrence. There exists a set B of positive measure and an integer N such that, for $x \in B$, $\mu_x^\infty(T_p) < \infty$ and $\mu_x^\infty(T_p^N) > (0.9)\mu_x^\infty(T_p)$. Let $x \in B$ and $\omega = a/p^K$ (with $K > 2N$) such that $y = x - \omega \in B$. Then μ_x^∞ is proportional to $(+\omega)_* \mu_y^\infty$, thus $\mu_x^\infty(T_p^N - \omega) > (0.9)\mu_x^\infty(T_p - \omega) = (0.9)\mu_x^\infty(T_p)$. This is absurd for T_p^N and $T_p^N - \omega$ are disjoint, and each has measure at least $(0.9)\mu_x^\infty(T_p)$. $\square$

9.4.4 *Invariant Measures*

Let $\alpha = f_2 : x \mapsto 2x$ on $\mathbf{S}^1$. A measure μ on the circle $\mathbf{S}^1$ is said to be α-invariant if for every continuous function f on $\mathbf{S}^1$,

$$\int_{\mathbf{S}^1} f \circ \alpha \, d\mu = \int_{\mathbf{S}^1} f d\mu.$$

If q is an odd integer, then 2 is invertible modulo q and generates a (finite order) subgroup in $(\mathbf{Z}/q\mathbf{Z})^\times$. The orbit of a rational number p/q, where q is odd, is thus periodic (and finite). The sum of the Dirac masses at each of the points in such an orbit is an invariant measure. Just as multiplication by 3 leaves the triadic Cantor set invariant, the multiplication by 2 map α leaves certain Cantor subsets invariant. Moreover, there exist invariant measures supported on these Cantor sets.

We see then that there exist many invariant probability measures which are left invariant by multiplication by 2. These are the analogs of the measures supported on compact B-orbits in Conjecture 9.10 and the following is a result in the style of Einsiedler, Katok, and Lindenstrauss.

Theorem 9.16 *Let $\alpha = f_2 : x \mapsto 2x$ on $\mathbf{S}^1$. Assume $p \neq 2$. Let μ be a probability measure invariant under α and T_p-recurrent. Then μ is the Lebesgue measure.*

Proof We show that for any $v \in \mathbf{Z}^*$ we have $\int_{\mathbf{S}^1} e(vx) \, d\mu(x) = 0$.
For $m > 0$, write $g_m(x) = \frac{1}{m} \sum_{k=0}^{m-1} e(v2^k x)$. Then

$$g_m = \frac{1}{m} \sum_0^{m-1} e(v\cdot) \circ \alpha^k,$$

thus $\int_{\mathbf{S}^1} g_m d\mu = \int_{\mathbf{S}^1} e(vx) \, d\mu(x)$ by invariance, and it therefore suffices to show that $\int_{\mathbf{S}^1} g_{m_n} \to 0$ along a subsequence $m_n \to \infty$.

The idea is to estimate $\int_{\mathbf{S}^1} |g_m|^2$: when we expand $|g_m|^2$ we'll end up with exponential sums having significant inner cancellation, to the point that very few terms will contribute and the integral will be small. Unfortunately, to make use of these algebraic symmetries, we'll need μ to give the same weight to each of the points $x + \omega$, as ω varies in T_p^N, which is not the case in general. We shall thus amplify the weight of the points having a small probability of being chosen (i.e., with $\mu_x^N(\{0\})$ small), calculating rather $\int_{\mathbf{S}^1} \frac{|g_m|^2}{\mu_x^N(\{0\})} d\mu(x)$.

Lemma 9.17 *Let N be such that, for $0 \leqslant k < \ell < m$, $v2^k \not\equiv v2^\ell \mod p^N$. Then*

$$\int_{\mathbf{S}^1} \frac{|g_m|^2}{\mu_x^N(\{0\})} \, d\mu(x) \leqslant \frac{p^N}{m}.$$

Proof Let $h(x) = |g_m(x)|^2/\mu_x^N(\{0\})$ if $\mu_x^N(\{0\}) \neq 0$ (which is satisfied almost everywhere), and 0 elsewhere. Then, again using the notation $\xi = x + T_p^N$, we have

$$\int_\xi h d\mu_\xi = \sum_{\omega \in T_p^N} \mu_x^N(\{\omega\}) h(x + \omega)$$

$$= \sum_{\omega \in T_p^N} \mu_{x+\omega}^N(\{0\}) h(x + \omega) \leqslant \sum_{\omega \in T_p^N} |g_m(x + \omega)|^2.$$

Thus,

$$\int_{\mathbf{S}^1} h(x) \, d\mu(x) = \int_{\mathbf{S}^1} \left(\int_\xi h d\mu_\xi \right) d\mu(x) \leqslant \int_{\mathbf{S}^1} \sum_{a=0}^{p^N-1} |g_m(x + a/p^N)|^2 d\mu(x).$$

But

$$\sum_{a=0}^{p^N-1} |g_m(x+a/p^N)|^2$$

$$= \frac{1}{m^2} \sum_{a=0}^{p^N-1} \sum_{0 \leqslant k,\ell < m} e(v2^k(x + a/p^N)) e(-v2^\ell(x + a/p^N))$$

$$= \frac{1}{m^2} \sum_{0 \leqslant k,\ell < m} e(v(2^k - 2^\ell)x) \sum_{a=0}^{p^N-1} e(v(2^k - 2^\ell)a/p^N).$$

If $k \neq \ell$ then $v(2^k - 2^\ell)$ is non-zero modulo p^N by hypothesis, thus

$$\sum_{a=0}^{p^N-1} e(v(2^k - 2^\ell)a/p^N) = 0.$$

If $k = \ell$, the sum equals p^N, and we obtain the claimed estimate.   □

Under the hypotheses of the lemma, we obtain by Cauchy-Schwarz

$$\left| \int_{\mathbf{S}^1} g_m d\mu \right|^2 \leqslant \int_{\mathbf{S}^1} \frac{|g_m|^2}{\mu_x^N(\{0\})} d\mu \cdot \int_{\mathbf{S}^1} \mu_x^N(\{0\}) \, d\mu(x) \leqslant \frac{p^N}{m} \int_{\mathbf{S}^1} \mu_x^N(\{0\}) \, d\mu(x).$$

But $\mu_x^N(\{0\}) = 1/\mu_x^\infty(T_p^N)$. By T_p-recurrence, this function tends to 0 almost everywhere so that, by dominated convergence, $\int \mu_x^N(\{0\}) \, d\mu(x) \to 0$ as $N \to +\infty$.

Lemma 9.18 *Let p and q be primes. The sequence of $q^k v$ ($k \in \mathbb{N}$) in $\mathbf{Z}/p^N\mathbf{Z}$ is periodic of period $a_N \geq cp^N$ for a certain constant $c > 0$ independent of N.*

Proof The integer v is not important, its contribution will be incorporated into the constant. We shall in fact show that if p and q are distinct primes, there exists a constant c such that the order of q in the multiplicative group $(\mathbf{Z}/p^N\mathbf{Z})^*$ is greater than cp^N. Thus the period of the sequence (q^k) modulo p^N will be greater than cp^N.

The integer q is invertible in $\mathbf{Z}/p\mathbf{Z}$ and the multiplicative group $(\mathbf{Z}/p\mathbf{Z})^*$ is cyclic, so there exists an integer $k \geq 1$ such that $q^k \equiv 1 \pmod{p}$. We can therefore write

$$q^k = 1 + ap^\ell \quad \text{with } k, \ell \geq 1 \text{ and } (a, p) = 1.$$

Now we have

$$(1 + ap^\ell)^{p^j} \equiv 1 + ap^{j+\ell} \pmod{p^{j+\ell+1}}.$$

In particular,

$$(1 + ap^\ell)^{p^{N-\ell-1}} \equiv 1 + ap^{N-1} \pmod{p^N} \not\equiv 1 \pmod{p^N}.$$

Since the order of $1 + ap^\ell$ in $(\mathbf{Z}/p^N\mathbf{Z})^*$ divides $p^{N-\ell}$ it is necessarily equal to $p^{N-\ell}$. The integer q^k is therefore the order $p^{N-\ell}$ modulo p^N and q is of order $\geq \frac{1}{p^\ell}p^N$ modulo p^N. $\qquad\square$

Taking $p = 2$, $q = p$ and $m = a_N$, we get $\left| \int_{\mathbf{S}^1} g_{a_N} d\mu \right|^2 \leq c^{-1} \int_{\mathbf{S}^1} \mu_x^N(\{0\}) \, d\mu(x)$, which goes to 0 as N goes to $+\infty$. $\qquad\square$

9.4.5 Entropy

In practice the T_p-recurrence is verified via arguments making use of entropy. Let $f_p : x \mapsto px$ on $\mathbf{S}^1$. An f_p-invariant measure μ is said to be *ergodic* if for every f_p-invariant subset A in $\mathbf{S}^1$ either $\mu(A) - 0$ or $\mu(A) = \mu(\mathbf{S}^1)$. For $x \in \mathbf{S}^1$, denote by $I_n(x)$ the interval of the form $[i/p^n, (i+1)/p^n)$ containing x. We shall say that f_p is of *strong positive entropy* for μ if there exists a constant $\kappa > 0$ such that for $x \in \mathbf{S}^1$

and for all n,

$$\mu(I_n(x)) = O(e^{-\kappa n}), \tag{9.22}$$

where the implied constant is independent of x.

Proposition 9.19 *Let μ be an f_p-invariant ergodic probability measure on $\mathbf{S}^1$. If f_p is of strong positive entropy with respect to μ, then μ is T_p-recurrent.*

Proof Let I be one of the intervals of the form $[i/p, (i+1)/p)$ for $i = 0, 1, \ldots, p-1$. Let $\mu_I(A) = \mu(f_p(A \cap I))$. The restriction of μ to I is absolutely continuous with respect to μ_I (car $\mu \leqslant \mu_I$). We can therefore define a function φ on I by the Radon-Nikodym derivative:

$$\varphi = \frac{d\mu}{d\mu_I}.$$

Letting I vary, we obtain a function φ defined on the whole circle $\mathbf{S}^1$. We can interpret $\varphi(x)$ as the μ-probability of choosing x knowing the value of $f_p(x)$ (this is conditional probability). In particular, for every $y \in \mathbf{S}^1$ we have

$$\sum_{f_p(x)=y} \varphi(x) = 1.$$

Consider then the operator P which to a function g on the circle associates the function Pg defined by

$$(Pg)(y) = \sum_{f_p(x)=y} \varphi(x)g(x).$$

The operator P is the *transfer operator* associated with f_p. It allows one to write

$$\int_{\mathbf{S}^1} f \circ f_p(x)g(x)\, d\mu(x) = \int_{\mathbf{S}^1} f(y)(Pg)(y)\, d\mu(y). \tag{9.23}$$

Lemma 9.20 *Let $x \in \mathbf{S}^1$ and $N \in \mathbb{N}$, and again write $\xi = x + T_p^N$ for the leaf of F_p^N passing through x. Then*

$$\int_\xi f\, d\mu_\xi = (P^N f) \circ f_p^N(x).$$

Proof Let A be a measurable set for $\mathcal{F}_p^N$. One can write $A = f_p^{-N}(B)$ for a certain set B. Thus,

$$\int_A (P^N f) \circ f_p^N(x)\, d\mu(x) = \int_{\mathbf{S}^1} 1_B \circ f_p^N \cdot (P^N f) \circ f_p^N(x)\, d\mu(x)$$

$$= \int_{\mathbf{S}^1} 1_B(y) \cdot (P^N f)(y)\, d\mu(y) \ (\mu \text{ is invariant})$$

$$= \int_{S^1} 1_B \circ f_p^N(z) f(z) \, d\mu(z) \text{ (from (9.23))}$$

$$= \int_A f(z) \, d\mu(z).$$

As $(P^N f) \circ f_p^N$ is constant on each leaf of F_p^N, this concludes the proof of the lemma.

$\square$

Note that

$$(P^N f) \circ f_p^N(x) = \sum_{f_p^N(y) = f_p^N(x)} \varphi(y)\varphi(f_p(y)) \cdots \varphi(f_p^{N-1}(y)) f(y).$$

This allows us to identify the measure μ_x^N: letting $y = x + \omega$, we have $\mu_x^N(\{\omega\}) = \varphi(y) \cdots \varphi(f_p^{N-1}(y))$. In particular,

$$\mu_x^N(\{0\}) = \varphi(x) \cdots \varphi(f_p^{N-1}(x)).$$

The function φ being μ-almost everywhere non-zero, we can take its logarithm, which will be non-positive since $\varphi \leq 1$. The Birkhoff ergodic theorem asserts then that for μ-almost every x we have

$$\frac{1}{N} \log(\mu_x^N(\{0\})) \longrightarrow \int_{S^1} \log \varphi(x) \, d\mu(x) \tag{9.24}$$

as N goes to infinity. The limit integral is negative; to this, since $\log \varphi$ is non-positive, it suffices to show that φ is not equal to 1 almost everywhere. So let x_0 be a point of S^1 and consider the interval $I = I_N(x_0)$ of the form $[i/p^N, (i+1)/p^N)$ containing x_0. Now any point $x \in S^1$ has a unique preimage y in I so that, if χ_I denotes the characteristic function of I, we have

$$P^N(\chi_I)(x) = \varphi(y) \cdots \varphi(f_p^{N-1}(y)).$$

On the other hand (9.23) implies

$$\mu(I) = \int_{S^1} \chi_I(x) \, d\mu(x)$$

$$= \int_{S^1} P^N(\chi_I)(x) \, d\mu(x).$$

It then follows from (9.22) that φ is not μ-almost everywhere equal to the constant function 1.

Finally, the limit (9.24) is negative and, for almost all x in $\mathbf{S}^1$, the expression

$$\mu_x^N(\{0\}) = \frac{1}{\mu_x^\infty(T_p^N)}$$

tends to 0 with N. Thus, for almost all x in $\mathbf{S}^1$ we have $\mu_x^\infty(T_p) = \infty$ and the T_p-recurrence follows from Proposition 9.15. $\square$

To recap, we started from the multiplication by 2 on the circle, and the identification

$$\mathbf{Z}\backslash\mathbf{R} = \mathbf{Z}\,[1/p]\,\backslash(\mathbf{R}\times\mathbf{Q}_p)/\mathbf{Z}_p \quad \text{(where } \mathbf{Z}_p \text{ acts only on } \mathbf{Q}_p) \qquad (9.25)$$

allowed us to bring into play an additional action. This created a dynamical setting with two commuting degrees of freedom.

In the case of hyperbolic surfaces, we start with the dynamical action given by the geodesic flow. We must transpose (9.25) to this setting in order to adjoin to the geodesic flow a second dynamical action commuting with it.

9.5   Hecke Operators and Lindenstrauss's Theorem

Let a and b be two positive integers. As in Chap. 1, we shall consider the quaternion algebra $A = D_{a,b}(\mathbf{Q})$ and an order $\mathcal{O}$ in A. Let p be a prime not dividing the discriminant of A. According to § 8.1, the algebra $A_p = A \otimes_{\mathbf{Q}} \mathbf{Q}_p$ is then a matrix algebra. For almost all p it moreover holds that $\mathcal{O}_p$ is a maximal order in A_p. Denote by $\mathcal{P}$ the set of these "good" primes and assume henceforth that $p \in \mathcal{P}$. We then write $\Gamma = \mathcal{O} \otimes_{\mathbf{Z}} \mathbf{Z}\,[1/p]$. In the proof of Lemma 9.12 we made essential use of the fact that the set $\mathbf{Z}[1/p] \cdot (\mathbf{R} \times \mathbf{Z}_p)$ (diagonal action) is equal to the product $\mathbf{R} \times \mathbf{Q}_p$. Here, Theorem 8.6 implies that

$$A_\infty^* \times A_p^* = \Gamma \cdot \left(\mathrm{GL}^+(2,\mathbf{R}) \times \mathcal{O}_p^*\right).$$

We then denote $G = A_\infty^* \times A_p^* \cong \mathrm{GL}(2,\mathbf{R}) \times \mathrm{GL}(2,\mathbf{Q}_p)$, and write K for the compact maximal subgroup of G isomorphic to $\mathrm{O}(2) \times \mathrm{GL}(2,\mathbf{Z}_p)$ and C for the center of $\mathrm{GL}(2,\mathbf{R})$ viewed as a subgroup of G. The following lemma is analogous to Lemma 9.12.

Lemma 9.21 *The diagonal embedding of Γ in G is discrete and the double quotient $C \cdot \Gamma\backslash G/K$ is homeomorphic to the quotient $\mathcal{O}^1\backslash\mathcal{H}$.*

Let C_p be the intersection $A_p^* \cap C\Gamma$ in G. Then C_p is contained in the center of $A_p^* \cong \mathrm{GL}(2, \mathbf{Q}_p)$, and is nothing other than the set of matrices

$$\begin{pmatrix} \pm p^r & 0 \\ 0 & \pm p^r \end{pmatrix} \in \mathrm{GL}(2, \mathbf{Q}_p)$$

with $r \in \mathbf{Z}$. Define

$$T_p = \mathrm{GL}(2, \mathbf{Q}_p)/C_p\mathrm{GL}(2, \mathbf{Z}_p) = \mathrm{PGL}(2, \mathbf{Q}_p)/\mathrm{PGL}(2, \mathbf{Z}_p).$$

As in the previous subsection, we can think of the double quotient

$$C \cdot \Gamma \backslash G / K$$

as a T_p-foliated space. Before studying this "foliation" we describe, in the following subsection, the geometry of T_p, which is the p-adic analog of the hyperbolic plane $\mathcal{H} = \mathrm{PGL}(2, \mathbf{R})/\mathrm{PO}(2)$.

9.5.1 The Tree of $\mathrm{PGL}(2, \mathbf{Q}_p)$

Let $V = \mathbf{Q}_p \times \mathbf{Q}_p$, a 2-dimensional vector space over the field $\mathbf{Q}_p$. A $\mathbf{Z}_p$-lattice is a $\mathbf{Z}_p$-submodule of V generated by two linearly independent vectors. We denote by L_0 the lattice $\mathbf{Z}_p \times \mathbf{Z}_p$ and consider the equivalence relation $\sim$ defined on the set of lattices by

$$L_1 \sim L_2 \iff L_2 = \alpha L_1, \ \alpha \in \mathbf{Q}_p^*.$$

Write $[L]$ for the equivalence class of a $\mathbf{Z}_p$-lattice L.

We then define the graph $\mathcal{T}_p$ by the data of its vertices

$$\mathrm{vertices}(\mathcal{T}_p) = \{[L] \mid L \text{ is a } \mathbf{Z}_p\text{-lattice}\}$$

and its edges

$$\mathrm{edges}(\mathcal{T}_p) = \{\{[L_1], [L_2]\} \mid \exists L_i' \in [L_i] \ (i = 1, 2) \text{ such that } [L_2' : L_1'] = p\}.$$

We have written $[L_2' : L_1']$ for the index of L_1' in L_2' (in particular $L_1' \subset L_2'$).

Remark 9.22 If L is a $\mathbf{Z}_p$-lattice then the index $[L : pL]$ is equal to p^2. The definition of edges is therefore symmetric: the lattices pL_2' and L_2' are equivalent and $[L_1' : pL_2'] = p$ (in particular pL_2' is contained in L_1').

The group $\mathrm{GL}(2, \mathbf{Q}_p)$ acts naturally and transitively on the space of $\mathbf{Z}_p$-lattices and its center preserves equivalence classes; the group $\mathrm{PGL}(2, \mathbf{Q}_p)$ thus acts transitively on the graph $\mathcal{T}_p$. Clearly $\mathrm{PGL}(2, \mathbf{Z}_p)$ is the stabilizer of L_0 in $\mathrm{PGL}(2, \mathbf{Q}_p)$. The set of vertices of $\mathcal{T}_p$ is therefore in bijection with the quotient $\mathrm{PGL}(2, \mathbf{Q}_p)/\mathrm{PGL}(2, \mathbf{Z}_p)$. Note, moreover, that the action of $\mathrm{PGL}(2, \mathbf{Q}_p)$ preserves the adjacency relation; thus $\mathrm{PGL}(2, \mathbf{Q}_p)$ embeds in the automorphism group of $\mathcal{T}_p$ which is then necessarily a regular graph (i.e., the number of edges emanating from a vertex is the same for all vertices).

The number of edges emanating from the vertex $[L_0]$ is

$$|\{L \subset L_0 \mid [L_0 : L] = p\}| = |\{pL_0 \subset L \subset L_0 \mid [L_0 : L] = p\}|$$

$$= |\{\text{line} \subset L_0/pL_0 \simeq \mathbf{F}_p \times \mathbf{F}_p\}| \qquad (9.26)$$

$$= |\mathbf{P}^1(\mathbf{F}_p)| = p + 1.$$

Proposition 9.23 *The graph $\mathcal{T}_p$ is a $(p + 1)$-regular tree.*

Proof Since we already know that $\mathcal{T}_p$ is $(p + 1)$-regular, we only have to show that it is connected and contains no closed loop. This is equivalent to showing that there is a unique path in $\mathcal{T}_p$ between any two vertices.

Let $[L]$ be a vertex of $\mathcal{T}_p$. Let L be the unique representative of $[L]$ such that $L \subset L_0$ and $L \not\subset pL_0$. There exists then a basis (e_1, e_2) of L_0 and an integer $a > 0$ such that $(e_1, p^a e_2)$ is a basis of L. It follows that the quotient L_0/L is a finite cyclic p-group. The existence and uniqueness of a path between $[L_0]$ and $[L]$ in $\mathcal{T}_p$ can then be read off from the existence and uniqueness of the Jordan-Hölder series in a finite cyclic p-group.

We can give an alternative proof of the uniqueness (which is less classical) by showing by hand that there are no loops in $\mathcal{T}_p$. Let $[L_0], [L_1], \ldots, [L_n]$ be a sequence of vertices of a path without back-tracking in $\mathcal{T}_p$. It will be sufficient to show by induction that $[L_0] \neq [L_n]$.

We may assume that $L_{i+1} \subset L_i$ and $[L_i : L_{i+1}] = p$. We argue by induction that $L_n \not\subset pL_0$ (we will then necessarily have $[L_n] \neq [L_0]$). By the induction hypothesis, we have $L_{n-1} \not\subset pL_0$. The lattices L_n and pL_{n-2} are the inverse images of two lines of the $\mathbf{F}_p$-plane L_{n-1}/pL_{n-1}. These two lines are distinct, for otherwise $[L_{n-2}] = [L_n]$, in contradiction with the fact that the given path has no back-tracking. We thus have

$$L_{n-1} = L_n + pL_{n-2}$$

hence

$$L_{n-1} \equiv L_n \pmod{pL_0}$$

and we deduce that $L_n \not\subset pL_0$ as desired. $\qquad \square$

9.5.2 Hecke Operators

The analog of the Laplacian on the tree $\mathcal{T}_p$ is the operator

$$(p+1)\,\mathrm{Id} - \delta$$

where δ is the *Hecke operator* defined by

$$\delta f(x) = \sum_{y \sim x} f(y),$$

where f is a function on the vertices of $\mathcal{T}_p$ and $\sim$ is the adjacency relation between the vertices of $\mathcal{T}_p$. The Hecke operator thus evaluates the sum of the values of f over all neighboring vertices of the given vertex x.

Just like the hyperbolic Laplacian, these operators induce operators on the corresponding arithmetic surfaces. It is the latter operators that are more often referred to as Hecke operators. We now describe them more explicitly.

We have

$$C \backslash G / C_p \cdot K \cong \mathcal{H} \times T_p$$

where T_p is identified with the set of vertices of $\mathcal{T}_p$. The group $\Gamma \subset G$ acts on $\mathcal{H} \times T_p$, the action on T_p being induced by that on the tree $\mathcal{T}_p$. We can therefore think of the double quotient $C \cdot \Gamma \backslash G / K$ as a "discretely foliated space", the points of each leaf being identified with the vertices of $\mathcal{T}_p$. Note that z-almost all $\{z\} \times T_p$ ($z \in \mathcal{H}$) project injectively into the quotient $C \cdot \Gamma \backslash G / K$.[8] We deduce that through every point x of

$$\mathcal{O}^1 \backslash \mathcal{H} \cong C \cdot \Gamma \backslash G / K$$

there passes a "T_p-leaf". We shall write $x \sim_{T_p} y$ when two points $x, y \in \mathcal{O}^1 \backslash \mathcal{H}$ lie on the same leaf.

We can then define the *Hecke operators* on the surface $\mathcal{O}^1 \backslash \mathcal{H}$ by

$$(S_p f)(x) = \sum_{y \sim_{T_p} x} f(y),$$

where $f : \mathcal{O}^1 \backslash \mathcal{H} \to \mathbf{C}$ is any function.

A matrix $A \in \mathrm{GL}(2, \mathbf{Q}_p)$ sends the lattice L_0 to a sublattice L of index $[L_0 : L] = p$ if and only if the determinant of A lies in $p\mathbf{Z}_p^*$. From (9.26), there exist $p + 1$ classes of such matrices modulo $\mathrm{GL}(2, \mathbf{Z}_p)$; let $\mathcal{S}_p$ denote this set of classes.

---

[8] The fixed points of matrices in $\mathrm{SL}(2, \mathbf{Z}[1/p])$ form a countable set.

The $p + 1$ elements of $\mathcal{S}_p$ correspond to the $p + 1$ lines in $L_0/pL_0 \simeq \mathbf{F}_p \times \mathbf{F}_p$ which in the canonical basis (e_1, e_2) are generated by the vectors e_1 or $e_2 + be_1$ with $0 \leqslant b < p$. We can thus represent the elements of $\mathcal{S}_p$ by the matrices

$$\begin{pmatrix} 1 & 0 \\ 0 & p \end{pmatrix} \text{ and } \begin{pmatrix} p & b \\ 0 & 1 \end{pmatrix} \text{ with } 0 \leqslant b < p.$$

When $\mathcal{O}^1$ is a congruence subgroup $\Gamma_0(N) \subset \mathrm{SL}(2, \mathbf{Z})$, with p and N relatively prime, we obtain the following expression for the Hecke S_p operator

$$(S_p f)(z) = f(pz) + \sum_{b=0}^{p-1} f\left(\frac{z + b}{p}\right), \tag{9.27}$$

for any $f : \mathcal{O}^1 \backslash \mathcal{H} \to \mathbf{C}$. We recover in this way the expression for the Hecke operators considered in Chap. 7, up to a normalization factor.

In general we put

$$\mathcal{O}(p) = \{\alpha \in \mathcal{O} \mid \mathrm{N}_{\mathrm{red}}(\alpha) = p\}.$$

The group $\mathcal{O}^1$ then acts by left multiplication on $\mathcal{O}(p)$ with $p + 1$ orbits, and the Hecke operator S_p is expressed as

$$(S_p f)(z) = \sum_{\alpha \in \mathcal{O}^1 \backslash \mathcal{O}(p)} f(\alpha z), \tag{9.28}$$

for any $f : \mathcal{O}^1 \backslash \mathcal{H} \to \mathbf{C}$.

9.5.3 Lindenstrauss's Theorem

It is clear that the Hecke operators S_p and the Laplacian Δ commute. We then call any weak limit of $|\phi_i(x)|^2 d\mu(x)$, where the sequence of ϕ_i consists of joint Δ *and* S_p eigenfunctions for almost every prime p, *an arithmetic quantum limit*.

Theorem 9.24 *Let S be an arithmetic surface as above. Then every arithmetic quantum limit is equal to* $\mathrm{area}(S)^{-1} d\mu$.

In what follows we shall explain how Lindenstrauss is able to deduce Theorem 9.24 from the analog of the following analog of Theorem 9.11. We keep the above notational conventions; in particular $G = \mathrm{GL}(2, \mathbf{R}) \times \mathrm{GL}(2, \mathbf{Q}_p)$. Define $K_p = \{e\} \times \mathrm{GL}(2, \mathbf{Z}_p) \subset K$. Finally, we shall say that a probability measure ν on $X = \mathcal{O}^1 \backslash \mathrm{SL}(2, \mathbf{R})$ which is invariant under the action of the diagonal group is of *strongly positive entropy* if there exists $\tau_0, \kappa > 0$ such that for every compact subset

$\Omega \subset X$ and for all $x \in \Omega$,

$$v(xB(\varepsilon, \tau_0)) = O_\Omega(\varepsilon^\kappa), \quad \forall \varepsilon > 0,$$

where

$$B(\varepsilon, \tau) = a((-\tau, \tau))u^-((-\varepsilon, \varepsilon))u^+((-\varepsilon, \varepsilon)), \tag{9.29}$$

$$u^+(x) = \begin{pmatrix} 1 & 0 \\ x & 1 \end{pmatrix}, \quad u^-(x) = \begin{pmatrix} 1 & x \\ 0 & 1 \end{pmatrix} \quad \text{and} \quad a(t) = \begin{pmatrix} e^t & 0 \\ 0 & e^{-t} \end{pmatrix}.$$

Henceforth the set $B(\varepsilon, \tau)$ will denote the above open neighborhood of the identity in $\mathrm{PSL}(2, \mathbf{R})$.

Theorem 9.25 *Let v be a probability measure on $X = C \cdot \Gamma \backslash G / K_p$ which is invariant under the action of the diagonal group*

$$A \times \{e\} = \left\{ \begin{pmatrix} * & 0 \\ 0 & * \end{pmatrix} \right\} \times \{e\} \subset G.$$

Assume that v is

1. *of strongly positive entropy;*
2. *T_p-recurrent.*

Then v is $\mathrm{GL}(2, \mathbf{R})$-invariant.

The proof of Theorem 9.25 is difficult and technical (it forms the heart of the proof of Theorem 9.24) but has nothing to do with spectral theory or arithmetic; it is purely ergodic theoretic along the lines of Theorem 9.16. We shall content ourselves here with the more modest goal of showing how the underlying arithmetic, realized by the action of the Hecke operators, allows one to verify entropy hypotheses 1 and 2, which are otherwise hard to get a hold of.

9.6 Use of Hecke Operators

In this section we explain why any arithmetic quantum limit μ satisfies the hypotheses of Theorem 9.25. We begin by checking the strongly positive entropy hypothesis. We must show that the density $|\psi|^2 d\mu$ associated with a Hecke eigenfunction is sufficiently "diffuse". We begin by showing that on the tree $\mathcal{T}_p$, the distributional density defined by a Hecke eigenfunction cannot give too much mass to a vertex or a geodesic axis. In what follows we shall express this property by saying that *an eigenfunction cannot be too concentrated on a vertex or along a geodesic axis*. The idea is to find enough Hecke translates of the balls $B(\varepsilon, \tau)$ which are pairwise "almost" disjoint.

9.6.1 Local Contributions

Let p be a prime belonging to $\mathcal{P}$ (see §9.5). Denote by e the base point in the tree $\mathcal{T}_p = \mathrm{PGL}(2, \mathbf{Q}_p)/\mathrm{PGL}(2, \mathbf{Z}_p)$ corresponding to the class of the lattice L_0. The Hecke operators arise in the following way: an eigenfunction $f : \mathcal{T}_p \to \mathbf{C}$ of the Hecke operator δ_p cannot be too concentrated on a vertex (Proposition 9.26); moreover, if f is the restriction of a Maaß form, we show that it cannot be too concentrated along a geodesic axis of $\mathcal{T}_p$ passing through e (Proposition 9.27).

Local Contribution I

In this first subsection we prove that an eigenfunction for the Hecke operator δ_p on the tree $\mathcal{T}_p$ cannot be too concentrated on a vertex. This property is true more generally for eigenfunctions of the Laplace operator on a regular tree.

Let $f : \mathcal{T}_p \to \mathbf{C}$. Define

$$\delta_{p^k} f(x) = \sum_{d_{\mathcal{T}_p}(x,y)=k} f(y),$$

where $d_{\mathcal{T}_p}$ is the combinatorial distance in the tree $\mathcal{T}_p$. Finally, write $B_n^{\mathcal{T}_p}$ for the ball of radius n centered at e in the tree $\mathcal{T}_p$ equipped with the distance function $d_{\mathcal{T}_p}$.

Proposition 9.26 *Assume that $\delta_p f = \lambda f$ for some $\lambda \in \mathbf{R}$. Then there exists a constant c independent of λ and p such that for all $n \geqslant 0$,*

$$\sum_{y \in B_n^{\mathcal{T}_p}} |f(y)|^2 \geqslant cn|f(e)|^2. \tag{9.30}$$

Proof It is clear that

$$\delta_{p^k} \circ \delta_p = \delta_{p^{k+1}} + p\delta_{p^{k-1}} \quad \text{if } k \geqslant 2$$

and

$$\delta_p \circ \delta_p = \delta_{p^2} + (p+1)\,\mathrm{Id}.$$

Thus f is an eigenfunction for all the δ_{p^k}. Write λ_{p^k} for the corresponding eigenvalue. The sequence (λ_{p^k}) satisfies the linear recurrence relation associated with the degree 2 equation $X^2 - \lambda X + p$. We then distinguish two cases, according to whether $|\lambda| > 2\sqrt{p}$ or $|\lambda| \leqslant 2\sqrt{p}$.

We begin by assuming that $|\lambda| > 2\sqrt{p}$. We then write $\cosh\alpha = |\lambda/2\sqrt{p}|$. The two roots of the polynomial $X^2 - \lambda X + p$ are then $\mathrm{sgn}(\lambda)\sqrt{p}\,e^{\alpha}$ and $\mathrm{sgn}(\lambda)\sqrt{p}\,e^{-\alpha}$ and a simple calculation gives

$$\lambda_{p^k} = \left[\frac{1 - pe^{2\alpha}}{p(1 - e^{2\alpha})}\,e^{k\alpha} + \frac{1 - pe^{-2\alpha}}{p(1 - e^{-2\alpha})}\,e^{-k\alpha}\right]\mathrm{sgn}(\lambda)p^{k/2}.$$

With the convention that $\lambda_{p^0} = 1$ we thus obtain

$$\sum_{k=0}^{n}\lambda_{p^{2k}} = p^n\,\frac{\sinh(2n+1)\alpha}{\sinh\alpha} \geq (2n+1)p^n.$$

In other words,

$$\left|\sum_{d_{T_p}(e,y)\in\{0,2,\dots,2n\}} f(y)\right| \geq (2n+1)p^n f(e).$$

The Cauchy-Schwarz inequality then implies

$$\sum_{d_{T_p}(e,y)\in\{0,2,\dots,2n\}} |f(y)|^2 \geq n^2|f(e)|^2,$$

yielding (9.30).

Now assume that $|\lambda| \leq 2\sqrt{p}$ and write $\cos\theta = \lambda/2\sqrt{p}$. In the same way as above one may show that

$$\sum_{k=0}^{n}\lambda_{p^{2k}} = p^n\,\frac{\sin(2n+1)\theta}{\sin\theta}. \qquad (9.31)$$

On the other hand, the Cauchy-Schwarz inequality implies

$$\sum_{d_{T_p}(e,y)=2k} |f(y)|^2 \geq \frac{\left|\sum_{d(e,y)=2k} f(y)\right|^2}{(p+1)p^{2k-1}}$$

$$= \frac{|\lambda_{p^{2k}}|^2|f(e)|^2}{(p+1)p^{2k-1}}.$$

Subtracting (9.31) with $n = k - 1$ from (9.31) with $n = k$ we get

$$\lambda_{p^{2k}} = p^k\left[\frac{\sin(2k+1)\theta}{\sin\theta} - \frac{\sin(2k-1)\theta}{p\sin\theta}\right].$$

Thus

$$\sum_{d(x,y)=2k} |f(y)|^2 \geq c|f(e)|^2$$

if $(2k+1)\theta \in [2\pi/5, 3\pi/5] \pmod{\pi}$.

If n is sufficiently large $(n > c_1/\theta)$, we can assume that

$$|\{k \mid 1 \leq k \leq n, \ (2k+1)\theta \in [2\pi/5, 3\pi/5] \pmod{\pi}\}| > c_2 n$$

so that (9.30) is satisfied for some constant c. On the other hand, if n is sufficiently small $(n \leq c_3/\theta)$ we have $\frac{\sin(2n+1)\theta}{\sin\theta} \geq n$ so that (9.31) implies

$$\left| \sum_{k=0}^{n} \lambda_{p^{2k}} \right| \geq np^n.$$

This gives

$$\sum_{y \in B_{2n}^T} |f(y)|^2 \geq cn^2 |f(e)|^2 \geq cn|f(e)|^2.$$

We can thus choose a constant c so that (9.30) always holds. $\qquad\square$

In order to determine the local contribution away from a geodesic in $\mathcal{T}_p$ we shall use the hypothesis that f is the restriction of a Maaß form. In this way we can appeal to the approximation to the Ramanujan-Petersson conjecture established in § 7.6.

Ramanujan-Petersson Conjecture

Suppose firstly that $\mathcal{O}^1 = \mathrm{SL}(2, \mathbf{Z})$ and write Γ for this group. Let $\phi : \Gamma \backslash \mathcal{H} \to \mathbf{C}$,

$$\phi(z) = \sum_{r \in \mathbf{Z}} a_r \sqrt{y} \, K_\nu(2\pi|r|y) e(rx),$$

be a non-constant Maaß form, which we assume to be an eigenfunction for all Hecke operators and normalized so that $a_r = 1$. Then ϕ is primitive and cuspidal, and we have

$$S_p \phi = \sqrt{p} \, a_p \phi.$$

Let $\lambda = \sqrt{p} \, a_p$.

Fix a "T_p-leaf" embedded in $\Gamma \backslash \mathcal{H}$. The restriction of ϕ to this leaf defines a function $f : \mathcal{T}_p \to \mathbf{C}$ which is a δ_p-eigenfunction with eigenvalue λ. The

Ramanujan-Petersson conjecture implies that for any $\varepsilon > 0$,

$$|\lambda| = O_\varepsilon\big(p^{1/2+\varepsilon}\big),$$

where the implied constant is independent of p. The Ramanujan-Petersson conjecture is still open. Nevertheless, Theorem 7.42 implies that there exists a universal constant $\delta > 0$ such that

$$|\lambda| = O\left(p^{1-\delta}\right). \tag{9.32}$$

Neither δ nor the implied constant depends on p.

This result remains true for more general $\mathcal{O}^1$ and $p \notin \mathcal{P}$. Just as for Theorem 8.25, the Jacquet-Langlands correspondence (sufficiently refined) allows us to reduce to the case of a Hecke-Maaß newform for a certain group $\Gamma_0(N)$, where p does not divide N.

Local Contribution II

Let $\mathcal{A}_p$ be the geodesic axis in the metric space $\mathcal{T}_p$ of the element

$$\begin{pmatrix} p & 0 \\ 0 & 1 \end{pmatrix} \in \mathrm{PGL}(2, \mathbf{Q}_p).$$

Then $\mathcal{A}_p$ is the line in $\mathcal{T}_p$ passing through the vertices $[(p^n \mathbf{Z}_p) \times \mathbf{Z}_p] \in \mathcal{T}_p$, $n \in \mathbf{Z}$. Figure 9.1 – realized by U. Goertz – represents an axis in the tree $\mathcal{T}_2$.

Fig. 9.1 An axis in the tree $\mathcal{T}_2$

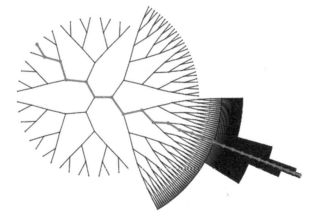

We introduce two subsets of "transversal" vertices to the axis $\mathcal{A}_p$. First let $\mathcal{S}_1$ be the set of $p-1$ neighbors of e in $\mathcal{T}_p$ not belonging to $\mathcal{A}_p$. Next let $\mathcal{S}_2$ be the union over all the $x \in \mathcal{S}_1$ of the p neighbors of x different from e. We have

$$|\mathcal{S}_1| = p - 1 \quad \text{and} \quad |\mathcal{S}_2| = p^2 - p,$$

$$\mathcal{S}_1 = \left\{ \begin{pmatrix} p & b \\ 0 & 1 \end{pmatrix} \cdot e \,\middle|\, 0 < b < p \right\}$$

and

$$\mathcal{S}_2 = \left\{ \begin{pmatrix} p & c + bp^{-1} \\ 0 & p^{-1} \end{pmatrix} \cdot e \,\middle|\, 0 < b < p \text{ and } 0 \leqslant c < p \right\}.$$

Let $\mathcal{S} = \mathcal{S}_1 \cup \mathcal{S}_2$.

When f is a Hecke eigenfunction on $\mathcal{T}_p$ obtained as a restriction of a Maaß form, we have at our disposal – according to the preceding subsection – the bound (9.32) on the eigenvalue λ. This allows us to prove the following proposition.

Proposition 9.27 *Let f and δ be as above. Then there exists a constant c, independent of f, λ, and p, such that*

$$\sum_{x \in \mathcal{S}} |f(x)|^2 \geqslant \frac{c}{p^{1-2\delta}} |f(e)|^2. \tag{9.33}$$

Proof Let $x \in \mathcal{S}_1$. Denote by $\mathcal{S}_2(x)$ the set of p neighbors of x belonging to $\mathcal{S}_2$; the point $e \notin \mathcal{S}_2$ is the remaining neighbor x. We then have

$$\sum_{y \in \mathcal{S}_2(x)} f(y) + f(e) = \lambda f(x).$$

The sets $\mathcal{S}_2(x)$ ($x \in \mathcal{S}_1$) are pairwise disjoint. Summing over $x \in \mathcal{S}_1$, we find

$$\left(\sum_{y \in \mathcal{S}_2} f(y) \right) + (p-1) \cdot f(e) = \lambda \cdot \left(\sum_{x \in \mathcal{S}_1} f(x) \right).$$

It then follows from the Cauchy-Schwarz inequality that

$$(p-1)^2 |f(e)|^2 \leqslant 2(p^2 - p) \left(\sum_{y \in \mathcal{S}_2} |f(y)|^2 \right) + 2\lambda^2 (p-1) \left(\sum_{x \in \mathcal{S}_1} |f(x)|^2 \right)$$

and thus that

$$|f(e)|^2 \leq (4 + 8\lambda^2 p^{-1}) \cdot \left(\sum_{y \in \mathcal{S}} |f(y)|^2 \right).$$

Inequality (9.32) finally allows us to conclude the proof of the proposition.   □

9.6.2 Intersections of Hecke Translates

In this section we aim to construct many Hecke translates of the balls $B(\varepsilon, \tau)$ – the open neighborhoods of the identity in $SL(2, \mathbf{R})$ defined by (9.29) – which are pairwise disjoint.

Let $\tau_0 > 0$ be sufficiently small and $0 < \varepsilon < 1/100$. To simplify we shall henceforth assume that $A = D_{a,b}(\mathbf{Q})$ is a division algebra; the group $\mathcal{O}^1$ is then cocompact in $SL(2, \mathbf{R})$. Fix an element x in $SL(2, \mathbf{R})$ and denote by $\underline{\alpha}$ the image in $SL(2, \mathbf{R})$ of the element $N_{\mathrm{red}}(\alpha)^{-1/2}\alpha \in A$.

Lemma 9.28 *There exists a constant c depending only on τ_0 and a constant c' depending only on τ_0 and x such that the set*

$$\{\alpha \in \mathcal{O} \mid N_{\mathrm{red}}(\alpha) \leq c\varepsilon^{-1} \text{ and } \underline{\alpha}x \in xB(4\varepsilon, 3\tau_0)\}$$

is contained in a quadratic subfield $F \subset A$ of discriminant less than $c'\varepsilon^{-1}$.

Proof Let $\alpha, \beta \in \mathcal{O}$ be such that

$$\underline{\alpha}x, \ \underline{\beta}x \in xB(4\varepsilon, 3\tau_0). \tag{9.34}$$

We can then write

$$\underline{\alpha}x \in xa(t_\alpha)B(4\varepsilon, 0) \quad \text{and} \quad \underline{\beta}x \in xa(t_\beta)B(4\varepsilon, 0).$$

As a consequence of Lemma 9.29, which follows, the image of x by the commutator $\rho := \underline{\alpha}\underline{\beta}^{-1}\underline{\alpha}^{-1}\underline{\beta}$ is contained in $xB(c_1\varepsilon, c_1\varepsilon^2)$. Here c_1 is a constant which, according to (9.34), depends only on τ_0; it can can be explicitly calculated with the help of Lemma 9.29.

Note that there exists a constant c_2, depending only on c_1, such that for all $z \in xB(c_1\varepsilon, c_1\varepsilon^2)x^{-1}$,

$$|\operatorname{tr}(z) - 2| \leq c_2\varepsilon^2.$$

The commutator ρ is equal to

$$\frac{1}{N_{red}(\alpha)N_{red}(\beta)}\alpha\bar{\beta}\bar{\alpha}\beta.$$

We thus have

$$tr(\rho) \in \frac{1}{N_{red}(\alpha)N_{red}(\beta)}\cdot\mathbf{Z}.$$

As long as

$$c_2\varepsilon^2 \leq [N_{red}(\alpha)N_{red}(\beta)]^{-1}$$

it must hold that $tr(\rho) = 2$ and, since $\mathcal{O}$ contains no unipotent elements, ρ is equal to the identity element. Putting $c = c_2^{-1/2}$, it follows from all of this that the elements of the set

$$\{\alpha \in \mathcal{O} \mid N_{red}(\alpha) \leq c\varepsilon^{-1} \text{ and } \underline{\alpha}x \in xB(\varepsilon, \tau_0)\}$$

commute. They generate a subfield of A which, since $[A : \mathbf{Q}] = 4$, is either $\mathbf{Q}$ or a quadratic extension of $\mathbf{Q}$. In the latter case the field, write the field as $\mathbf{Q}(\alpha)$ for some primitive element α belonging to the above set. Then α is a root of the polynomial $t^2 - tr(\alpha)t + N_{red}(\alpha) = 0$ and the discriminant of $\mathbf{Q}(\alpha)$ is equal to $D = tr(\alpha)^2 - 4N_{red}(\alpha)$. But by definition

$$\frac{|tr(\alpha)|}{N_{red}(\alpha)^{1/2}} = |tr(\underline{\alpha})| \in |tr(xB(4\varepsilon, 3\tau_0)x^{-1})|$$

is bounded from above by a constant depending only on τ_0 and $N_{red}(\alpha) \leq c\varepsilon^{-1}$. Thus there exists a constant c' depending only on τ_0 and x such that $D \leq c'\varepsilon^{-1}$. □

Lemma 9.29 *For all $\tau > 0$ and $\varepsilon \in (0, 1/10)$ we have*

$$B(\varepsilon, \tau)B(\varepsilon, \tau) \subset B(O_\tau(\varepsilon), 2\tau, O_\tau(\varepsilon^2))$$

and

$$B(\varepsilon, \tau)^{-1} \subset B(O_\tau(\varepsilon), \tau + O_\tau(\varepsilon^2)).$$

Proof We shall show only the first inclusion, the proof of the second one being similar. Let

$$g_1 = a(t_1)u^-(a_1)u^+(b_1) \quad \text{and} \quad g_2 = a(t_2)u^-(a_2)u^+(b_2).$$

Then

$$g_1g_2 = a(t_1)u^-(a_1)u^+(b_1)a(t_2)u^-(a_2)u^+(b_2)$$
$$= a(t_1 + t_2)u^-(e^{-2t_2}a_1)u^+(e^{2t_2}b_1)u^-(a_2)u^+(b_2).$$

Put $\tilde{b}_1 = e^{2t_2}b_1$. Writing

$$u^+(\tilde{b}_1)u^-(a_2) = \begin{pmatrix} 1 & a_2 \\ \tilde{b}_1 & 1 + a_2\tilde{b}_1 \end{pmatrix}$$

in the form

$$u^- \left(a_2[1 + a_2\tilde{b}_1]^{-1} \right) a \left(-\log \left[1 + \tilde{b}_1 a_2 \right] \right) u^+ \left(\tilde{b}_1[1 + a_2\tilde{b}_1]^{-1} \right)$$

we find that

$$g_1g_2 = a(t_1 + t_2 + O_\tau(\varepsilon^2))u^-(O_\tau(\varepsilon))u^+(O_\tau(\varepsilon)).$$

The lemma then follows. □

Denote by $F = \mathbf{Q}(\alpha) \subset A$ the quadratic field and $R = \mathbf{Z}[\alpha] \subset \mathcal{O}_F$ the order given by Lemma 9.28. Let

$$\mathcal{P}_\varepsilon = \{p \in \mathcal{P} \mid (p, D_\varepsilon) = 1\}.$$

Fix a prime p in $\mathcal{P}_\varepsilon$. Then the algebra $F_p = F \otimes_\mathbf{Q} \mathbf{Q}_p$ is unramified and R_p is maximal. Then the subalgebra F_p of A_p is isomorphic either to the unique unramified quadratic field extension of $\mathbf{Q}_p$ or to the product $\mathbf{Q}_p \times \mathbf{Q}_p$. In the first case, $F_p^* = \mathbf{Q}_p^* \cdot F_p^1$, where F_p^1 denotes the group of elements of norm $N_{F_p/\mathbf{Q}_p}$ equal to 1; these elements of norm 1 are contained in the ring of integers $\mathcal{O}_{F_p} = R \otimes \mathbf{Z}_p \subset \mathcal{O}_p = K_p$. In the latter case let x_1 (resp. x_2) denote the element of F_p corresponding to $(1,0)$ (resp. $(0,1)$) in $\mathbf{Q}_p \times \mathbf{Q}_p$. Then $x_1, x_2 \in \mathcal{O}_{F_p} = R \otimes \mathbf{Z}_p \subset \mathcal{O}_p \cong M_2(\mathbf{Z}_p)$ are commuting idempotents and can therefore be simultaneously diagonalized under conjugation by a matrix $GL(2, \mathbf{Z}_p)$. Moreover, F_p^* is conjugate by a matrix in K_p to the subgroup of diagonal matrices in $A_p^* \cong GL(2, \mathbf{Q}_p)$. In either case, there exists $k_p \in K_p$ such that F_p^* is contained in

$$k_p \cdot \left\{ \begin{pmatrix} a & 0 \\ 0 & d \end{pmatrix} \, \middle| \, a, d \in \mathbf{Q}_p^* \right\} \cdot K_p.$$

In particular, the orbit of the base point $e = [K_p]$ in $\mathcal{T}_p$ under the action of F_p^* is contained in the geodesic $k_p \cdot \mathcal{A}_p$. Recall that we defined in § 9.6.1 a subset $\mathcal{S} = \mathcal{S}_1 \cup \mathcal{S}_2$ of $\mathcal{T}_p$ consisting of vertices "transversal" to the axis $\mathcal{A}_p$. Let

$$\mathcal{R}_p = \{\alpha_s \mid s \in \mathcal{S}\}$$

be a system of representatives $\alpha_s \in \mathcal{O}(p)$, if $s \in \mathcal{S}_1$, and $\alpha_s \in \mathcal{O}(p^2)$, if $s \in \mathcal{S}_2$, such that $k_p \cdot \mathcal{S} = \{[\alpha_s^{-1} K_p] \in \mathcal{T}_p\}$. Given a point $x \in X = \mathcal{O}^1 \backslash \mathrm{SL}(2, \mathbf{R})$, the elements of $\{\alpha_s x \mid s \in \mathcal{S}\}$ project onto the points of the T_p-leaf passing through the projection of x in $\mathcal{O}^1 \backslash \mathcal{H}$. These points correspond – via the identification of the leaf with the tree $\mathcal{T}_p$ – to the points of $k_p \cdot \mathcal{S}$. Finally, let

$$\mathcal{R}_\varepsilon = \bigcup_{p \in \mathcal{P}_\varepsilon, \, p^4 \leq c\varepsilon^{-1}} \mathcal{R}_p,$$

where c is the constant given in Lemma 9.28.

Proposition 9.30 *There exists a constant c_0 such that for any $\varepsilon \in (0, 1/100)$ and for all $y \in \mathrm{SL}(2, \mathbf{R})$, the set*

$$\{\beta \in \mathcal{R}_\varepsilon \mid y \in \underline{\beta} x B(\varepsilon, \tau_0)\}$$

contains at most c_0 elements.

Proof Let β_1 and $\beta_2 \in \mathcal{R}_\varepsilon$ be two distinct elements such that

$$\underline{\beta_1} x B(\varepsilon, \tau_0) \cap \underline{\beta_2} x B(\varepsilon, \tau_0) \neq \varnothing.$$

There then exists a primitive element $\beta \in \mathcal{O}$ such that $\overline{\beta}_1 \beta_2 \in \mathbf{Z}\beta$. The reduced norm of β divides $p_1^2 p_2^2$, where $\beta_i \in \mathcal{R}_{p_i}$ $(i = 1, 2)$. Moreover, by the definition of β, we have

$$\underline{\beta} x \in x B(4\varepsilon, 3\tau_0).$$

It then follows from Lemma 9.28 that β belongs to the field F.

Lemma 9.31 *We have $p_1 = p_2$.*

Proof We assume $p_1 \neq p_2$ and argue by contradiction. For notational convenience we write $p = p_2$. The image of $\beta^{-1} \in F$ in $A^*(\mathbf{Q}_p)$ acts on the tree $\mathcal{T}_p$ by sending e to a vertex on the geodesic $k_p \cdot \mathcal{A}_p \subset \mathcal{T}_p$. Since $p_1 \neq p$, the image of β_1 in $A^*(\mathbf{Q}_p)$ lies in K_p. It follows that $[\beta_2^{-1} K_p]$ lies on the geodesic $k_p \cdot \mathcal{A}$, which contradicts the definition of $\mathcal{R}_p$. $\qquad \square$

We thus have $\beta_1, \beta_2 \in \mathcal{R}_p$ for some $p \in \mathcal{P}_\varepsilon$ and the reduced norm of β is a power of p. Viewed as an element of $A^*(\mathbf{Q}_p)$, β^{-1} sends $[\beta_1^{-1} K_p]$ to $[\beta_2^{-1} K_p]$. These two points lie in $k_p \cdot \mathcal{S}$ and project onto the same element e of the geodesic $k_p \cdot \mathcal{A}_p$. But $\beta^{-1} \in F$ sends e to the geodesic $k_p \cdot \mathcal{A}_p$. All of this forces β^{-1} to stabilize e. Thus β, being primitive, lies in $\mathcal{O}^1$. We finish the proof of the proposition by using the properness of the action of $\mathcal{O}^1$ on $\mathrm{SL}(2, \mathbf{R})$. $\qquad \square$

9.6.3   Strongly Positive Entropy

We again write $X = \mathcal{O}^1 \backslash \mathrm{SL}(2, \mathbf{R})$ and assume that X is compact. For a prime $p \in \mathcal{P}$ we defined the Hecke operator S_p acting on functions on the surface $\mathcal{O}^1 \backslash \mathcal{H}$, see (9.28). We can lift the S_p to operators $C^\infty(X) \to C^\infty(X)$ by putting, for $\psi \in C^\infty(X)$,

$$(S_p \psi)(x) = \sum_{\alpha \in \mathcal{O}^1 \backslash \mathcal{O}(p)} \psi(\underline{\alpha} x).$$

Theorem 9.32 *There exists constants $\kappa, c > 0$ such that for any $\psi \in C^\infty(X)$ of L^2-norm 1 which is an eigenfunction for all Hecke operators S_p, for all $x \in X$ and for all $\varepsilon > 0$, we have*

$$\mu_\psi(xB(\varepsilon, \tau_0)) \leqslant c\varepsilon^\kappa,$$

where $d\mu_\psi(x) = |\psi(x)|^2 dm(x)$ and dm is a fixed Haar measure on $\mathrm{SL}(2, \mathbf{R})$.

Proof Fix $x \in X$. It follows from Proposition 9.30 that the function

$$\sum_{\beta \in \mathcal{R}_\varepsilon} 1_{\underline{\beta} xB(\varepsilon, \tau_0)}$$

is everywhere less than or equal to c_0. Multiplying this inequality by $|\psi|^2$ and integrating over a fundamental domain for the action of $\mathcal{O}^1$ on $\mathrm{SL}(2, \mathbf{R})$, we obtain

$$\sum_{\beta \in \mathcal{R}_\varepsilon} \mu_\psi(\underline{\beta} xB(\varepsilon, \tau_0)) \leqslant c_0.$$

Now recall that the set $\mathcal{R}_\varepsilon$ is a union of sets $\mathcal{R}_p$ for p running through $\mathcal{P}_\varepsilon$. For every $p \in \mathcal{P}_\varepsilon$, we have

$$\sum_{\beta \in \mathcal{R}_p} \mu_\psi(\underline{\beta} xB(\varepsilon, \tau_0)) = \int_{B(\varepsilon, \tau_0)} \sum_{\beta \in \mathcal{R}_p} |\psi(\underline{\beta} xb)|^2 dm(b).$$

But it follows from Proposition 9.27 that there is a universal constant c_1 such that for every b,

$$\sum_{\beta \in \mathcal{R}_p} |\psi(\underline{\beta} xb)|^2 \geqslant \frac{c_1}{p^{1-2\delta}} |\psi(xb)|^2.$$

We deduce from this that

$$\sum_{\beta \in \mathcal{R}_p} \mu_\psi(\underline{\beta} xB(\varepsilon, \tau_0)) \geqslant \frac{c_1}{p^{1-2\delta}} \mu_\psi(xB(\varepsilon, \tau_0)).$$

Summing over the set of $p \in \mathcal{P}_\varepsilon$, we find

$$\sum_{\beta \in \mathcal{R}_\varepsilon} \mu_\psi \left(\underline{\beta} x B(\varepsilon, \tau_0) \right) \geq c_1 \left(\sum_{p \in \mathcal{P}_\varepsilon} \frac{1}{p^{1-2\delta}} \right) \mu_\psi \left(x B(\varepsilon, \tau_0) \right),$$

whence

$$\mu_\psi \left(x B(\varepsilon, \tau_0) \right) \leq \frac{c_0}{c_1} \left(\sum_{p \in \mathcal{P}_\varepsilon} \frac{1}{p^{1-2\delta}} \right)^{-1}. \qquad (9.35)$$

Note that the set of primes $p \notin \mathcal{P}$ is finite and that the sum

$$\sum_{p | D_\varepsilon} \frac{1}{p^{1-2\delta}},$$

being less than or equal to $\log D_\varepsilon$, is uniformly bounded from above by $\log c' + \log \varepsilon^{-1}$, where c' is given by Lemma 9.28. Moreover, from the prime number theorem we see that

$$\sum_{p^4 \leq c\varepsilon^{-1}} \frac{1}{p^{1-2\delta}} \leq \int_1^{c^{1/4}\varepsilon^{-1/4}} \frac{d \left(x/\log x \right)}{x^{1-2\delta}} = \varepsilon^{-\kappa}$$

for a certain constant $\kappa > 0$ depending only on c and δ. Inserting these estimates into (9.35) we deduce the theorem. All of the above constants depend a priori on x but, by the compactness of X, we can choose them uniformly. $\qquad \square$

Corollary 9.33 *The microlocal lift of any quantum arithmetic limit is of strongly positive entropy.*

Proof Let ν be a quantum arithmetic weak limit of a sequence $|\phi_i(x)|^2 d\mu(x)$, where each $\phi_i \in C^\infty(\mathcal{O}^1 \backslash \mathcal{H})$ is of L^2-norm 1, is a Laplacian eigenfunction $\Delta \phi_i = \lambda_i \phi_i$ $(\lambda_i = 1/4 + r_i^2 \to +\infty)$ and an eigenfunction of all Hecke operators S_p for $p \in \mathcal{P}$. From Theorem 9.3 (and its proof), each measure $|\phi_i(x)|^2 d\mu(x)$ lifts to a measure $\widetilde{\nu}_{\phi_i} = |\Psi_i|^2 dm$ on $X = \mathcal{O}^1 \backslash \mathrm{SL}(2, \mathbf{R})$, where

$$\Psi_i = \frac{1}{\sqrt{2[r_i^{1/2}] + 1}} \sum_{n=-[r_i^{1/2}]}^{[r_i^{1/2}]} (\phi_i)_n.$$

Moreover, the sequence $\widetilde{\nu}_{\phi_i}$ converges weakly to the microlocal lift $\widetilde{\nu}$ of ν. The function Ψ_i being itself an eigenfunction of all Hecke operators, we can apply Theorem 9.32 to it. Passing to the weak limit we find that the microlocal lift of ν is of strongly positive entropy. $\qquad \square$

In particular, Corollary 9.33 implies the following result.

Corollary 9.34 *Let $S = \mathcal{O}^1 \backslash \mathcal{H}$ be as above. Let v be a quantum arithmetic limit on S and σ the singular support of v with respect to Lebesgue measure μ. Then if σ is contained in a finite union of points and closed geodesics, we have $\sigma = \emptyset$.*

9.6.4 T_p-Recurrence

To conclude we fix a prime $p \in \mathcal{P}$ and note that the proofs of Corollary 9.33 and Proposition 9.26 imply the following proposition, where c is the constant given in Proposition 9.26.

Proposition 9.35 *Let v be a quantum arithmetic limit on $S = \mathcal{O}^1 \backslash \mathcal{H}$ and $\tilde{v}$ its microlocal lift to X. Let $n \in \mathbb{N}$, $x \in X$. Let $r > 0$ be sufficiently small. Then*

$$\sum_{y \in t(x, B_n^T)} \tilde{v}(\overline{B_r(y)}) \geq cn\mu(B_r(x)), \tag{9.36}$$

where $t(x, B_n^T)$ denotes the set of points in the T_p-leaf passing through x and at distance $\leq n$ from x in T_p.

Similarly to § 9.2 we can "disintegrate" the measure $\tilde{v}$ along T_p-leaves. Proposition 9.35 then implies that for $\tilde{v}$-almost all $x \in X$,

$$\tilde{v}_{x,T}^\infty(T) = \infty.$$

This in turn is equivalent – again as in § 9.2 – to the T_p-recurrence of $\tilde{v}$.

From Corollary 9.33 we can thus apply Theorem 9.25 to the measure $\tilde{v}$ on X, concluding the proof of Theorem 9.24.

9.7 Commentary and References

The quantum unique ergodicity conjecture is due to Rudnick and Sarnak [108]. It is a remarkable conjecture: it posits that the chaotic aspect of the geodesic flow on S manifests itself at the quantum level. Indeed, we can think of the equation $\Delta \phi = \lambda \phi$ as the quantization of the Hamiltonian equation generating the geodesic flow, here $\lambda = 1/\hbar^2$ with $\hbar$ equal to the Planck constant (or quantization level).

Conjecture 9.1 followed the work of Šnirel'man [125], Zelditch [144] and Colin de Verdière [29], who proved a weak version: Corollary 9.8. The proof of their theorem is based on the notion of microlocal lift.

The quantum unique ergodicity conjecture is surely out of reach for the time being. Nevertheless, spectacular progress has recently been made by Nalini Anantharaman [2] and, in the case of arithmetic surfaces, by Elon Lindenstrauss [82] by ergodic theoretic methods. In this chapter we have described a part of Lindenstrauss's work by making the link with general ergodic theoretic results, for the proofs of which we refer the reader to the original articles.

§9.1

The fact that the geodesic flow on a finite volume hyperbolic surface is mixing is a theorem of E. Hopf; see for example [7]. This gives a precise sense in which the geodesic flow is chaotic.

The Weyl calculus and microlocal analysis are the subject of the books [90, 104]. The microlocal lift is usually defined via microlocal analysis. This is not the approach we have followed here. Our "by hand" construction of this lift then leaves in the background the term "microlocal".

§9.2

The presentation that we give here of the microlocal lift is due to Zelditch [144], Wolpert [142] and Lindenstrauss [82]. We have primarily followed the article [82] of Lindenstrauss.

§9.4

This subsection is based on a text written by Sébastien Gouëzel for a study group on the work of Lindenstrauss.

Theorem 9.13 is due to Rokhlin; for a proof, see for example [48].

We give four equivalent definitions of T_p-recurrence. The fourth is the definition that one can find in the article of Lindenstrauss [82], the first is the classical definition.

For the construction of invariant measures supported on Cantor sets on the circle the reader can consult [50, Th. 4.1.1].

Historically, Theorem 9.16 is the first result in the style of Einsiedler, Katok, Lindenstrauss. It is due to Rudolph [109] and is motivated by an important work of Furstenberg [40]. The proof that we reproduce here is due to Host [58].

We have not defined the entropy of a measure. For that the reader can consult, for example, [50]. The entropy of f_p is equal to

$$-\int_{\mathbb{S}^1} \log \varphi(x) \, d\mu(x).$$

In particular, this integral must be strictly negative. An invariant measure can be decomposed into ergodic components (see [50, Th. 4.1.12]). We can then verify that if μ satisfies (9.22), then almost all ergodic components of μ are of entropy $\geqslant \kappa$.

For the Birkhoff ergodicity theorem we refer the reader to [50, Cor. 4.1.9].

More generally than Proposition 9.19, we can show that a measure μ is T_p-recurrent if and only if all of its ergodic components are of positive entropy (see [82]).

Taking $p = 3$, Theorem 9.16 and Proposition 9.19 imply Rudolph's theorem in its original form:

Theorem 9.36 *Let μ be a measure on $\mathbf{S}^1$ which is invariant under multiplication by 2 and 3. Assume that every ergodic component of μ with respect to multiplication by 3 is of positive entropy (with respect to $\times 3$). Then μ is Lebesgue measure.*

The relation between Theorem 9.16 and the theorem of Einsiedler, Katok and Lindenstrauss is then obvious: one can think of the action of A_1 as multiplication by 3 and the action of A_2 as multiplication by 2.

§ 9.5

We have followed Serre's book [117] to describe the tree associated with PGL$(2, \mathbf{Q}_p)$.

Theorem 9.24 is proved in the case of compact hyperbolic surfaces by Lindenstrauss in [82]. The case of the modular surface presents an additional difficulty: the mass can escape to infinity. Soundararajan showed that this is not the case in [126]. One then finally obtains the general Theorem 9.24 which confirms the arithmetic version of the conjecture of Rudnick and Sarnak.

It is conjectured that the spectrum of the Laplacian on an arithmetic hyperbolic surface has bounded multiplicities. Under this hypothesis it is not difficult to verify – using the quantum ergodicity theorem – that the quantum unique ergodicity conjecture for arithmetic surfaces follows from Theorem 9.24. Nevertheless, bounding the multiplicity of the spectrum seems, at present, to be completely out of reach.

Compared with Theorems 9.11 or 9.16, we make an additional hypothesis on the measure v in the statement of Theorem 9.25. Indeed, in Theorem 9.11 we do not ask that v be T_p-recurrent and in Theorem 9.16 we do not ask that v have strongly positive entropy. This difference comes from the fact that, in the case of Theorem 9.11, if μ is an ergodic $A_1 \times A_2$-invariant measure with positive entropy with respect to A_1, then μ is automatically recurrent with respect to the foliation induced by the second factor of SL$(2, \mathbf{R})$. Similarly, in the case of Theorem 9.16, if v is ergodic and T_p-recurrent then v is of positive entropy with respect to multiplication by 2.

§ 9.6

The idea of the proof that arithmetic quantum limits are of positive entropy goes back to the article of Rudnick and Sarnak [108]. The general result is due to Bourgain and Lindenstrauss [17] and the proof that we give is taken from Silberman's thesis (issuing from joint work with Venkatesh, see [124] for a third approach).

Lemma 9.28 reproduces [17, Lcm. 3.3] and Lemma 9.29 reproduces [17, Lcm. 3.2].

Theorem 9.32 is the main result of [17]. It directly implies the strongly positive entropy of arithmetic quantum limits.

In the case of a general negatively curved Riemannian manifold, Nalini Anantharaman [2] showed that every quantum limit (not necessarily arithmetic!) is of positive entropy. This remarkable theorem implies that if v is a quantum limit, v has at least one ergodic component of positive entropy. Nevertheless, this does not allow one to deduce the statement of Corollary 9.33, for it does not rule out the possibility that other ergodic components of v are of zero entropy.

Corollary 9.34 is due to Rudnick and Sarnak [108]. It is historically the first result in the direction of their conjecture.

9.8 Exercises

Exercise 9.37 (Sensitivity to initial conditions)

1. Consider two geodesics $\gamma_1, \gamma_2 : \mathbf{R}^+ \to \mathbf{R}^2$ in the Euclidean plane. Let $\varepsilon > 0$ and $T > 0$. Assume that γ_1 and γ_2 coincide at $t = 0$ and are close to each other at time T, i.e., $\mathrm{dist}_{\mathrm{Euc}}(\gamma_1(T), \gamma_2(T)) \leq \varepsilon$. Show that $\mathrm{dist}_{\mathrm{Euc}}(\gamma_1(2T), \gamma_2(2T)) \leq 2\varepsilon$.

2. Now consider two geodesics $\gamma_1, \gamma_2 : \mathbf{R}^+ \to \mathcal{H}$ in the hyperbolic plane. Let $\varepsilon > 0$ and $T > 0$. Assume that γ_1 and γ_2 coincide at $t = 0$ and are close to each other at time T, i.e., $\rho(\gamma_1(T), \gamma_2(T)) \leq \varepsilon$. Show that if $\varepsilon > 0$ is fixed and T goes to infinity, we have[9]:

$$\rho(\gamma_1(T), \gamma_2(T)) \sim 2T.$$

Exercise 9.38 Show that the space of K-finite functions is dense in $C^\infty(\Gamma \backslash G)$ and that it corresponds to the restriction to the unit tangent bundle (viewed as the level hypersurface $|w|^2 = 1/2$ in $\mathcal{H} \times \mathbf{R}^2$) of functions in $C^\infty(\mathcal{H})[w, \overline{w}]$. Here $C^\infty(\mathcal{H})[w, \overline{w}]$ is the space of polynomials in w and $\overline{w}$ with coefficients in $C^\infty(\mathcal{H})$.

Exercise 9.39 Let $\gamma(\alpha, \beta)$ be the geodesic in $\mathcal{H}$ with endpoints $\alpha, \beta \in \mathbf{P}^1(\mathbf{R}) = \mathbf{R} \cup \{\infty\}$. We send the reader to § 4.4 for the notion of the *size of a geodesic excursion* with respect to the Farey triangulation of $\mathcal{H}$.

1. Let $\alpha \in [-1, 0)$ and $\beta > 1$. Show that the n-th excursion (beginning with that at infinity) of $\gamma = \gamma(\alpha, \beta)$ is associated with the rational number $p_{n-1}/q_{n-1} = [b_0, b_1, \ldots, b_{n-1}, 0, \ldots]$ obtained by truncating the continued fraction expansion of β to order $n - 1$.

2. Show that a geodesic of $S = \mathrm{PSL}(2, \mathbf{Z}) \backslash \mathcal{H}$ is compact if and only if the associated sequence is periodic, and that a geodesic in S is relatively compact if and only if its sequence of excursion sizes is bounded.

3. Deduce that there exist A-orbits in $\mathrm{PSL}(2, \mathbf{Z}) \backslash \mathrm{PSL}(2, \mathbf{R})$ which are relatively compact but not compact.

---

[9]Thus, even a precise control on the geodesic flow over an interval $[0, T]$ cannot lead to an estimation on $\rho(\gamma_1(2T), \gamma_2(2T))$ (other than that implied by the triangle inequality). The future is independent of the past; this is a very simple manifestation of deterministic chaos.

Appendix A
Three Coordinate Systems for $\mathcal{H}$

In this appendix we shall describe three systems of coordinates for the half-plane $\mathcal{H}$ that we use at various places in the text.

Polar Coordinates

Let $p_0 \in \mathcal{H}$ be an arbitrary base point. For any point $p \in \mathcal{H} - \{p_0\}$, there exists a unique geodesic parametrized by arc length $\gamma : [0, +\infty) \to \mathcal{H}$ with $\gamma(0) = p_0$ and passing through p. Let $r = r(p)$ be the distance from p_0 to p, so that $\gamma(r(p)) = p$. Furthermore, fix an arbitrary unit tangent vector v at p_0 and let $\theta = \theta(p) \in [-\pi, \pi)$ be the directly oriented angle from v to the tangent vector along γ at p_0. Then $(r, \theta) = (r(p), \theta(p))$ are the *polar coordinates* of p relative to the choice of the base point p_0 and the vector v. In polar coordinates, the hyperbolic metric admits the following expression:

$$ds^2 = dr^2 + \sinh^2 r d\theta^2. \tag{A.1}$$

Rather than taking $\theta(p)$ in the interval $[-\pi, \pi)$, we can view $\theta(p)$ as an element of the unit circle

$$\mathbf{S}^1 = \mathbf{R}/[s \longmapsto s + 2\pi].$$

Fermi Coordinates

By replacing the reference to the base point p_0 by a reference to a fixed geodesic, we obtain the Fermi coordinates. They are defined as follows. Let $\gamma_0 : t \mapsto \gamma_0(t) \in \mathcal{H}$ be

© Springer International Publishing Switzerland 2016
N. Bergeron, *The Spectrum of Hyperbolic Surfaces*, Universitext,
DOI 10.1007/978-3-319-27666-3

a fixed geodesic line in the hyperbolic plane, parametrized by arc length dt. Then γ_0 separates $\mathcal{H}$ into two half-planes, one to the right and one to the left of γ_0. For every point $p \in \mathcal{H}$ we thus have a *signed distance* r from p to γ_0: positive on one side, negative on the other, and zero if $p \in \gamma_0$. In these coordinates the hyperbolic metric becomes

$$ds^2 = dr^2 + \cosh^2 rdt^2. \tag{A.2}$$

We follow the following *sign convention*: if γ_0 is oriented, the signed distance r from a point p to γ_0 is negative if p is to the left of γ_0 and positive otherwise.

Horocyclic Coordinates

The last coordinates we introduce are useful in the study of cusps. Let p_0 be a point on the *boundary of* $\mathcal{H}$, i.e., a point of the extended complex plane $\widehat{\mathbf{C}}$ lying in $\widehat{\mathbf{R}} = \mathbf{R} \cup \{\infty\}$, the union of the real axis and the point at infinity. Euclidean circles (or lines if $p_0 = \infty$) contained in $\mathcal{H} \cup \{p_0\}$ and passing through p_0 are called *horocycles* with "center" p_0.

The elements in $\mathrm{SL}(2, \mathbf{R})$ send horocycles to horocycles. In the special case where $p_0 = \infty$, the horocycles with ∞ as their center are horizontal Euclidean lines, and the vertical lines are geodesics having ∞ as an endpoint. These vertical and horizontal lines intersect each other orthogonally. More generally, regardless of the position of the point p_0, the geodesics having p_0 as an endpoint and the horocycles having p_0 as their center form an orthogonal family.

Now let $h : t \mapsto h(t) \in \mathcal{H}$ be a horocycle parametrized by arc length. The hyperbolic plane is again separated into two half-planes by h. We choose the parametrization of h for which p_0 is to the left of h. By convention the signed distance r from a point p to h is negative if p is to the left of h, positive if p is to the right of h, and zero if $p \in h$.

There exists a unique real number t such that the geodesic passing through p and perpendicular to h meets h at $h(t)$. By definition, the *horocyclic coordinates* are (r, t). In horocyclic coordinates the hyperbolic metric is expressed as

$$ds^2 = dr^2 + e^{2r}dt^2. \tag{A.3}$$

Appendix B
The Gamma Function and Bessel Functions

The Gamma Function

The Euler Gamma function is defined for $\mathrm{Re}(s) > 0$ as the Mellin transform of the exponential function, see (1.26). Integration by parts produces the relation

$$\Gamma(s+1) = s\Gamma(s).$$

From this it follows, via a recurrence argument, that $\Gamma(s)$ admits a meromorphic continuation to all of $\mathbf{C}$, whose only poles are simple and located at $s = -k$ for integer $k \geqslant 0$. Moreover, we have $\Gamma(n) = (n-1)!$ for $n \geqslant 1$.

We refer the reader to the book of Titchmarsh [129, particularly §4.41 and 4.42] for the proofs of the following properties (the last three are referred to collectively as *Stirling's formula*):

1. We have $\Gamma(s) \neq 0$ for all $s \in \mathbf{C}$.
2. We have

$$|\Gamma(\sigma + it)| = (2\pi)^{1/2}|t|^{\sigma-1/2}e^{-\pi|t|/2}(1 + O_{a,b}(t^{-1})), \tag{B.1}$$

   uniformly for all $a \leqslant \sigma \leqslant b$ and $|t| \geqslant 1$.
3. We have

$$\Gamma(\sigma) \sim \sqrt{\frac{2\pi}{\sigma}}\,\sigma^\sigma e^{-\sigma} \tag{B.2}$$

   as $\sigma \in \mathbf{R}$ tends toward $+\infty$.

© Springer International Publishing Switzerland 2016
N. Bergeron, *The Spectrum of Hyperbolic Surfaces*, Universitext,
DOI 10.1007/978-3-319-27666-3

4. We have

$$\frac{\Gamma'}{\Gamma}(s) = \log|s| + O_\varepsilon(s^{-1}), \tag{B.3}$$

uniformly in every angle $|\arg(s)| \leqslant \pi - \varepsilon$

The book by Whittaker and Watson [138] contains a mine of information on special functions. The Gamma function forms the subject of Chapter XII. In the following paragraph we shall state a few properties of Bessel functions used in the text. One can find proofs of these properties in [138, Chap. XVII] or [135].

Bessel Functions

These are the solutions to the differential equation

$$z^2 f'' + z f' - (z^2 + v^2)f = 0, \tag{B.4}$$

where v is a complex number.

Equation (B.4) is singular at $z = 0$; we can eliminate this singularity by cutting the complex z-plane along the interval $(-\infty, 0]$. Equation (B.4) then admits two linearly independent solutions which are holomorphic in $z \in \mathbf{C} - (-\infty, 0]$. One of them is easily obtained as the sum of the series

$$\sum_{k=0}^{+\infty} \frac{1}{k!\Gamma(k+1+v)} \left(\frac{z}{2}\right)^{v+2k} \tag{B.5}$$

which converges absolutely on the entire complex plane. We denote it by $I_v(z)$ and v is called the *order of* I_v. As a function of v the function I_v is entire.

Replacing v with $-v$ leaves equation (B.4) invariant, so that the function $I_{-v}(z)$ is another solution. The solutions $I_v(z), I_{-v}(z)$ are linearly independent if and only if the Wronskian $W(I_v(z), I_{-v}(z)) = -2(\pi z)^{-1} \sin \pi v$ is not identically zero, which is to say that v is not an integer.

If $v = n$ is an integer we have the relation

$$I_n(z) = I_{-n}(z). \tag{B.6}$$

To obtain a suitable pair of linearly independent solutions for arbitrary v we form the linear combination

$$K_v(z) = \frac{\pi}{2}(\sin \pi v)^{-1}(I_{-v}(z) - I_v(z)), \tag{B.7}$$

while passing to the limit $\nu \to n$ to define K_n. The functions $I_\nu(z)$ and $K_\nu(z)$ are always linearly independent since the Wronskian is

$$W(I_\nu(z), K_\nu(z)) = -z^{-1}.$$

For $\nu = n$ integral, the function $I_n(z)$ is entire. If ν is not an integer there is a discontinuity along the negative real axis. More precisely, we have

$$I_\nu(-x + \varepsilon i) - I_\nu(-x - \varepsilon i) \sim 2i \sin(\pi \nu) I_\nu(x)$$

for $x > 0$ when ε tends toward 0.

The Bessel functions of different orders are linked by the following recurrence relations:

$$I_{\nu-1}(z) - I_{\nu+1}(z) = 2\nu z^{-1} I_\nu(z),$$
$$I_{\nu-1}(z) + I_{\nu+1}(z) = 2 I_\nu'(z),$$
$$(z^\nu I_\nu(z))' = z^\nu I_{\nu-1}(z),$$
$$(z^{-\nu} I_\nu(z))' = z^{-\nu} I_{\nu+1}.$$

(For the K-Bessel functions the above formulae remain valid as long as we change the sign of the right-hand side.)

The Bessel functions of order $1/2$ are elementary functions:

$$I_{1/2}(z) = (2/\pi z)^{1/2} \sinh z, \quad K_{1/2}(z) = (\pi/2z)^{1/2} e^{-z}.$$

With the help of the above recurrence formulae we can find elementary expressions for the Bessel functions of order half an odd integer.

For the K-Bessel functions we have the following asymptotic behavior:

$$K_\nu(x) \sim \frac{\Gamma(\nu)}{2} \left(\frac{x}{2}\right)^{-\nu} \quad \text{as } x \longrightarrow 0. \tag{B.8}$$

For $0 < x < 1$,

$$K_\nu(x) = \frac{2^{\nu-1} \Gamma(\nu)}{x^\nu} + O(x^{2-\nu}), \tag{B.9}$$

where the constant in the O is uniform for $\nu \in (\nu_0, 1/2)$ and $x \in (0, 1/2)$.

For $x > 1 + \nu^2$

$$K_\nu(x) = \left(\frac{\pi}{2x}\right)^{1/2} e^{-x} \left(1 + O\left(\frac{1 + \nu^2}{x}\right)\right), \tag{B.10}$$

where the constant in the O is uniform for $v \in (v_0, 1/2)$ and $x > 2$. More precisely, with $\mu = 4v^2$

$$K_v(x) \sim \sqrt{\frac{\pi}{2x}} e^{-x} \left\{ 1 + \frac{v-1}{8x} + \frac{(v-1)(v-9)}{2!(8x)^2} + \cdots \right\}.$$

There are various integral representations for Bessel functions. Here are a few for the K-Bessel functions:

$$K_v(z) = \pi^{1/2} \Gamma\left(v + \frac{1}{2}\right)^{-1} \left(\frac{z}{2}\right)^v \int_1^{+\infty} (t^2 - 1)^{v-1/2} e^{-tz} dt$$

$$= \pi^{-1/2} \Gamma\left(v + \frac{1}{2}\right) \left(\frac{z}{2}\right)^{-v} \int_0^{+\infty} (t^2 + 1)^{-v-1/2} \cos(tz) dt$$

$$= \left(\frac{\pi}{2z}\right)^{1/2} \Gamma\left(v + \frac{1}{2}\right)^{-1} e^{-z} \int_0^{\infty} e^{-t} (t(1 + t/2z))^{v-1/2} dt$$

$$= \frac{1}{2} \int_0^{+\infty} \exp\left(-\frac{z}{2}(t + 1/t)\right) t^{-v-1} dt$$

$$= \int_0^{+\infty} e^{-z\cosh t} \cosh(vt) dt,$$

where $\mathrm{Re}(z) > 0$ and $\mathrm{Re}(v) > -1/2$. From the last integral representation we easily deduce the two following properties:

1. For fixed v and $x > 0$ the function $K_v(x)$ is positive and decreasing.
2. For fixed $x > 0$ and $v > 0$ the function $K_v(x)$ is positive and increasing.

Appendix C
Elementary Bounds on Hyper-Kloosterman Sums

by Valentin Blomer and Farrell Brumley

Let q and n be integers such that $q, n \geqslant 2$, and let r be an integer prime to q. Define

$$S_n(r, q) = \sum_{\chi \ (\mathrm{mod}\ q)}^{*} \overline{\chi}(r) \tau(\chi)^n, \tag{C.1}$$

the star indicating that the sum is taken over primitive characters. For a character χ mod q we denote by

$$\tau(\chi) = \sum_{x \ (\mathrm{mod}\ q)} \chi(x) e(x/q) \tag{C.2}$$

the Gauß sum, where $e(x) = e^{2\pi i x}$. The aim of this appendix is to prove the following elementary bound.

Theorem C.1 *Let p be a prime not dividing n and $q = p^{2\alpha}$ with $\alpha \geqslant 2$. If r admits no n-th root $r^{1/n}$ in $(\mathbf{Z}/p^{\alpha}\mathbf{Z})^{\times}$ then $S_n(r, q) = 0$; otherwise*

$$S_n(r, q) = \phi(q) q^{(n-1)/2} \sum_{\zeta^n \equiv 1 \ (\mathrm{mod}\ p^{\alpha})} e\big(nr^{1/n}\zeta/q\big), \tag{C.3}$$

where ϕ is the Euler totient function.

(V. Blomer): Universiät Göttingen, Mathematisches Institut, Bunsenstr. 3-5, 37073 Göttingen, Allemagne.
e-mail: blomer@uni-math.gwdg.de
(F. Brumley): Institut Galilée, Université Paris 13, 99 avenue J.-B. Clément, 93430 Villetaneuse, France.
e-mail: brumley@math.univ-paris13.fr

© Springer International Publishing Switzerland 2016
N. Bergeron, *The Spectrum of Hyperbolic Surfaces*, Universitext,
DOI 10.1007/978-3-319-27666-3

Analogous formulae can be obtained for $q = p^{2\alpha+1}$ with $\alpha \geq 1$ and p fixed. For simplicity, we only treat the case of even powers here. Note that the condition $p \nmid n$ can be eliminated if one is content with the upper bound $S_n(r, q) \leq nq^{(n-1)/2}$.

We begin by relating $S_n(r, q)$ to the hyper-Kloosterman sum

$$K_n(r, q) = \sum_{\substack{x_1, \ldots, x_n \,(\mathrm{mod}\, q) \\ x_1 \cdots x_n \equiv r \,(\mathrm{mod}\, q)}} e\left(\frac{x_1 + \cdots + x_n}{q}\right). \tag{C.4}$$

Since r is prime to q, the above sum runs over all invertible residue classes mod q.

Lemma C.2 *For all $q \geq 2$ we have $\phi(q)^{-1} S_n(r, q) = K_n(r, q)$.*

Proof The idea is to use Fourier inversion on the group $G = (\mathbf{Z}/q\mathbf{Z})^\times$. For primitive characters χ, we interpret the Gauß sum $\tau(\chi)$ as the Fourier transform, evaluated at χ, of the function $x \mapsto e(x/q)$ defined on G. On the other hand, $\phi(q)^{-1} S_n(r, q)$ can be seen as the inverse Fourier transform, evaluated at r, of the function $\chi \mapsto \tau(\chi)^n$ defined on the group $\widehat{G}$ of *all* characters of G. The convolution theorem (or just a simple change of summation order) then states that $\phi(q)^{-1} S_n(r, q)$ is the convolution of the function $e(x/q)$ iterated n times with itself. This multi-convolution of exponentials is nothing other than $K_n(r, q)$.

For the above argument to go through, one must check that (C.1) can be extended to all characters. This is possible since $\tau(\chi) = 0$ unless χ is primitive. Indeed, if the conductor of χ is p^β with $0 \leq \beta < \alpha$, then writing $x = y + p^\beta z$ with $y \,(\mathrm{mod}\, p^\beta)$ and $z \,(\mathrm{mod}\, p^{\alpha-\beta})$ we find

$$\tau(\chi) = \sum_{y\,(\mathrm{mod}\, p^\beta)} \sum_{z\,(\mathrm{mod}\, p^{\alpha-\beta})} \chi(y + p^\beta z) e\left(\frac{y + p^\beta z}{p^\alpha}\right)$$

$$= \sum_{y\,(\mathrm{mod}\, p^\beta)} \chi(y) e\,(y/p^\alpha) \sum_{z\,(\mathrm{mod}\, p^{\alpha-\beta})} e\,(z/p^{\alpha-\beta}) = 0,$$

since the inner sum vanishes. $\square$

To prove Theorem C.1 it therefore suffices to show that if r has an n-th root $r^{1/n}$ in $(\mathbf{Z}/p^\alpha\mathbf{Z})^\times$ then

$$K_n(r, q) = q^{(n-1)/2} \sum_{\zeta^n \equiv 1 \,(\mathrm{mod}\, p^\alpha)} e\left(nr^{1/n}\zeta/q\right) \qquad (q = p^{2\alpha}); \tag{C.5}$$

otherwise $K_n(r, q) = 0$.

Remark C.3 There are $n - 1$ free variables in (C.4), each running through the q elements of $\mathbf{Z}/q\mathbf{Z}$. As each term is of size 1 in absolute value, a trivial bound on $K_n(r, q)$ is given by q^{n-1}. Thus (C.5) expresses square-root cancellation in the number of free variables.

When $q = p$ is prime, the optimal bound $|K_n(r,p)| \leqslant np^{(n-1)/2}$ is a deep result of Deligne [35]. Other the other hand, when $q = p^\alpha$ with $\alpha > 1$, the exponential sums can be explicitly calculated, and in an elementary way. This is a very general principle that applies in a large variety of situations (see [64, Section 12]).

Finally, note that we can easily obtain a bound on $K_n(r,q)$ which is intermediate between the trivial bound q^{n-1} and the optimal bound $q^{(n-1)/2}$. Indeed, the Gauß sum $\tau(\chi)$ of a primitive character $\chi \bmod q$ satisfies $|\tau(\chi)| = \sqrt{q}$. The expression (C.1) of $S_n(r,q)$ then implies $|S_n(r,q)| \leqslant \phi(q)q^{n/2}$, and from Lemma C.2 we see that

$$|K_n(r,q)| \leqslant q^{n/2}.$$

To simplify the notation we write

$$f(x) = f(x_1, \ldots, x_{n-1}) = x_1 + \cdots + x_{n-1} + \frac{r}{x_1 \cdots x_{n-1}}.$$

Then

$$K_n(r,q) = \sum_{x \,(\mathrm{mod}\ q)}^{*} e(f(x)/q), \tag{C.6}$$

where the star indicates that the sum is taken over invertible residue classes. The gradient of f is given by the column vector of rational functions

$$\nabla f(x) = \begin{pmatrix} 1 - \frac{r}{(x_1 \cdots x_{n-1})x_1} \\ \vdots \\ 1 - \frac{r}{(x_1 \cdots x_{n-1})x_{n-1}} \end{pmatrix}. \tag{C.7}$$

Lemma C.4 (Taylor expansion) *For $j = 1, \ldots, n-1$, let x_j be an invertible class* $(\mathrm{mod}\ p^{2\alpha})$. *Let $x_j = y_j + p^\alpha z_j$ where y_j and z_j are determined mod p^α, and y_j is invertible. We put $x = (x_j)_j$ and similarly for y and z. Then*

$$f(x) = f(y + p^\alpha z) \equiv f(y) + p^\alpha \nabla f(y) \cdot z \pmod{p^{2\alpha}}, \tag{C.8}$$

where $a \cdot b$ denotes the usual scalar product of vectors.

Proof The congruence (C.8) is valid for any rational function f as long as x, y are distinct from the poles of f, but we can also verify it directly for the particular function f by a direct computation. Indeed, by definition of f,

$$f(y + p^\alpha z) = \sum_{j=1}^{n-1} y_j + p^\alpha z_j + r \prod_{j=1}^{n-1} (y_j + p^\alpha z_j)^{-1}. \tag{C.9}$$

Now

$$(y_j + p^\alpha z_j)(y_j^{-1} - p^\alpha z_j y_j^{-2}) = 1 - p^{2\alpha}(z_j/y_j)^2 \equiv 1 \pmod{p^{2\alpha}},$$

so that

$$(y_j + p^\alpha z_j)^{-1} \equiv y_j^{-1} - p^\alpha z_j y_j^{-2} \pmod{p^{2\alpha}}$$

and

$$\prod_{j=1}^{n-1}(y_j + p^\alpha z_j)^{-1} \equiv \prod_{j=1}^{n-1}(y_j^{-1} - p^\alpha z_j y_j^{-2})$$

$$\equiv (y_1 \cdots y_{n-1})^{-1} - p^\alpha (y_1 \cdots y_{n-1})^{-1} \sum_{j=1}^{n-1} y_j^{-1} z_j \pmod{p^{2\alpha}}.$$

Inserting this last expression in (C.9) we obtain (C.8).   □

We now return to the proof of (C.5). We shall see that only the critical points of f contribute to the final expression for $K_n(r, q)$. These are the points where the exponential $e(f(x)/q)$ becomes "stationary". This phenomenon is similar to that for oscillatory integrals over the reals, where only a small neighborhood around the stationary points contribute to the asymptotic behavior of the integral. The p-adic case which is the subject of this appendix is much simpler, and we obtain here not just an asymptotic formula but an exact one.

Proof of Theorem C.1 We write

$$K_n(r, p^{2\alpha}) = \sideset{}{^*}\sum_{y \,(\text{mod } p^\alpha)} \sum_{z \,(\text{mod } p^\alpha)} e\left(\frac{f(y) + p^\alpha \nabla f(y) \cdot z}{p^{2\alpha}}\right)$$

$$= \sideset{}{^*}\sum_{y \,(\text{mod } p^\alpha)} e\left(f(y)/p^{2\alpha}\right) \sum_{z \,(\text{mod } p^\alpha)} e\left(\frac{\nabla f(y) \cdot z}{p^\alpha}\right).$$

The inner sum vanishes unless $\nabla f(y)$ is identically 0 mod p^α, in which case it evaluates to $p^{\alpha(n-1)}$. In other words,

$$K_n(r, p^{2\alpha}) = p^{\alpha(n-1)} \sideset{}{^*}\sum_{\substack{y \,(\text{mod } p^\alpha) \\ \nabla f(y) \equiv 0 \,(\text{mod } p^\alpha)}} e(f(y)/p^{2\alpha}). \tag{C.10}$$

At this point, the proof already shows that this expression is independent of the choice of representatives y (mod p^α).

We now look more closely at the solutions to $\nabla f(\mathbf{y}) = 0 \pmod{p^{\alpha}}$. According to Definition (C.7) these are precisely the invertible classes $\mathbf{y}$ such that

$$(y_1 \cdots y_{n-1})y_j \equiv r \pmod{p^{\alpha}} \quad j = 1, \ldots, n-1.$$

This implies $y_1 \equiv \cdots \equiv y_{n-1} \pmod{p^{\alpha}}$, which gives $y_j^n \equiv r \pmod{p^{\alpha}}$ for $1 \leqslant j \leqslant n-1$. If r admits no n-th root $r^{1/n}$ in $(\mathbf{Z}/p^{\alpha}\mathbf{Z})^{\times}$, then f has no critical points; the sum in (C.10) is therefore empty and $K_n(r, q)$ vanishes. Otherwise the solutions to $y^n \equiv r$ $\pmod{p^{\alpha}}$ are given by $r^{1/n}\zeta$, where ζ runs over a complete set of representatives of solutions to the equation $y^n = 1 \pmod{p^{\alpha}}$. We are free to make a suitable choice of such representatives, and as $p \nmid n$, Hensel's lemma [47, Proposition 3.4.2] assures us that we can take ζ to satisfy $\zeta^n = 1 \pmod{p^{2\alpha}}$. Thus

$$f(y, \ldots, y) = f(r^{1/n}(\zeta, \ldots, \zeta)) \equiv nr^{1/n}\zeta \pmod{p^{2\alpha}}.$$

Inserting this into (C.10) we obtain (C.5). $\square$

References

1. Norbert A'Campo and Marc Burger, *Réseaux arithmétiques et commensurateur d'après G.A. Margulis*, Invent. Math. **116** (1994), no. 1–3, 1–25.
2. Nalini Anantharaman, *Entropy and the localization of eigenfunctions*, Ann. of Math. (2) **168** (2008), no. 2, 435–475.
3. V. I. Arnol'd, *Mathematical methods of classical mechanics*, Graduate Texts in Mathematics, vol. 60, Springer-Verlag, New York, 1989, Translated from the 1974 Russian original by K. Vogtmann and A. Weinstein.
4. James Arthur, *An introduction to the trace formula*, Harmonic analysis, the trace formula, and Shimura varieties, Clay Math. Proc., vol. 4, American Mathematical Society, Providence, RI, 2005, pp. 1–263.
5. A. O. L. Atkin and J. Lehner, *Hecke operators on $\Gamma_0(m)$*, Math. Ann. **185** (1970), 134–160.
6. Alan F. Beardon, *The geometry of discrete groups*, Graduate Texts in Mathematics, vol. 91, Springer-Verlag, New York, 1995, Corrected reprint of the 1983 original.
7. Tim Bedford, Michael Keane, and Caroline Series (eds.), *Ergodic theory, symbolic dynamics, and hyperbolic spaces (Trieste, April 17–28, 1989)*, Oxford Science Publications, New York, The Clarendon Press Oxford University Press, 1991.
8. Pierre H. Bérard, *Spectral geometry: direct and inverse problems*, Lecture Notes in Mathematics, vol. 1207, Springer-Verlag, Berlin, 1986.
9. Nicolas Bergeron and Laurent Clozel, *Spectre automorphe des variétés hyperboliques et applications topologiques*, Astérisque, vol. 303, Société mathématique de France, Paris, 2005.
10. Joseph Bernstein and Andre Reznikov, *Analytic continuation of representations and estimates of automorphic forms*, Ann. of Math. (2) **150** (1999), no. 1, 329–352.
11. Valentin Blomer and Farrell Brumley, *On the Ramanujan conjecture over number fields*, Ann. of Math. (2) **174** (2011), no. 1, 581–605.
12. Jens Bolte and Stefan Johansson, *A spectral correspondence for Maaß waveforms*, Geom. Funct. Anal. **9** (1999), no. 6, 1128–1155.
13. _____, *Theta-lifts of Maaß waveforms*, Emerging applications of number theory (Minneapolis, MN, 1996), IMA Vol. Math. Appl., vol. 109, Springer, New York, 1999, pp. 39–72.
14. Andrew R. Booker and Andreas Strömbergsson, *Numerical computations with the trace formula and the Selberg eigenvalue conjecture*, J. Reine Angew. Math. **607** (2007), 113–161.
15. Andrew R. Booker, Andreas Strömbergsson, and Akshay Venkatesh, *Effective computation of Maass cusp forms*, Int. Math. Res. Not. (2006), Art. ID 71281, 34.
16. Armand Borel, *Linear algebraic groups*, second ed., Graduate Texts in Mathematics, vol. 126, Springer-Verlag, New York, 1991.

© Springer International Publishing Switzerland 2016
N. Bergeron, *The Spectrum of Hyperbolic Surfaces*, Universitext,
DOI 10.1007/978-3-319-27666-3

17. Jean Bourgain and Elon Lindenstrauss, *Entropy of quantum limits*, Comm. Math. Phys. **233** (2003), no. 1, 153–171.

18. Haïm Brezis, *Analyse fonctionnelle. Théorie et applications*, Collection Mathématiques Appliquées pour la Maîtrise, Masson, Paris, 1983.

19. D. Bump, J. W. Cogdell, E. de Shalit, D. Gaitsgory, E. Kowalski, and S. S. Kudla, *An introduction to the Langlands program*, Birkhäuser Boston Inc., Boston, MA, 2003, Lectures presented at the Hebrew University of Jerusalem, Jerusalem, March 12–16, 2001, Edited by Joseph Bernstein and Stephen Gelbart.

20. Daniel Bump, *Automorphic forms and representations*, Cambridge Studies in Advanced Mathematics, vol. 55, Cambridge University Press, Cambridge, 1997.

21. Daniel Bump, W. Duke, Jeffrey Hoffstein, and Henryk Iwaniec, *An estimate for the Hecke eigenvalues of Maass forms*, Internat. Math. Res. Notices (1992), no. 4, 75–81.

22. Peter Buser, *Riemannsche Flächen mit Eigenwerten in* (0, 1/4), Comment. Math. Helv. **52** (1977), no. 1, 25–34.

23. _____, *A note on the isoperimetric constant*, Ann. Sci. École Norm. Sup. (4) **15** (1982), no. 2, 213–230.

24. _____, *Geometry and spectra of compact Riemann surfaces*, Progress in Mathematics, vol. 106, Birkhäuser Boston Inc., Boston, MA, 1992.

25. Isaac Chavel, *Eigenvalues in Riemannian geometry*, Pure and Applied Mathematics, vol. 115, Academic Press Inc., Orlando, FL, 1984, Including a chapter by Burton Randol, With an appendix by Jozef Dodziuk.

26. Jeff Cheeger, *A lower bound for the smallest eigenvalue of the Laplacian*, Problems in analysis (Papers dedicated to Salomon Bochner, 1969), Princeton Univ. Press, Princeton, N.J., 1970, pp. 195–199.

27. James W. Cogdell and Ilya I. Piatetski-Shapiro, *Remarks on Rankin-Selberg convolutions*, Contributions to automorphic forms, geometry, and number theory, Johns Hopkins Univ. Press, Baltimore, MD, 2004, pp. 255–278.

28. Paul Cohen and Peter Sarnak, *Notes on the trace formula*, available at: http://web.math. princeton.edu/sarnak/, 1980.

29. Yves Colin de Verdière, *Ergodicité et fonctions propres du laplacien*, Comm. Math. Phys. **102** (1985), no. 3, 497–502.

30. Françoise Dal'Bo, *Geodesic and horocyclic trajectories*, Universitext, Springer-Verlag London, Ltd., London; EDP Sciences, Les Ulis, 2011, Translated from the 2007 French original.

31. Harold Davenport, *Multiplicative number theory*, third ed., Graduate Texts in Mathematics, vol. 74, Springer-Verlag, New York, 2000.

32. Giuliana Davidoff, Peter Sarnak, and Alain Valette, *Elementary number theory, group theory, and Ramanujan graphs*, London Mathematical Society Student Texts, vol. 55, Cambridge University Press, Cambridge, 2003.

33. E. B. Davies, *Heat kernels and spectral theory*, Cambridge Tracts in Mathematics, vol. 92, Cambridge University Press, Cambridge, 1990.

34. Henri Paul de Saint-Gervais, *Uniformisation des surfaces de Riemann. Retour sur un théorème centenaire*, ENS Éditions, Lyon, 2010.

35. Pierre Deligne, *La conjecture de Weil. I*, Inst. Hautes Études Sci. Publ. Math. (1974), no. 43, 273–307.

36. J.-M. Deshouillers and H. Iwaniec, *The nonvanishing of Rankin-Selberg zeta-functions at special points*, The Selberg trace formula and related topics (Brunswick, Maine, 1984), Contemp. Math., vol. 53, American Mathematical Society, Providence, RI, 1986, pp. 51–95.

37. Régine et Adrien Douady, *Algèbre et théories galoisiennes*, Cassini, 2005.

38. Nelson Dunford and Jacob T. Schwartz, *Linear operators.*, Wiley Classics Library, John Wiley & Sons Inc., New York, 1988.

39. Lester R. Ford, *Automorphic functions*, Chelsea, 2004.

40. Harry Furstenberg, *Disjointness in ergodic theory, minimal sets, and a problem in Diophantine approximation*, Math. Systems Theory **1** (1967), 1–49.

41. Sylvestre Gallot, Dominique Hulin, and Jacques Lafontaine, *Riemannian geometry*, third ed., Universitext, Springer-Verlag, Berlin, 2004.

42. Alex Gamburd, *On the spectral gap for infinite index "congruence" subgroups of* $SL_2(\mathbf{Z})$, Israel J. Math. **127** (2002), 157–200.

43. Stephen S. Gelbart, *Automorphic forms on adèle groups*, Princeton University Press, Princeton, N.J., 1975, Annals of Mathematics Studies, No. 83.

44. Étienne Ghys, *Poincaré and his disk*, The scientific legacy of Poincaré (Éric Charpentier, Étienne Ghys, and Annick Lesne, eds.), Hist. Math., vol. 36, American Mathematical Society, Providence, RI, 2010, Translated from the French original (2006), pp. 17–46.

45. Dorian Goldfeld, *Explicit formulae as trace formulae*, Number theory, trace formulas and discrete groups (Oslo, 1987), Academic Press, Boston, MA, 1989, pp. 281–288.

46. _____, *Automorphic forms and L-functions for the group* $GL(n, \mathbf{R})$, Cambridge Studies in Advanced Mathematics, vol. 99, Cambridge University Press, Cambridge, 2006, With an appendix by Kevin A. Broughan.

47. Fernando Q. Gouvêa, *p-adic numbers. An introduction*, second ed., Universitext, Springer-Verlag, Berlin, 1997.

48. Paul R. Halmos, *Measure Theory*, D. Van Nostrand Company, Inc., New York, N.Y., 1950.

49. Gergely Harcos, *Uniform approximate functional equation for principal L-functions*, Int. Math. Res. Not. (2002), no. 18, 923–932.

50. Boris Hasselblatt and Anatole Katok, *A first course in dynamics*, Cambridge University Press, New York, 2003.

51. E. Hecke, *Eine neue Art von Zetafunktionen und ihre Beziehungen zur Verteilung der Primzahlen*, Math. Z. **1** (1918), no. 4, 357–376.

52. _____, *Eine neue Art von Zetafunktionen und ihre Beziehungen zur Verteilung der Primzahlen*, Math. Z. **6** (1920), no. 1–2, 11–51.

53. Dennis A. Hejhal, *The Selberg trace formula and the Riemann zeta function*, Duke Math. J. **43** (1976), no. 3, 441–482.

54. _____, *The Selberg trace formula for* $PSL(2, \mathbf{R})$. *Vol. I*, Lecture Notes in Mathematics, vol. 548, Springer-Verlag, Berlin, 1976.

55. _____, *A classical approach to a well-known spectral correspondence on quaternion groups*, Number theory (New York, 1983–84), Lecture Notes in Mathematics, vol. 1135, Springer, Berlin, 1985, pp. 127–196.

56. Sigurdur Helgason, *Topics in harmonic analysis on homogeneous spaces*, Progress in Mathematics, vol. 13, Birkhäuser Boston, Mass., 1981.

57. _____, *Differential geometry, Lie groups, and symmetric spaces*, Graduate Studies in Mathematics, vol. 34, American Mathematical Society, Providence, RI, 2001, Corrected reprint of the 1978 original.

58. Bernard Host, *Nombres normaux, entropie, translations*, Israel J. Math. **91** (1995), no. 1–3, 419–428.

59. Heinz Huber, *Zur analytischen Theorie hyperbolischer Raumformen und Bewegungsgruppen. II*, Math. Ann. **142** (1960/1961), 385–398.

60. M. N. Huxley, *Cheeger's inequality with a boundary term*, Comment. Math. Helv. **58** (1983), no. 3, 347–354.

61. Aleksandar Ivić, *An approximate functional equation for a class of Dirichlet series*, J. Anal. **3** (1995), 241–252.

62. Henryk Iwaniec, *Prime geodesic theorem*, J. Reine Angew. Math. **349** (1984), 136–159.

63. _____, *Spectral methods of automorphic forms*, second ed., Graduate Studies in Mathematics, vol. 53, American Mathematical Society, Providence, RI, 2002.

64. Henryk Iwaniec and Emmanuel Kowalski, *Analytic number theory*, American Mathematical Society Colloquium Publications, vol. 53, American Mathematical Society, Providence, RI, 2004.

65. H. Jacquet and R. P. Langlands, *Automorphic forms on* $GL(2)$, Lecture Notes in Mathematics, vol. 114, Springer-Verlag, Berlin, 1970.

66. Hervé Jacquet, *Automorphic forms on* GL(2). *Part II*, Lecture Notes in Mathematics, vol. 278, Springer-Verlag, Berlin, 1972.

67. Svetlana Katok, *Fuchsian groups*, Chicago Lectures in Mathematics, University of Chicago Press, Chicago, IL, 1992.

68. Henry H. Kim, *Functoriality for the exterior square of* GL$_4$ *and the symmetric fourth of* GL$_2$, J. Amer. Math. Soc. **16** (2003), no. 1, 139–183 (electronic), With Appendix 1 by Dinakar Ramakrishnan and Appendix 2 by Kim and Peter Sarnak.

69. Anthony W. Knapp, *Representation theory of semisimple groups. An overview based on examples*, Princeton Landmarks in Mathematics, Princeton University Press, Princeton, NJ, 2001, Reprint of the 1986 original.

70. Neal Koblitz, *Introduction to elliptic curves and modular forms*, second ed., Graduate Texts in Mathematics, vol. 97, Springer-Verlag, New York, 1993.

71. Emmanuel Kowalski, *Un cours de théorie analytique des nombres*, Cours Spécialisés, vol. 13, Société Mathématique de France, Paris, 2004.

72. N. V. Kuznetsov, *Arifmeticheskaya forma formuly sleda Selberga i raspredelenie norm primitivnykh giperbolicheskikh klassov modulyarnoi gruppy*, Akad. Nauk SSSR Dal'nevostoč. Naučn. Centr, 1978, Presented at the Far Eastern Mathematical School.

73. T. Y. Lam, *Introduction to quadratic forms over fields*, Graduate Studies in Mathematics, vol. 67, American Mathematical Society, Providence, RI, 2005.

74. E. Landau, *Über die Anzahl der Gitterpunkte in gewissen Bereichen, II*, Nachr. v. d. Gesellschaft d. Wiss. zu Göttingen, math.-phys. Klasse (1915), 209–243.

75. Serge Lang, *Algebraic number theory*, second ed., Graduate Texts in Mathematics, vol. 110, Springer-Verlag, New York, 1994.

76. _____, *Algebra*, third ed., Graduate Texts in Mathematics, vol. 211, Springer-Verlag, New York, 2002.

77. Robert P. Langlands, *Base change for* GL(2), Annals of Mathematics Studies, vol. 96, Princeton University Press, Princeton, N.J., 1980.

78. A. F. Lavrik, *Functional and approximate functional equations for Dirichlet functions*, Mat. Zametki **3** (1968), 613–622.

79. Peter D. Lax and Ralph S. Phillips, *Scattering theory for automorphic functions*, Princeton Univ. Press, Princeton, N.J., 1976, Annals of Mathematics Studies, No. 87.

80. J. Lewis and D. Zagier, *Period functions for Maass wave forms. I*, Ann. of Math. (2) **153** (2001), no. 1, 191–258.

81. Wen Ch'ing Winnie Li, *L-series of Rankin type and their functional equations*, Math. Ann. **244** (1979), no. 2, 135–166.

82. Elon Lindenstrauss, *Invariant measures and arithmetic quantum unique ergodicity*, Ann. of Math. (2) **163** (2006), no. 1, 165–219.

83. Elon Lindenstrauss and Akshay Venkatesh, *Existence and Weyl's law for spherical cusp forms*, Geom. Funct. Anal. **17** (2007), no. 1, 220–251.

84. W. Luo, Z. Rudnick, and P. Sarnak, *On Selberg's eigenvalue conjecture*, Geom. Funct. Anal. **5** (1995), no. 2, 387–401.

85. Wenzhi Luo, *Nonvanishing of L-values and the Weyl law*, Ann. of Math. (2) **154** (2001), no. 2, 477–502.

86. Wenzhi Luo, Zeév Rudnick, and Peter Sarnak, *On the generalized Ramanujan conjecture for* GL(n), Automorphic forms, automorphic representations, and arithmetic (Fort Worth, TX, 1996), Proc. Sympos. Pure Math., vol. 66, American Mathematical Society, Providence, RI, 1999, pp. 301–310.

87. Hans Maass, *Über eine neue Art von nichtanalytischen automorphen Funktionen und die Bestimmung Dirichletscher Reihen durch Funktionalgleichungen*, Math. Ann. **121** (1949), 141–183.

88. Colin Maclachlan and Alan W. Reid, *The arithmetic of hyperbolic 3-manifolds*, Graduate Texts in Mathematics, vol. 219, Springer-Verlag, New York, 2003.

89. G. A. Margulis, *Discrete subgroups of semisimple Lie groups*, Ergebnisse der Mathematik und ihrer Grenzgebiete (3), vol. 17, Springer-Verlag, Berlin, 1991.

90. André Martinez, *An introduction to semiclassical and microlocal analysis*, Universitext, Springer-Verlag, New York, 2002.

91. William S. Massey, *A basic course in algebraic topology*, Graduate Texts in Mathematics, vol. 127, Springer-Verlag, New York, 1991.

92. Toshitsune Miyake, *Modular forms*, Springer Monographs in Mathematics, Springer-Verlag, Berlin, 2006, Translated from the 1976 Japanese original by Yoshitaka Maeda.

93. Rached Mneimné and Frédéric Testard, *Introduction à la théorie des groupes de Lie classiques*, Collection Méthodes, Hermann, Paris, 1986.

94. Jean-Pierre Otal, *Three topological properties of small eigenfunctions on hyperbolic surfaces*, Geometry and dynamics of groups and spaces, Progress in Mathematics, vol. 265, Birkhäuser, Basel, 2008, pp. 685–695.

95. Jean-Pierre Otal and Eulalio Rosas, *Pour toute surface hyperbolique de genre g,* $\lambda_{2g-2} > 1/4$, Duke Math. J. **150** (2009), no. 1, 101–115.

96. Daniel Perrin, *Cours d'algèbre*, Collection CAPES/AGREG Mathématiques, Ellipses, Paris, 1996.

97. ———, *Algebraic geometry. An introduction*, Universitext, Springer-Verlag London, Ltd., London; EDP Sciences, Les Ulis, 2008, Translated from the 1995 French original by Catriona Maclean.

98. R. S. Phillips and P. Sarnak, *On cusp forms for co-finite subgroups of* PSL(2, **R**), Invent. Math. **80** (1985), no. 2, 339–364.

99. Vladimir Platonov and Andrei Rapinchuk, *Algebraic groups and number theory*, Pure and Applied Mathematics, vol. 139, Academic Press Inc., Boston, MA, 1994, Translated from the 1991 Russian original.

100. R. A. Rankin, *Contributions to the theory of Ramanujan's function* $\tau(n)$ *and similar arithmetical functions. I. The zeros of the function* $\sum_{n=1}^{\infty} \tau(n)/n^s$ *on the line* Re$s = 13/2$. *II. The order of the Fourier coefficients of integral modular forms*, Proc. Cambridge Philos. Soc. **35** (1939), 351–372.

101. I. Reiner, *Maximal orders*, London Mathematical Society Monographs. New Series, vol. 28, The Clarendon Press Oxford University Press, Oxford, 2003, Corrected reprint of the 1975 original.

102. Georg Friedrich Bernhard Riemann, *Œuvres mathématiques de Riemann*, Traduites de l'allemand par L. Laugel, avec une préface de C. Hermite et un discours de Félix Klein. Nouveau tirage, Librairie Scientifique et Technique Albert Blanchard, Paris, 1968.

103. Frigyes Riesz and Béla Sz.-Nagy, *Functional analysis*, Dover Books on Advanced Mathematics, Dover Publications Inc., New York, 1990, Translated from the second French edition by Leo F. Boron, Reprint of the 1955 original.

104. Didier Robert, *Autour de l'approximation semi-classique*, Progress in Mathematics, vol. 68, Birkhäuser Boston Inc., Boston, MA, 1987.

105. Walter Roelcke, *Über die Wellengleichung bei Grenzkreisgruppen erster Art*, S.-B. Heidelberger Akad. Wiss. Math.-Nat. Kl. **1953/1955** (1953/1955), 159–267 (1956).

106. Walter Rudin, *Real and complex analysis*, third ed., McGraw-Hill Book Co., New York, 1987.

107. ———, *Functional analysis*, second ed., International Series in Pure and Applied Mathematics, McGraw-Hill Inc., New York, 1991.

108. Zeév Rudnick and Peter Sarnak, *The behaviour of eigenstates of arithmetic hyperbolic manifolds*, Comm. Math. Phys. **161** (1994), no. 1, 195–213.

109. Daniel J. Rudolph, $\times 2$ *and* $\times 3$ *invariant measures and entropy*, Ergodic Theory Dynam. Systems **10** (1990), no. 2, 395–406.

110. Peter Sarnak, *Class numbers of indefinite binary quadratic forms*, J. Number Theory **15** (1982), no. 2, 229–247.

111. ———, *Nonvanishing of L-functions on* re$(s) = 1$, available at: http://web.math.princeton.edu/sarnak/ShalikaBday2002.pdf, 2002.

112. ———, *Spectra of hyperbolic surfaces*, Bull. Amer. Math. Soc. (N.S.) **40** (2003), no. 4, 441–478 (electronic).

113. Peter Sarnak and Xiao Xi Xue, *Bounds for multiplicities of automorphic representations*, Duke Math. J. **64** (1991), no. 1, 207–227.

114. Atle Selberg, *Bemerkungen über eine Dirichletsche Reihe, die mit der Theorie der Modulformen nahe verbunden ist*, Arch. Math. Naturvid. **43** (1940), 47–50.

115. _____, *Harmonic analysis and discontinuous groups in weakly symmetric Riemannian spaces with applications to Dirichlet series*, J. Indian Math. Soc. (N.S.) **20** (1956), 47–87.

116. _____, *On the estimation of Fourier coefficients of modular forms*, Proc. Sympos. Pure Math., Vol. VIII, American Mathematical Society, Providence, R.I., 1965, pp. 1–15.

117. Jean-Pierre Serre, *Arbres, amalgames*, SL_2, Astérisque, vol. 46, Société Mathématique de France, Paris, 1977.

118. _____, *Cours d'arithmétique*, Le Mathématicien, vol. 2, Presses Universitaires de France, Paris, 1977, Deuxième édition revue et corrigée.

119. _____, *Représentations linéaires des groupes finis*, Hermann, Paris, 1978.

120. Hideo Shimizu, *On zeta functions of quaternion algebras*, Ann. of Math. (2) **81** (1965), 166–193.

121. Goro Shimura, *Introduction to the arithmetic theory of automorphic functions*, Publications of the Mathematical Society of Japan, vol. 11, Princeton University Press, Princeton, NJ, 1994, Reprint of the 1971 original, Kano Memorial Lectures, 1.

122. Carl Ludwig Siegel, *The average measure of quadratic forms with given determinant and signature*, Ann. of Math. (2) **45** (1944), 667–685.

123. _____, *Advanced analytic number theory*, second ed., Tata Institute of Fundamental Research Studies in Mathematics, vol. 9, Tata Institute of Fundamental Research, Bombay, 1980.

124. Lior Silberman and Akshay Venkatesh, *On quantum unique ergodicity for locally symmetric spaces*, Geom. Funct. Anal. **17** (2007), no. 3, 960–998.

125. A. I. Šnirel'man, *Ergodic properties of eigenfunctions*, Uspehi Mat. Nauk **29** (1974), no. 6(180), 181–182.

126. Kannan Soundararajan, *Quantum unique ergodicity for* $SL_2(\mathbb{Z})\backslash\mathbb{H}$, Ann. of Math. (2) **172** (2010), no. 2, 1529–1538.

127. Michael E. Taylor, *Partial differential equations*, Texts in Applied Mathematics, vol. 23, Springer-Verlag, New York, 1996, Basic theory.

128. Richard Taylor, *Galois representations*, Ann. Fac. Sci. Toulouse Math. (6) **13** (2004), no. 1, 73–119.

129. E. C. Titchmarsh, *The theory of the Riemann zeta-function*, second ed., The Clarendon Press Oxford University Press, New York, 1986.

130. Jerrold Tunnell, *Artin's conjecture for representations of octahedral type*, Bull. Amer. Math. Soc. (N.S.) **5** (1981), no. 2, 173–175.

131. A. B. Venkov, *Spectral theory of automorphic functions*, Trudy Mat. Inst. Steklov. **153** (1981), 172.

132. Marie-France Vignéras, *Arithmétique des algèbres de quaternions*, Lecture Notes in Mathematics, vol. 800, Springer, Berlin, 1980.

133. _____, *Variétés riemanniennes isospectrales et non isométriques*, Ann. of Math. (2) **112** (1980), no. 1, 21–32.

134. _____, *Quelques remarques sur la conjecture* $\lambda_1 \geq 1/4$, Seminar on number theory, Paris 1981–82 (Paris, 1981/1982), Progr. Math., vol. 38, Birkhäuser Boston, Boston, MA, 1983, pp. 321–343.

135. G. N. Watson, *A treatise on the theory of Bessel functions*, Cambridge Mathematical Library, Cambridge University Press, Cambridge, 1995, Reprint of the second (1944) edition.

136. André Weil, *Basic number theory*, Classics in Mathematics, Springer-Verlag, Berlin, 1995, Reprint of the second (1973) edition.

137. Edwin Weiss, *Algebraic number theory*, Dover Publications Inc., Mineola, NY, 1998, Reprint of the 1963 original.

138. E. T. Whittaker and G. N. Watson, *A course of modern analysis*, Cambridge Mathematical Library, Cambridge University Press, Cambridge, 1996, An introduction to the general theory

of infinite processes and of analytic functions; with an account of the principal transcendental functions, Reprint of the fourth (1927) edition.

139. David Vernon Widder, *The Laplace Transform*, Princeton Mathematical Series, vol. 6, Princeton University Press, Princeton, N.J., 1941.

140. Franck Wielonsky, *Séries d'Eisenstein, intégrales toroïdales et une formule de Hecke*, Enseign. Math. (2) **31** (1985), no. 1–2, 93–135.

141. Scott A. Wolpert, *Disappearance of cusp forms in special families*, Ann. of Math. (2) **139** (1994), no. 2, 239–291.

142. _____, *Semiclassical limits for the hyperbolic plane*, Duke Math. J. **108** (2001), no. 3, 449–509.

143. D. Zagier, *Eisenstein series and the Riemann zeta function*, Automorphic forms, representation theory and arithmetic (Bombay, 1979), Tata Inst. Fund. Res. Studies in Math., vol. 10, Tata Inst. Fundamental Res., Bombay, 1981, pp. 275–301.

144. Steven Zelditch, *Uniform distribution of eigenfunctions on compact hyperbolic surfaces*, Duke Math. J. **55** (1987), no. 4, 919–941.

145. Claude Zuily and Hervé Queffélec, *Analyse pour l'agrégation*, third ed., Sciences Sup, Dunod, Paris, 2007.

Index of notation

© Springer International Publishing Switzerland 2016
N. Bergeron, *The Spectrum of Hyperbolic Surfaces*, Universitext,
DOI 10.1007/978-3-319-27666-3

Index

© Springer International Publishing Switzerland 2016
N. Bergeron, *The Spectrum of Hyperbolic Surfaces*, Universitext,
DOI 10.1007/978-3-319-27666-3

Index of names

© Springer International Publishing Switzerland 2016
N. Bergeron, *The Spectrum of Hyperbolic Surfaces*, Universitext,
DOI 10.1007/978-3-319-27666-3

Printed in the United States
By Bookmasters